Eurotech Direct '91
Thermofluids Engineering

Conference Planning Panel

V Cole (Chairman)
Department of Energy

E Fisher
Department of Mechanical, Materials and Manufacturing Engineering
University of Newcastle upon Tyne

A Gilchrist
Department of Mechanical and Process Engineering
University of Strathclyde

F Mills
Frank Mills Associates
Leyland
Lancashire

M Moore
Consultant
Guildford
Surrey

J H Weaving
ERA Limited
Dunstable
Bedfordshire

D E Winterbone
Department of Mechanical Engineering
University of Manchester Institute of Science and Technology

Proceedings of the Institution of Mechanical Engineers

Eurotech Direct '91
Thermofluids Engineering

European Engineering Research and Technology Transfer Congress

2–4 July 1991
International Congress Centre, Birmingham

Co-sponsoring Bodies
Commission of the European Communities under the SPRINT programme
Department of Trade and Industry
Science and Engineering Research Council
Fellowship of Engineering
Institution of Electrical Engineers
Institution of Manufacturing Engineers
British Gear Association
British Computer Society
American Society of Mechanical Engineers
Société des Ingenieures et Scientifiques de France
Norwegian Society of Chartered Engineers
Institution of Engineers of Ireland
Verein Deutscher Ingenieure
Royal Flemish Association of Engineers
Svenska Mekanisters Riksforening
Japan Society of Mechanical Engineers
International Association for Continuing Engineering Education
European Association for the Transfer of Technologies, Innovation and Industrial Information

IMechE 1991–8

Published for IMechE by
Mechanical Engineering Publications Limited

First Published 1991
This publication is copyright under the Berne Convention and the International Copyright Convention. Apart from any fair dealing for the purpose of private study, research, criticism or review, as permitted under the Copyright, Designs and Patents Act, 1988, no part may be reproduced, stored in a retrieval system, or transmitted in any form or by any means, electronic, electrical, chemical, mechanical, photocopying, recording or otherwise, without the prior permission of the copyright owners. Reprographic reproduction is permitted only in accordance with the terms of licences issued by the Copyright Licensing Agency, 90 Tottenham Court Road, London W1P 9HE. *Unlicensed multiple copying of the contents of this publication is illegal.* Inquiries should be addressed to: The Managing Editor, Mechanical Engineering Publications Limited, Northgate Avenue, Bury St. Edmunds, Suffolk, IP32 6BW.

Authorization to photocopy items for personal or internal use, is granted by the Institution of Mechanical Engineers for libraries and other users registered with the Copyright Clearance Center (CCC), provided that the fee of $0.50 per page is paid direct to CCC, 21 Congress Street, Salem, Ma 01970, USA. This authorization does not extend to other kinds of copying such as copying for general distribution for advertising or promotional purposes, for creating new collective works, or for resale, 085298 $0.00 + .50.

© The Institution of Mechanical Engineers 1991

ISBN 0 85298 775 7

A CIP catalogue record for this book is available from the British Library.

The Publishers are not responsible for any statement made in this publication. Data, discussion and conclusions developed by authors are for information only and are not intended for use without independent substantiating investigation on the part of potential users.

Printed by Waveney Print Services Ltd, Beccles, Suffolk

Contents

C413/012	**A two-dimensional heat diffusion code incorporating one- and two-phase convection and radiation** T D A Kennedy and R L Moss	1
C413/014	**Numerical investigation of modified K-ε turbulence models for the simulation of a backward facing step flow** T Kobayashi, N Taniguchi, H-K Myong and M-A Ohmachi	9
C413/077	**A proposal for improving the modelling of turbulence** R K Duggins	15
C413/022	**Computational study on three-dimensional flow around a heavy-duty truck body** S Nishikawa, Y Watanabe, K Fujitani, R Himeno and M Takagi	21
C413/041	**Turbulent flow and heat transfer predictions for cross-corrugated rotary regenerators** M Ciofalo, M W Collins and G Perronne	27
C413/046	**Numerical simulation of transitions in axial-rotary flows in an annulus** T J Lockett, S M Richardson and W J Worraker	35
C413/043	**The rotating matrix swirling gas burner: performance and computational design** W A Abd Al-Masseeh, D Bradley, P H Gaskell, A Ishikawa and A K C Lau	43
C413/054	**Modelling gas-fired furnace flow and combustion using the PCOC code** S A Beltagui, P J Stopford, R N Fuggle, A M A Kenbar, T Ralston and N Marriott	51
C413/076	**Exploitation of modern flow computation techniques in the design of turbine blading** C T J Scrivener and J Grant	59
C413/067	**Prediction of smoke movement using CFD** N Rhodes, I W Clark and K Else	73
C413/059	**Meeting vehicle pollution regulations by combustion technology** J Chapman, A C Cole, S Wallace and J H Weaving	83
C413/039	**Experimental and theoretical study of the radiation of gases containing dust particles** J Stasiek and M W Collins	95
C413/065	**Swirl centre precession under steady flow conditions** S Nadarajah, M J Tindal and M Yianneskis	103
C413/009	**Estimation of the differences of swirl and charging characteristics between cylinders in a multi-cylinder diesel engine** Y Isshiki, Y Shimamoto and T Wakisaka	109
C413/079	**Numerical simulation of fluid flow in poppet valves** N D Vaughan, D N Johnston and K A Edge	119
C413/023	**Some developments in the application of computational fluid dynamics to flows in internal combustion engine cylinders** A P Watkins, C Diomataris, M Z Gül, H Khaleghi, C J Lea and D M Wang	129

C413/049	**Local void fraction measurement in flow channels of irregular shape** I K Smith, Z G Xu and C A Aldis	147
C413/044	**Pressure drop in foam flow** J R Calvert	153
C413/020	**Effect of confined geometry, using grids, on pool boiling curves** I Rajab and R H S Winterton	159
C413/036	**Electro-hydrodynamic enhancement of two-phase heat transfer** T G Karayiannis and P H G Allen	165
C413/038	**Numerical analysis of transient two-dimensional bubbly flow with size and number density distribution** Y Matsumoto, J Matsui and H Ohashi	183
C413/006	**A transient technique for determining average heat transfer coefficients in finned surfaces** N Lawrence and T Cowell	189
C413/062	**Flow resistance in circular tubes rotating about a parallel axis** A R Johnson	197
C413/003	**Biofilm formation and control in flowing aqueous systems** T R Bott	213
C413/040	**Liquid crystal mapping of local heat transfer in crossed corrugated geometrical elements for air heat exchangers** J Stasiek, M W Collins and P E Chew	223
C413/061	**Simulation: a tool for improving the design process** B Murphy, M A P Murray and V I Hanby	229
C413/011	**A study of flow oscillation and heating in a Hartmann-Sprenger tube—a literature survey** J Iwamoto	237
C413/017	**Thermal insulation for optimum design of annular cavity industrial chimneys** M M A Shahin	243
C413/028	**The design and testing of heat exchangers for a Stirling cycle heat pump** D H Rix	249

C413/012

A two-dimensional heat diffusion code incorporating one - and two - phase convection and radiation

T D A KENNEDY, MSc, C Eng, MIMechE, MBNES
Consultant, Pangbourne, Berkshire
R L MOSS, MSc, PhD
Joint Research Centre, Commission of the European Communities,
Petten, The Netherlands

SYNOPSIS A 2-D finite-difference heat transfer code has been developed for use with personal computers. It combines the solution of the diffusion equation (for heat) with analysis of the boundary('coolant') conditions, which may include convection and/or radiation of heat to or from each boundary.

The code solves transient problems, but can, of course, be used for finding steady-state conditions. It has been validated against a number of experimental results, as well as several simple test cases which can be solved analytically

Problems can be handled which include single and/or two-phase fluids at the boundaries, time dependent radiation to or from boundaries and internal heat generation rates which vary with time and position.

NOTATION

A	Surface area (m^2)
A_x	Flow channel cross-section (m^2)
B	Parameter in equation (10)
C_1	Factor in equation (7)
C_p	Specific heat (J/kg.K)
d_e	Equivalent diameter of channel (m)
g	Gravitational acceleration (m/s^2)
Gr	Grashof No. $(l^3.\rho^2.g.\beta.\Delta T)/\eta^2$
h_{lv}	Latent heat (J/kg)
I	Internal heat generation(W/m^3)
l	Length dimension (m)
m	Mass flux ($kg/m^2.s$)
m	Mass flow (kg/s)
Nu	Nusselt No. $(\alpha.d_e/\lambda)$
P	Pressure (N/m^2)
Pr	Prandtl No. $Cp.\eta/\lambda)$
q	Heat flux (W/m^2)
Qrad	Heat radiated (W)
r	Radius(m)
Re	Reynold's No
t	Time (s)
T	Temperature (°C)
z	Axial position relative to datum(m)

Subscripts

an	Annulus
b	Bulk property
cr	Critical
e	Equivalent
eff	Effective
l	Liquid
s	Surface
sat	Saturation
sub	Sub-cooled
v	Vapour
w	Wall

Greek

α	Heat transfer coeff ($W/m^2.K$)
β	Coeff. th. expansion (K^{-1})
Δ	Difference
η	Absolute viscosity (kg/m.s)
ϵ	Total thermal emissivity
λ	Th.conductivity (W/m.λ)
ρ	Density (kg/m^3)
σ	Surface tension (N/m)
σ_{sB}	Stefan Boltzmann constant

1. INTRODUCTION

Accurate prediction of temperatures is an important requirement for many designers and safety analysts. Many large finite element codes exist for thermal analysis. Some problems occur which require more detailed handling of boundary conditions than that which is available in typical large f.e. codes.

A two-dimensional finite difference code has been developed for easy use on I.B.M. compatible P.C.s. This uses a two-step iteration method in which the solution of the set of heat diffusion equations alternates with the solution of the heat balance equations for Newtonian cooling at the boundaries. Radiant heat transfer is also computed at both internal and external boundaries.

The alternating direction method is used for the solution of the diffusion equations for each real time step supplied by the code User. Transient and steady-state solutions can be provided; the latter are found by running the problem until steady-state is reached.

The paper describes the extension of the methods used in a well proven mainframe 2-D code(Kennedy[3]) to handle problems with 2-Phase cooling and time-dependent radiation.

The particular problem of coolants in 2-phase flow, where heat transfer coefficients may undergo step changes in magnitude is handled by smoothing the changes over a number of time steps. Where the radiation at the outside of the model envelope is significant, the temperature of the surface which receives/transmits that radiation is generally not the same as that of the local convection boundary. Examples are containers stored in the open air and buildings. To allow for this effect, separate temperatures are allowed for convection and radiation at all outside boundaries.

Predictions have been verified against results from experiments in the High Flux Reactor at the Joint Research Centre at Petten and against a range of simple test cases for which analytic solutions are available.

2. THE COMPUTER CODE 2DT-BOIL

In the code, the equation for the diffusion of heat

$$\nabla \lambda . \nabla T + I = \rho . C_p . \delta T / \delta t \quad - - - - - (1)$$

is replaced by an equivalent set of finite difference equations, which are solved by an alternating direction method described by Douglas [1], for each time step, t.

After each time step, the boundary conditions are re-calculated. Boundaries may be internal or external to the model envelope. They may provide adiabatic or isothermal conditions, or they may represent real coolants or heating fluids. Radiation from and to boundary surfaces is also allowed. Transient or steady-state conditions can be predicted; the latter are found by running the problem to equilibrium conditions.

Radiation is handled by inclusion in the heat generation term, I, in equation (1). The heat radiated from or to each boundary is calculated for conditions at the start of each time step, and the consequent heat gain or loss is added to the heat generation. Where an external boundary is concerned, the temperature of the region to or from which the heat is radiating is usually not the same as coolant boundary temperature. The code allows separate boundary temperatures for convection and radiation.

The model geometry may be cylindrical or rectangular. Cylinders may be solid or hollow. The model envelope must be completely filled. If the boundary of the real problem is irregular, 'dummy' (e.g. adiabatic or isothermal) material must be provided to complete the envelope. Up to 1577 mesh points are allowed for a standard PC (1887 for a PC AT). The code is normally set up with 30 radial X 50 axial mesh lines for a PC or 30 X 60 for a PC-AT. Mesh disposition is easily changed because the Common blocks holding the arrays are held in a separate file which is called up by each subroutine when the code is compiled. e.g. the 30 X 60 mesh for a PC-AT may be replaced by a 20 X 90 mesh.

Internal coolants are allowed inside the model envelope. These become effectively internal boundries, using the method of Clark and Troost [2]. This method 'freezes' the coolant temperatures during each iteration, which implies that the coolant is, during the duration of that time step, in equilibrium with its adjacent surfaces. Consequently, the code should not be used to study the transient behaviour of the coolant over time steps comparable to or greater than the coolant transit time.

A coolant can take up the outlet temperature of another coolant as its inlet temperature. This feature allows simple modelling of natural convection cells and of some heat exchanger situations. At the same time, conduction in adjacent solid materials and, for gaseous coolants, radiation between channel surfaces, are accounted for.

2.1 General Logic Flow.

The basic logic flow is shown below.

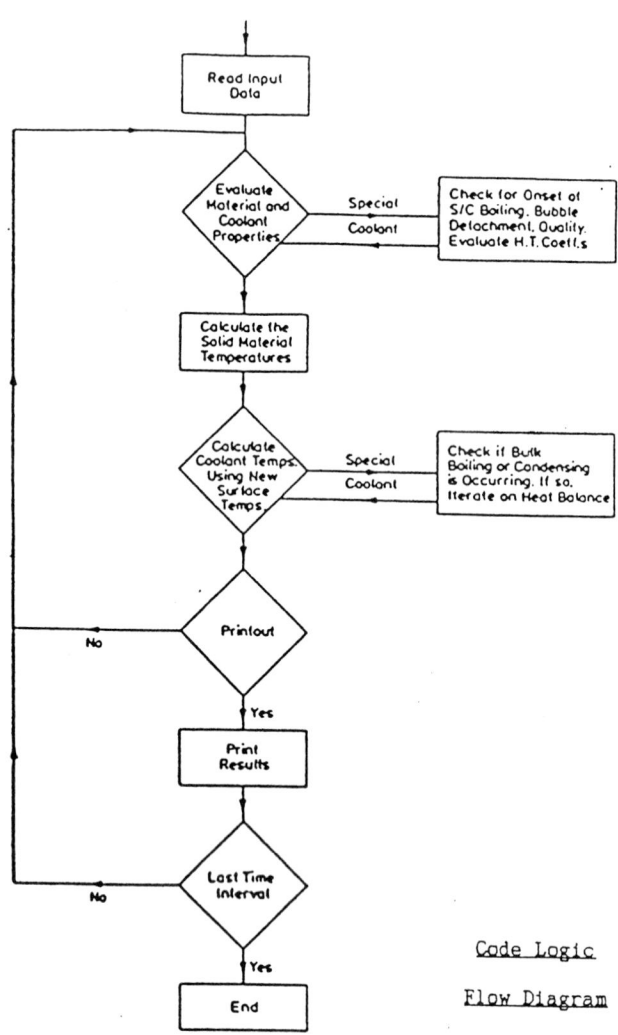

Code Logic Flow Diagram

As described earlier, the boundary conditions for the solution of the finite difference equations are controlled by the coolants at the boundaries. Thus, to the code, the terms 'coolant' and 'boundary' are synonomous. Whether a coolant in the model is a real fluid, cooling or heating, or a 'dummy' (e.g. adiabatic boundary) depends on the physical situation to be modelled and the code User's preference.

2.2 Solution of the Diffusion Equations.

The set of finite difference equations for the diffusion of heat over a discrete time are solved by an alternating direction method (Douglas [1]). It provides a stable solution up to reasonably large time steps where there are no large changes in conditions between successive iterations (e.g. order of magnitude increases in heat generation or coolant heat transfer coefficients at the onset of boiling or condensation).

The code follows the method used by Clark and Troost[2], with a modification to the equations, described by Kennedy[3]. The equations are not described here, as the method is well known.

2.3   Boundary Conditions.

After calculation of the material temperatures at the end of a time step, coolant temperatures are recalculated from the heat transferred into or out of each coolant. When the boundary is adiabatic, it is described as a 'dummy' coolant, which is assigned a very high heat capacity and a very low heat transfer co-efficient: in consequence no significant heat is transferred between the material in the model and the boundary. In similar fashion, when the boundary is isothermal, the associated 'dummy' coolant is assigned very high heat capacity and heat transfer co-efficient, so that the boundary keeps the adjacent material at the same constant temperature.

Real coolants are assigned specific heats, mass flows and heat transfer coefficients appropriate to the fluid in use. The calculation is described by Kennedy[3] for single-phase fluids. If saturation temperature is reached, either by a cooling fluid heating up, or by a heating fluid cooling down, the situation is changed. A cooling fluid remains at saturation temperature until it has all boiled off (mass quality, x=1) and a heating fluid remains isothermal until it has all condensed (x=0). The heat removed from or added to the adjacent material is no longer fluid sensible heat, but latent heat of vaporisation. Conservation of the overall model heat balance requires an additional routine which is called while coolant fluid mass quality is between 0 and 1. Calculation of coolant temperatures completes an iteration(time step). The code then returns to re-evaluate material and coolant physical properties and coolant heat transfer coefficients, based on the temperatures calculated and on the current time (Some properties may be time-dependant, e.g.coolant inlet temp.s).

When the temperature of the surface adjacent to a liquid coolant reaches a certain degree of superheat (ΔT_{sat}), sub-cooled boiling starts in the liquid, and the heat transfer between the liquid and surface increases very rapidly. This can result in large step changes in the heat transfer co-efficient with consequent instabilities in the solution of the diffusion equations. Stability over successive iterations is maintained by time averaging the computed heat transfer co-efficients over eight successive iterations. For these conditions, small time steps are used to avoid the introduction of significant errors in the solution. Similar considerations apply to a heating vapour cooling down below saturation temperature.

2.4   Radiation.

Heat is allowed to radiate from and to all surfaces adjacent to a gas or vacuum, at both internal and external boundaries. Because the mechanism of radiation is independant of conduction and convection, a 'source and sink' method is used. At the start of each iteration, The radiant heat transfer is found for each mesh space adjacent to a boundary:-

$$Q_{rad_{i,j}} = \frac{\sigma_{SB} A_1 \varepsilon_1 \varepsilon_2 (T_{m_1}^4 - T_{s_2}^4)}{(\varepsilon_2 + \varepsilon_1 (1-\varepsilon_2)(A_1/A_2))} \quad - - - (2)$$

Equation(2) uses a simplified gap model with heat radiating normally at each surface. Errors are likely to be introduced with wide and irregular gas spaces. The radiant heat terms are added to the internal heat generation terms, $I_{i,j}$ for each relevant mesh space. The combined heat generation and radiated heat terms are then used in the solution of the heat diffusion equations.

2.5   Coolant heat transfer coefficients.

Input options for these coefficients are :
(a) User supplied constants or functions of fluid bulk or film temperature or time.
(b) Code calculation of single-phase co-efficient with User supplied fluid physical properties.
(c) Code calculation of single- or two-phase co-eff. for liquid/vapour, using physical property data held with the code.

For option (b) the code determines whether flow is laminar or turbulent, and then evaluates the local co-efficient accordingly. For laminar flow, equations derived from analytic solutions(Kays[4]) are used. For turbulent flow, one of four correlations is used, depending on the fluid Prandtl No :

$(Pr)<0.1$ (metals) $(Nu)=5.55+0.003(Pr)(Re)$ --- (3)

$0.1<(Pr)<1$ (gases) $(Nu)=0.023(Re)^{0.8}(Pr)^{0.4}$ -- (4)

$1>(Pr)>20$ (water etc) $(Nu)=0.0155(Re)^{.83}(Pr)^{.5}$ - (5)

$20>(Pr)$ (oils etc) $(Nu)=0.0118(Re)^{0.9}(Pr)^{.5}$ --- (6)

For option(c), the code checks if there is enough superheating at each position along the heated surface to start sub-cooled boiling, using the correlation of Rohsenow[5] :

$$\Delta T_{sat}=C_1 h_{lv}/Cp_l [q/_l h_{lv}(\sigma/g/(\rho_l-\rho_v))^{.5}]^{.33}[Cp_l l/\lambda_l] \quad - - - (7)$$

With insufficient superheating, a single-phase co-efficient is evaluated. For forced convection flow, the procedure described for option(b) is followed. For natural convection problems, the correlation of Niemann[6] is used to evaluate the heat transfer:

$$\lambda_{eff}/\lambda_l = 1+[0.119(Gr_{an}Pr)^{1.27}/(Gr_{an}Pr+14500)] - (8)$$

With sufficient superheating, the sub-cooled boiling coefficient is evaluated, using the equation due to Forster/Zuber[7].

$$\alpha_{nb}=\frac{0.00122[(\lambda_l^{.79}Cp_l^{.45}\rho_l^{.49})/(\sigma^{.5}\mu_l^{.49}h_{lv}^{.24}\rho_v^{.24})]}{\Delta T_{sat}^{.24} \cdot \Delta P_{sat}^{.75}} \quad --- (9)$$

A check is made to ensure that the critical heat flux for sub-cooled burnout is not exceeded, using the correlation of Ivey and Morris[8] :-

$q_{cr} = 0.13 h_{lv}^{0.5}[\sigma.g(\rho_l-\rho_v)]^{0.25}[1+B.\Delta T_{sub}]$ -- (10)

where $B = 0.065(\rho_l-\rho_v)^{0.8}.Cp_v/h_{lv}$ ----- (11)

2.6   Radial Thermal Expansion.

Thermal expansion of solid materials surrounding a gas space can cause significant changes in the size of the space. In cylindrical geometry, the code calculates radial thermal expansion to determine the width of a gas space at operating temperatures, often very different from the width measured at ambient conditions. The code assumes that the User has supplied dimensions taken at 20°C.

2.7   Heat Balance Consistency.

With computer codes which provide temperature distributions over a large number of mesh points it is important to be able to check the overall and local heat balances; this helps the User to validate the results. For this purpose, results printed out include an overall heat balance. For a steady state solution, the sum of the heat flows at the model envelope boundaries must equal the heat generated within the model. In addition, optional extra output gives details of local heat flows across the length of each each internal or external coolant channel, and of radiated heat at all internal and external surfaces.

2.8 Material Physical Properties

This section refers to solid materials or to static fluids, which are treated in the same way as solids. i.e heat is transferred across the material by diffusion only. Up to 30 different materials are allowed. The User may select any of 22 standard materials, whose properties, taken from Touloukian[9], are stored with the code, in tabular form, as functions of temperature. Alternatively, he may write in his own properties. The two methods may be combined in any way the User wishes. e.g., if he wishes to scale the thermal conductivity of a standard material in one direction only, to model a physical shape which does not conform to the regular geometry required by the finite difference method used, or because the material is anisotropic, this can be done.

2.9 Coolant Fluid Physical Properties

As described above, coolants are handled separately from solids or static fluids. The User can choose the option of telling the code to evaluate heat transfer coefficients at specified coolant boundaries. In this case he must supply the necessary fluid physical properties. Up to 30 coolants are allowed. Coolants nos 1-15 are assumed to be gases and 16-24 to be liquids. Nos 25-30 are 'special' coolants which can be single- or two-phase. Physical properties of water and steam on the saturation line are stored with the code. The methods used for two-phase conditions are relevant to fluids other than water, but the code has only been checked for water/steam. Gas coolants are assumed transparent to thermal radiation while liquids are considered to be opaque.

3. COMPARISON OF PREDICTION WITH EXPERIMENT.

The code has been used to assist the design of experiments in the High Flux Reactor at Petten. Two examples are given which compare predictions with measurements.

3.1 Irradiation of UO_2 fuel pellets at specified temperatures for material testing.

The code forms an integral part of the design analysis of irradiation experiments at the High Flux Reactor at Petten. Whilst most of the experiments consist of a central specimen, encapsulated in 2 or 3 concentric cylindrical tubes, which considerably simplifies the analysis, certain aspects of the prevailing nuclear and thermal conditions in the reactor, make it essential that a well-defined experiment should be modelled.

As an example of this, an experiment was recently requested where the specimen consisted of 10 pellets (7mm diameter, 0.6mm thickness) of uranium dioxide (UO_2), separated by high-temperature resistant(TZM) molybdenum drums, and encapsulated in tubing of the same material. The design required that the ten UO_2 pellets should be irradiated at a peak nuclear heating of 40 W/g, temperatures varying linearly from 1800°C at the bottom to 1000°C at the top.

The flux conditions, as with other materials testing reactors, vary in a sinusoidal form, with the peak just below the core centre line. The proposed design consisted of the fuel capsule in a TZM tube, which is stepped along its outer diameter, immersed in sodium, and contained in concentric molybdenum and stainless tubes. The experiment, which is contained in a standard carrier, is shown schematically in fig.1. The model of the experiment used for analysis by the code is shown in fig.2. The analysis, taking into account the fission heating in the fuel and the gamma heating in all materials, was used to predict the width of gas gaps needed to provide the correct fuel pellet temperatures for the experiment.

The resulting temperature history of the experiment during the short 2 hour irradiation period is shown in fig.3. In Table 1, the temperature values of the averaged thermocouples at each axial level are given. The table compares the predicted values with these mean measured values. A similar comparison is given for fuel pellet temperatures. It can be seen that prediction is in excellent agreement with results.

Table 1. Measured and Predicted Temperatures

thermocouples			fuel pellet temperatures		
no	pred.n	measured	required	achieved	error(%)
9,10	730	751	1000	1005	0.5
			1089	1100	1.0
7,8	910	930	1178	1190	1.0
			1267	1280	1.0
			1356	1375	1.4
5,6	1120	1124	1445	1465	1.4
			1533	1555	1.4
3,4	1490	1512	1622	1640	1.1
			1711	1695	-0.9
1,2	1720	1671	1800	1750	-2.7

3.2 Two-phase heat transfer in an irradiation capsule for testing water reactor fuel.

Irradiation testing of BWR and PWR fuel rods has been carried out in the High Flux Reactor(HFR) at the Joint Research Centre(JRC) Petten for a number of years, using special irradiation capsules. The capsules rely on vigorous sub-cooled boiling as the heat transfer mechanism at the fuel rod surface. Heat transfer from this boiling water to the outer wall of the capsule then occurs by re-condensation of the vapour bubbles and by natural convection, enhanced by the turbulence of the boiling process. This aspect of the code and its validation against experiment have been described by Kennedy et al[10]

Knowledge of the upper operational limits for fuel irradiation capsules is essential for the safe operation of experiments in any materials testing reactor. Experimental determination of these limits is difficult, due to the sudden and irreversible nature of film boiling. The use of a reliable code as a route to deriving the limits is a considerable advantage.

Because the irradiation capsule relies on natural convection of the high pressure water inside, the code calculates the flow in the capsule, assuming a single convection cell. It is also assumed that the flow will not be affected by the sub-cooled boiling occurring at the surface. The simple expression used to calculate the mass flux is:-

$$m = 0.001(Gr)_{an}^{0.45} \quad - - - - - - - - - (10)$$

An electrically heated test capsule was constructed at JRC Petten to check the code predictions for sub-cooled boiling conditions against experimental results. A central electrical heater rod provided power up to 9.5 kW (380 W/cm linear rating). The heater has an active length of 25 cm and an outer diameter of 10.75 mm. Fig.4 gives the schematic layout of the test capsule, with the positions of the measuring thermocouples. A small sampling water flow of 1 g/s is passed through the capsule during the experiments - this simulates the conditions which exist in the in-reactor experiments. The out-of-reactor tests were carried under normal BWR conditions of 70 bar internal water pressure. During the tests, the power was increased in steps from 0 to 6.23 kW (249 W/cm); temperatures were allowed to stabilise for each step before increasing power to the next level. Results of all thermocouple measurements, taken at varying heater powers, are shown in Table 2.

Table 2.

Measured Heater, Shield & Water Temperatures.

Power (kW)	S1	S2	S3	S4	S5	S6	W1	W2	H4	H3	H2	H1
0.0	20	20	20	20	20	20	20	20	20	20	20	20
0.01	24	22	27	21	20	20	36	21	63	74	80	72
0.17	26	24	31	22	20	20	43	23	76	90	98	88
0.36	35	29	44	25	22	20	61	27	107	128	139	120
0.49	46	35	49	29	24	21	73	30	125	149	163	118
0.69	53	41	56	32	25	21	94	33	146	174	180	131
0.99	63	53	60	38	28	22	105	38	173	208	195	153
1.33	69	64	72	45	31	22	115	43	198	239	217	172
2.02	81	82	83	60	37	23	132	53	238	292	258	199
2.66	92	92	97	78	44	25	150	63	275	326	286	229
3.45	104	130	111	96	54	26	170	75	317	352	327	263
4.21	116	114	120	108	65	28	184	88	350	374	359	289
4.80	124	123	130	116	75	29	194	102	372	385	367	311
5.47	132	135	136	126	85	31	215	119	386	397	378	333
6.23	146	142	149	133	94	33	228	136	398	410	388	347

Fig.5 shows the physical system to be modelled and the code representation of the system used for the prediction of conditions in the test capsule. The re-circulating water convection cell between the heater and the aluminium shield is modelled by the two coolants, -25 representing the heated upflow past the heater and -28 representing the cooled down flow past the shield. Each block takes its inlet conditions from the outlet conditions of the other. The special water blocks (28) between the upflow and downflow legs are included to handle enhanced radial heat transfer resulting from sub-cooled boiling when it occurs. The sampling water flow through the capsule, which is used in the in-reactor experiment to check for any fission leakage, is represented by coolant nos -6, -8, and -5; coolant -8 takes its inlet conditions from the outlet conditions of -6 and -5 takes its inlet conditions from the outlet of -8. The secondary coolant, which flows outside the capsule at 158 g/s, is modelled by coolant -7.

The predictions by the code are compared with the measurements shown in Table 2 above. Fig.6 shows the measured and predicted temperatures of the capsule water and aluminium shield at 3 heater powers, 2.66, 4.21 and 6.23 kW. The onset of sub-cooled boiling in the capsule water was determined from experiment to start, at the top of the heater, at a power of above 3.75 kW. The code predicted that this would occur at 4.12 kW. Fig.7 shows the progression of the onset of sub-cooled boiling from the top to the bottom of the heater with increasing power, as predicted by the code.

4. DISCUSSION

The code was originally developed to assist in the design of irradiation experiments in materials testing reactors, and to aid safety calculations required for the operation of these reactors. Early proving and validation came from such experiments. However, the code has been found to be useful in many other applications. The ability to model complex and rapidly changing boundary conditions such as occur when a liquid coolant starts to boil, is a considerable asset. This feature was developed to assist the design of experimental fuel capsules which rely on natural convection and sub-cooled boiling of water; it could be useful in analysing conditions in other areas where boiling may occur, either by design or by default.

The main problem in incorporating boiling heat transfer coefficients at the boundaries in a transient f.d. code arises from the step changes in conditions which occur as sub-cooled boiling is initiated where the fluid was previously in single-phase. To overcome the instabilities which are introduced, the coefficients calculated by the code are smoothed over eight iterations. By this means, stability is maintained. To avoid distorting the solution of transient problems, which arise with large time steps between iterations, small time steps are used. However, the reductions in acceptable length of time step have not been found to result in excessive run times.

A particular feature of the code is the ability to accept internal heat generation which varies with both time and position. This facility was originally introduced to allow fissile and gamma heating curves to be used in models for in-reactor experiments. A typical example of the use of this feature outside the nuclear field was the thermal analysis of a solid-state battery to be used in a space satellite in geo-stationary orbit. Such a satellite is subject to a 24-hour cycle of varying insolation - all surfaces receive no insolation for 12 hours on the dark side of the Earth: the surfaces receive rates which vary, according to their orientation, between zero and the solar constant while the satellite is on the bright side of the Earth. This insolation was simulated in the model by thin layers at each surface, which were assigned appropriate time-dependant internal heat generation rates.

As the code uses a finite difference method, it is not suitable for analysing temperature distributions in complex solid shapes. For such problems a finite element code is appropriate. The advantages of this code lie in its ability to analyse situations where boundary conditions are complex and variable, and may not be known at the start of a problem. Single- and two-phase heat transfer coefficients can be evaluated at each iteration during the solution of a problem.

5. CONCLUSIONS

The extension of a well tried mainframe heat transfer code to include the rapidly changing boundary conditions, which occur with transition from single-phase liquid to sub-cooled boiling water in the boundary coolant when the water reaches saturation temperature, has been carried out successfully. The modified code (2DTBOIL) has been written for use with IBM or compatible PC's.

The 2DTBOIL code provides a versatile and well-proven tool for analysing a wide variety of thermal problems in cylindrical or rectangular geometry. It is easily loaded on and operated from a standard IBM or compatible PC.

Transient and steady-state problems can be accepted. Complex boundary conditions, including two-phase coolants and time varying radiation can be handled. Variation of heat generation rates in time and position for each material used in model can be defined easily by the User at input.

The code has been validated against a number of test cases for which analytic solutions are available. It has also been proved against experimental results. Extension to three dimensions is straightforward, but use would be limited to a comparatively small number of mesh-lines in each direction with a standard PC.

REFERENCES

(1) DOUGLAS, J. - 'Alternating-direction methods 3-space variables' Numer. Math 4, 41-63 (1962)

(2) CLARK, S.S. & TROOST, M. _ 'RAT-3D, A general three-dimensional heat transfer code' - GAMD-7346 (1966)

(3) KENNEDY, T.D.A. - '2DT & 3DT - two- and three-dimensional heat-transfer computer programs' - AERE-R9210, Harwell Lab., Oxon, England (1978)

(4) KAYS, W.M. -'Convective heat and mass transfer' pp 109-117, McGraw-Hill Book Co. New York, U.S.A. (1966).

(5) ROHSENOW, W.M. - 'A method of correlating heat transfer data for surface of boiling liquids - Trans. Am. Soc. Mech. Engrs, Vol. 74, p.969 (1952)

(6) NIEMANN, H. -'Die Warmeubertragung durch naturliche konvektion in spaltformigen hohlraumen' -Gesundheit Ingenieur, Vol. 69, pp 224-228 (1948)

(7) FORSTER, H.K. & ZUBER, N. - 'Dynamics of vapour bubble and boiling heat transfer' - Am. Inst. Chem. Eng. Journal 1(4), pp 531-535 (1955).

(8) IVEY, H.J. & MORRIS, D.J. -'On the relevance of the vapour/liquid exchange mechanism for sub-cooled boiling heat transfer at high pressure' -AEEW-R137, A.E.E.W. Winfrith, England (1962)

(9) TOULOUKIAN, Y.S. -'Thermophysical properties of high temperature solid materials.' - Purdue University, U.S.A.

(10) KENNEDY, T.D.A., McALLISTER, S., MARKGRAF, J. & RUYTER, I.A. -'Development of 2-D Computer Code for the Prediction of 2-Phase Heat Transfer in an Experimental LWR Irradiation Capsule' - Nuclear Energy - Volume 30 No.3, (1991).

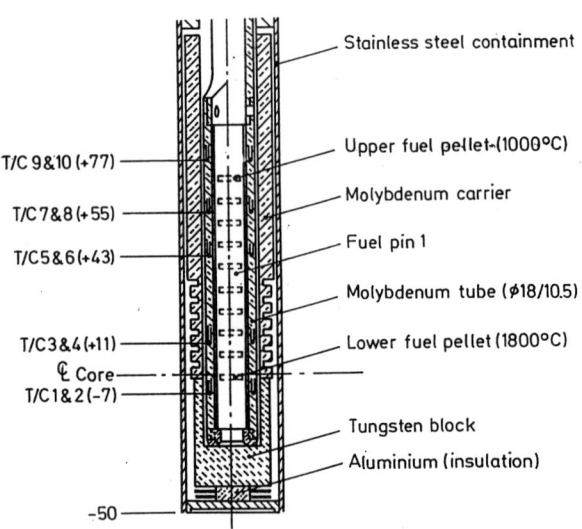

Fig 1: Sketch of experimental carrier (3.1)

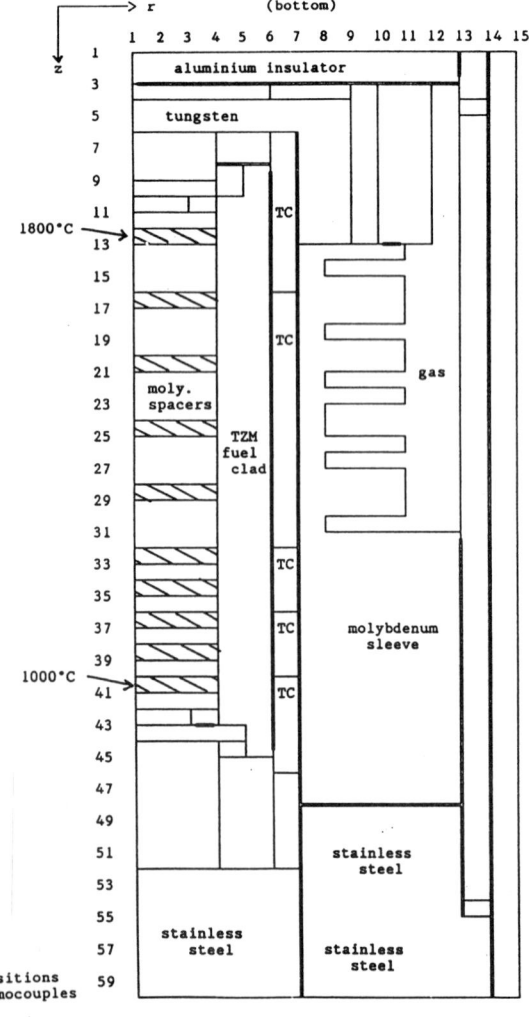

Fig 2: 2DT model of experiment (3.1)

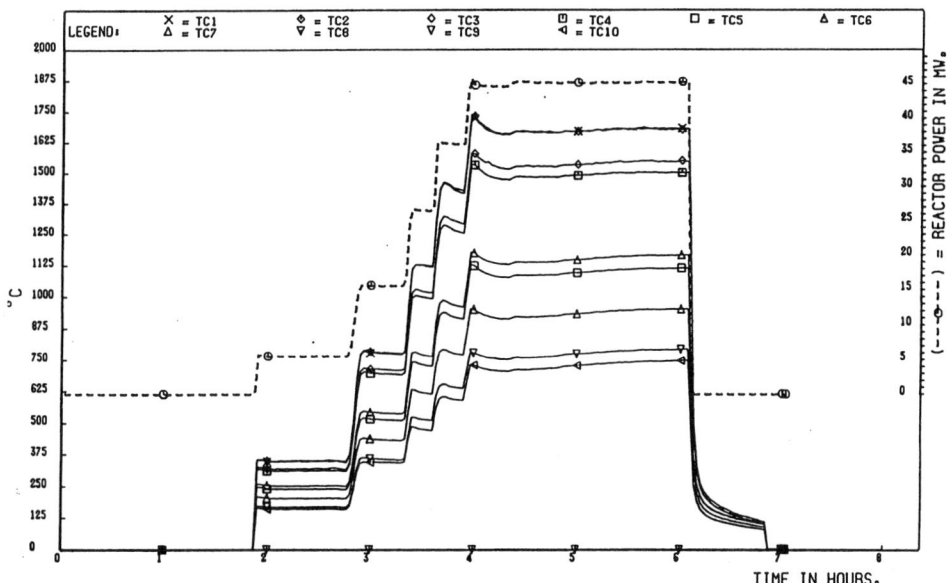

Fig 3: Temperature history of experiment (3.1)

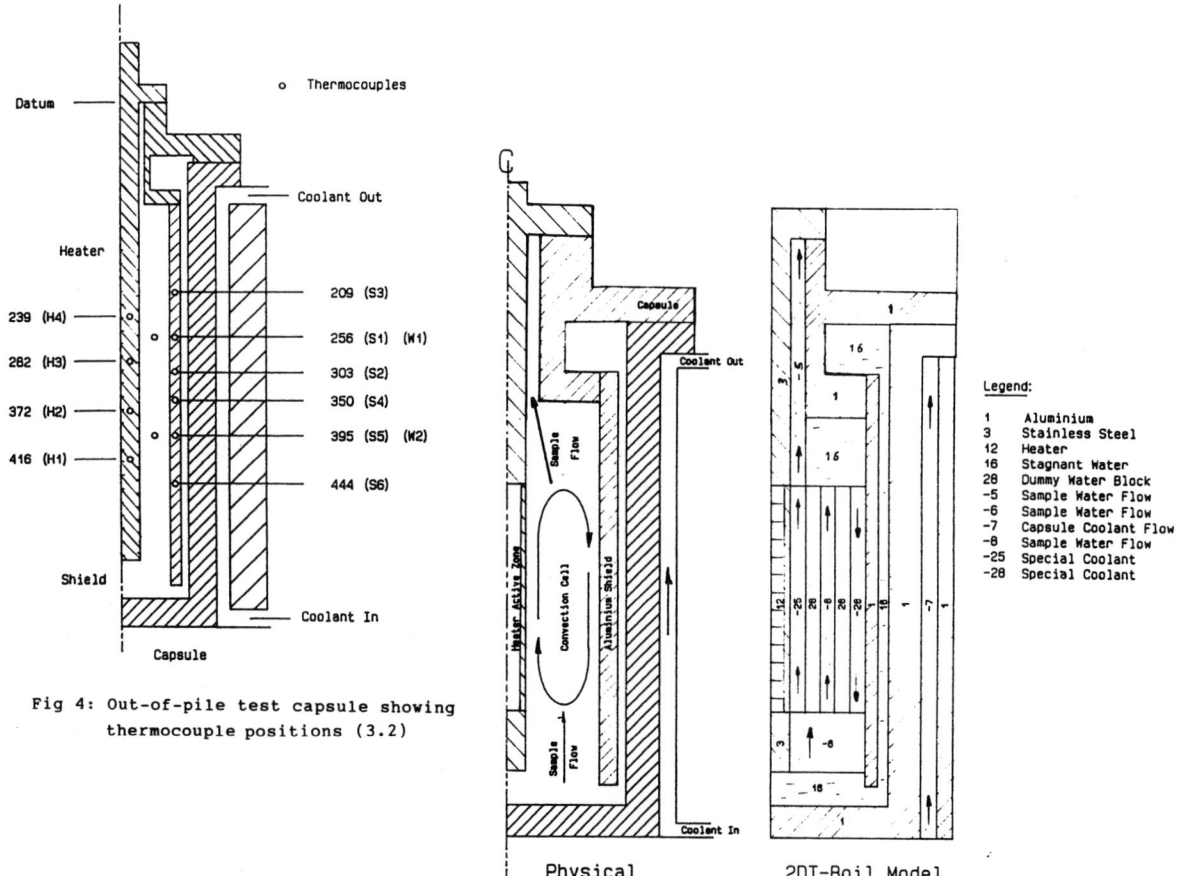

Fig 4: Out-of-pile test capsule showing thermocouple positions (3.2)

Fig 5: 2DT-Boil representation of out-of-pile capsule (3.2)

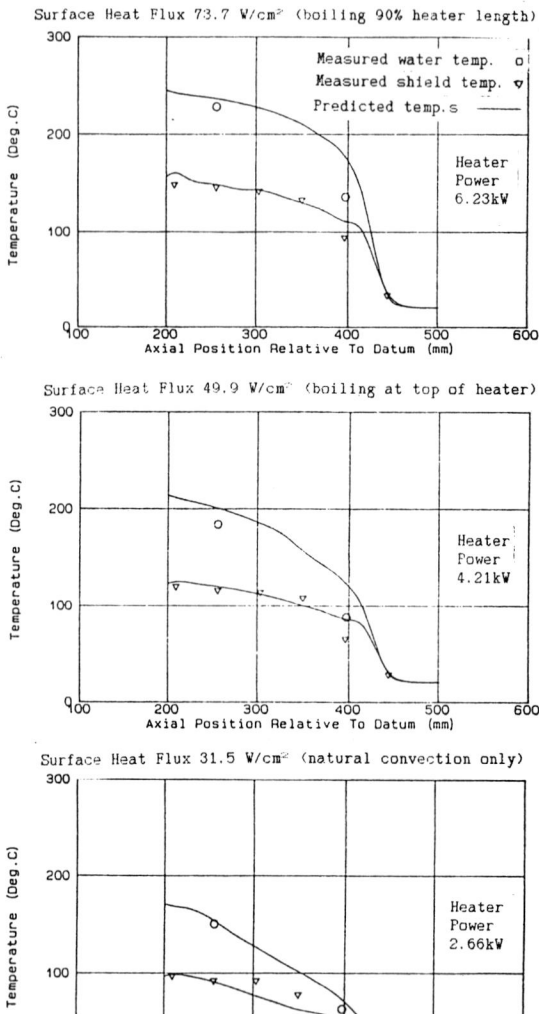

Fig 6: Shield and convection cell bulk water temperatures as a function of axial position and heater power (3.2)

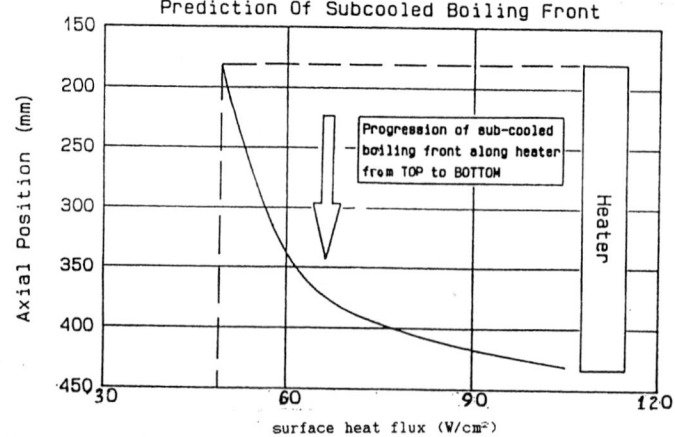

Fig 7: 2DT-Boil prediction of the progression of the sub-cooled boiling front as a function of surface heat flux (3.2)

C413/014

Numerical investigation of modified K-Σ turbulence models for the simulation of a backward facing step flow

T KOBAYASHI, PhD, MJSME and N TANIGUCHI, PhD, MJSME
Institute of Industrial Science, University of Tokyo, Japan
H-K MYONG, PhD, MJSME
Korea Institute of Science and Technology, Korea
M-A OHMACHI,
Japanese Patent Office, Tokyo, Japan

A k-ε turbulence model is expanded for the anisotropic expression of Reynolds stresses. Compared with the results of a isotropic standard model and the experiment data, the new model is verified in the separated flow past a backward facing step. The anisotropic model can more accurately predict the normal Reynolds stresses and the pressure recovery. Influences of the model constant values and the scheme for the convective terms are also estimated.

1. INTRODUCTION

The widely used k-ε model predicts well the turbulent shear flow which is mainly dominated by one component of turbulent shear stresses. However, for the types of turbulent flow where separation or recirculation exists, the k-ε model gives rise to considerable inaccuracy. Before now, it was tried to improve the accuracy by consideration of streamline curvature effect. Recently, however, it has become known that the isotropic turbulent eddy viscosity relationship itself on which the k-ε model is based yields inaccuracies in description of such types of flows. Therefore, anisotropic model of Reynolds stress is proposed to predict the flow problems with separation and recirculation [1,4].

In this study, the turbulent flow past a backward facing step is taken as a benchmark for the estimation of turbulent models in the prediction of separation and recirculation flows. Two-dimensional calculation results of the anisotropic k-ε model are compared, respectively, with those of isotropic k-ε model and with the experimental results (Kim et.al.[5]). Moreover, the influence of model constants and schemes for the convection terms in the momentum equation are also investigated.

2. CLOSURE EQUATIONS

The governing equations to be solved are the continuity equation and Reynolds equation:

(1) $\quad \dfrac{\partial U_i}{\partial x_i} = 0$

(2) $\quad U_j \dfrac{\partial U_i}{\partial x_j} = -\dfrac{1}{\rho}\dfrac{\partial P}{\partial x_i} + \dfrac{\partial}{\partial x_j}\left(\nu \dfrac{\partial U_i}{\partial x_j} - \overline{u_i u_j}\right)$

where U_i is the mean velocity field and u_i is the fluctuation part of the velocity in the i-direction.

In the anisotropic turbulence model, the component is the Reynolds stress tensor $\overline{u_i u_j}$ in the Eq.2 are given by:

(3) $\quad \overline{u_i u_j} = \dfrac{2}{3}k\delta_{ij} - \nu_t\left(\dfrac{\partial U_i}{\partial x_j} + \dfrac{\partial U_j}{\partial x_i}\right) + \dfrac{k}{\varepsilon}\nu_t \sum_{m=1}^{3} C_m\left(S_{mij} - \dfrac{1}{3}S_{m\alpha\alpha}\delta_{ij}\right)$

The third term in the Eq.3 is a nonlinear term superimposed in the isotropic model. Here C_m (m=1,2,3) are model constants, δ_{ij} is Kronecker's delta function, and,

$k = \dfrac{1}{2}\overline{u_i u_i}$

$\varepsilon = \nu \overline{\dfrac{\partial u_i}{\partial x_j}\dfrac{\partial u_i}{\partial x_j}}$

$S_{1ij} = \dfrac{\partial U_i}{\partial x_m}\dfrac{\partial U_j}{\partial x_m}$

$S_{2ij} = \dfrac{1}{2}\left(\dfrac{\partial U_m}{\partial x_i}\dfrac{\partial U_m}{\partial x_j} + \dfrac{\partial U_m}{\partial x_j}\dfrac{\partial U_i}{\partial x_m}\right)$

$S_{3ij} = \dfrac{\partial U_m}{\partial x_i}\dfrac{\partial U_m}{\partial x_j}$

The eddy viscosity is expressed by

$\nu_t = C_\mu \dfrac{k^2}{\varepsilon}$

The turbulent kinetic energy k and its dissipation rate ε are obtained from their transport equations as follows,

(5) $\quad U_j \dfrac{\partial k}{\partial x_j} = \dfrac{\partial}{\partial x_j}\left\{\left(\nu + \dfrac{\nu_t}{\sigma_k}\right)\dfrac{\partial k}{\partial x_j}\right\} - \overline{u_i u_j}\dfrac{\partial U_i}{\partial x_j} - \varepsilon$

(6) $\quad U_j \dfrac{\partial \varepsilon}{\partial x_j} = \dfrac{\partial}{\partial x_j}\left\{\left(\nu + \dfrac{\nu_t}{\sigma_\varepsilon}\right)\dfrac{\partial \varepsilon}{\partial x_j}\right\} - C_{\varepsilon 1}\dfrac{\varepsilon}{k}\overline{u_i u_j}\dfrac{\partial U_i}{\partial x_j} - C_{\varepsilon 2}\dfrac{\varepsilon^2}{k}$

In this study, the condition of the calculation corresponding to five cases are shown in Table 1. Two sets of model constants are used in these calculations; one is proposed ly Launder and Spalding [7] (LS) which would be optimized mainly in the boundary layer flows; the other by Myong and Kasagi [3,6] (MK) which was adopted to the duct flows, the streamwize and vertical directions are indicated by X and Y, the origin of which is at the bottom corner of step, and X_r is the reattachment point. The height of step, h and the inlet centerline mean velocity, U_0 are taken to be characteristics length and velocity respectively, and the Reynolds number is 46,000.

3. CALCULATION RESULTS

The streamline diagrams from the calculations of the 5 cases are shown in Fig.1. In the case 1 which is the situation of the standard isotropic k-ε model, the separation length, Xr/h is predicted as 5.8, which is 20% smaller than the experimental value of Xr/h=7.0. In the cases 2,3,4 and 5, the separation length are respectively Xr/h=6.0,6.2,6.3 and 6.2.

According to this results, it is shown that for the different sets of model constants in the isotropic k-ε models, MK's constants is superior to LS's. In the same time, between the isotropic and anisotropic models, better prediction is obtained by the use of the anisotropic models. In the cases 4 and 5 of the anisotropic model with MK's constants, the predicted value is closest to the experimental value. Concerning the improvement of the separation length prediction, the influence of changing the model constants in the isotropic k-ε model seems a little grater then that of the model improvement by the anisotropic terms. Furthermore, with the comparison between case 4 and 5, the difference of calculation schemes for the convection term makes little effect on the streamline pattern and the separation length.

Next, the calculation results of mean velocity field U/U_0 are shown in Fig.2. In order to observe the differences of velocity distribution generated from the isotropic and anisotropic models at several locations in the flow field, a comparison is presented for the case 2 and case 4 in which the same set of model constants (MK) is used. Only in the vicinity of the reattachment point X/h=6.22, the results of all cases are demonstrated for comparison.

As general tendency, the difference among each case is small, and almost same velocity distribution is shown except for the vicinity of reattachment point. As the same as the observation of streamline, there exists little influence from the calculation schemes of the convection terms.

With observation of the velocity profiles at X/h=2.67 and 6.22, anisotropic model produces more reduction of velocity in the strong shear layer around the recirculation zone. In the near wall region, the predictions of both models have not so good agreement with the experimental data. At X/h=6.22, near the reattachment point, the reversed flow is not appeared because the separation length predicted in the case of the standard isotropic k-ε model is shorter than this value. A relatively correct prediction of the separation length is presented in the anisotropic cases 4 and 5, where a reversed flow region can be recognized.

At the locations behind the reattachment point X/h=10.66 and 13.33, the calculation results are not always consistent with the experimental data, and similar tendency can be found in the calculations of other researchers. It is usually explained that the experimental results include three dimensional effect whereas the calculation is a two-dimensional one. At these location, the difference between the both cases becomes smaller as the flow goes downstream; at the X/h=13.33, it is almost indistinguishable. This is consistent with the fact that the anisotropic term in the Eq.3 dose not effect the velocity distribution in the fully developed turbulent channel flow.

Next, the distribution of turbulent

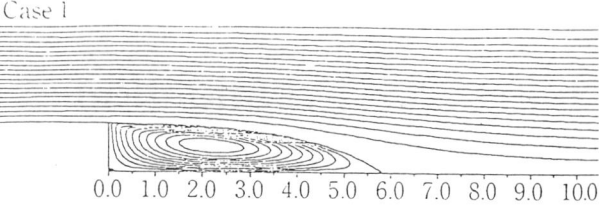
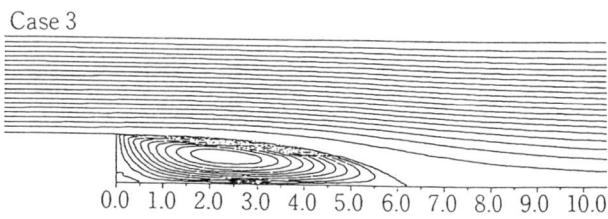
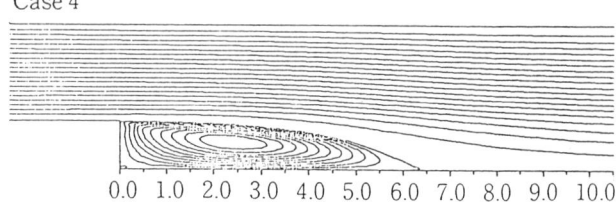
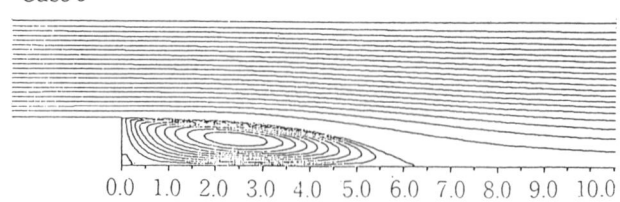

Fig.1 Stream line

quantities will be investigated. Unfortunately, because the experimental data is lack in the recirculation region, they are plotted only behind the reattachment point to the downstream in the figures.

Figure 3 shows turbulence intensities $\overline{uu}/U0$. In the anisotropic case, due to the normal Reynolds stress difference generated from the nonlinear term in Eq.3; with respect to the streamwize turbulence intensity, at any locations, the peak value of the anisotropic cases are higher and more upward shifted towards the up-wall than that of isotropic ones. Compared with the experimental data, there is a tendency of an overestimation of turbulence intensity in the anisotropic case; otherwise, underestimation of that in the isotropic case. Concerning the influence of the model constants, the predicted turbulence intensities by LS's constants is relatively higher than that by MK's. A similar tendency can be observed in the turbulent shear stress $\overline{uv}/U_0^2$ and the turbulent kinetic energy k/U_0^2.

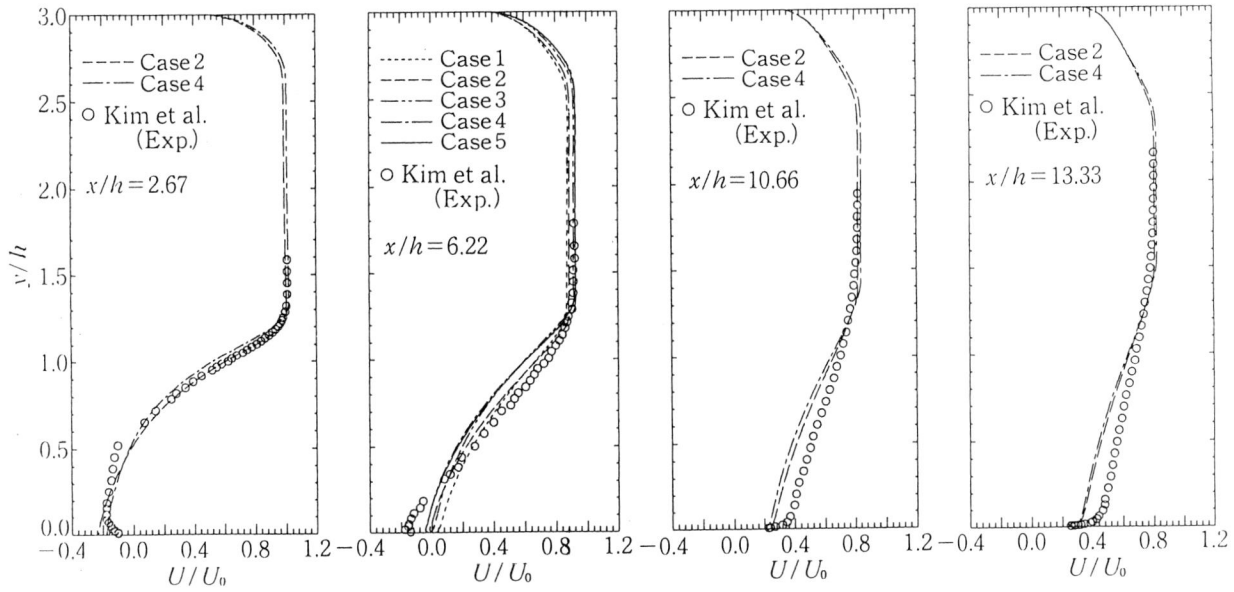

Fig.2 Mean velocity distribution

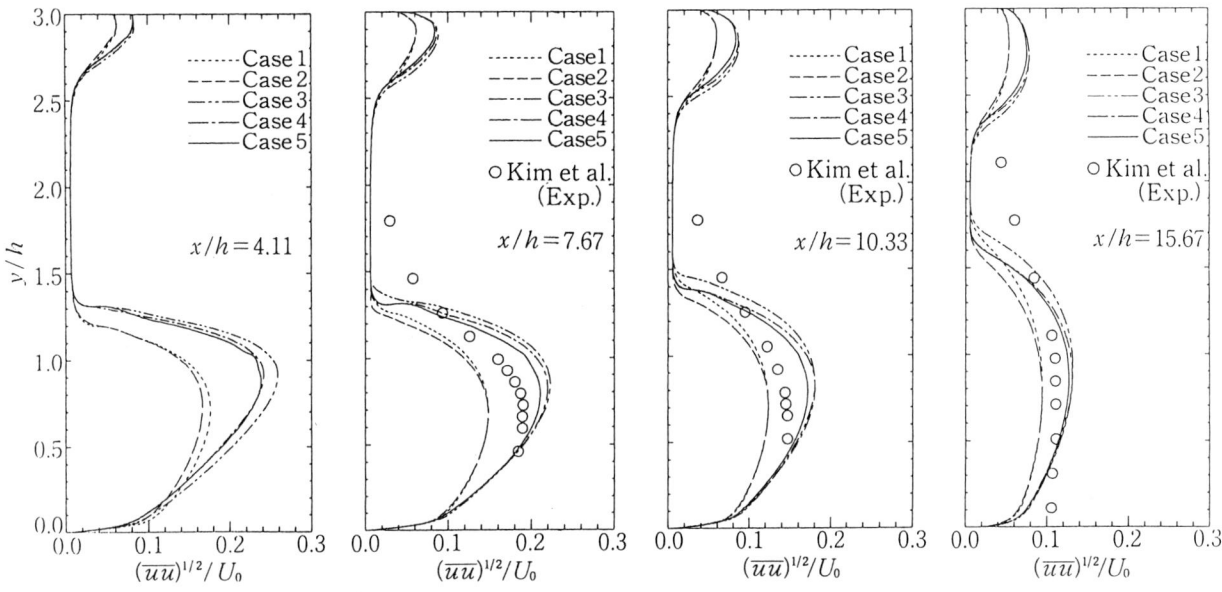

Fig.3 The distribution of the turbulence intensity in the main stream direction

Next, $\overline{uv}/U_0^2$ distributions are shown in Fig.4. Here we can see that the difference among the models is small. Therefore, with respect to the Reynolds shear stress, the prediction can be obtained by the isotropic model as reliably as by the anisotropic model. This also implies that the anisotropic model hold the characteristics of the isotropic ones when the problem is reduce to a simple turbulent boundary shear flow without the influence of normal Reynolds stresses.

In the prediction of Reynolds stresses, the calculation schemes for the convection term has some influences, which are clearly appear in the vicinity of reattachment point X/h=7.67. Generally, the case 5 using skew upwind scheme predicts the smaller values, which is closer to the experimental data than the case 4. Concerning the turbulent shear stress, the result of using skew upwind scheme is closer to the experimental value, but it is not so effective to improve the reattachment length.

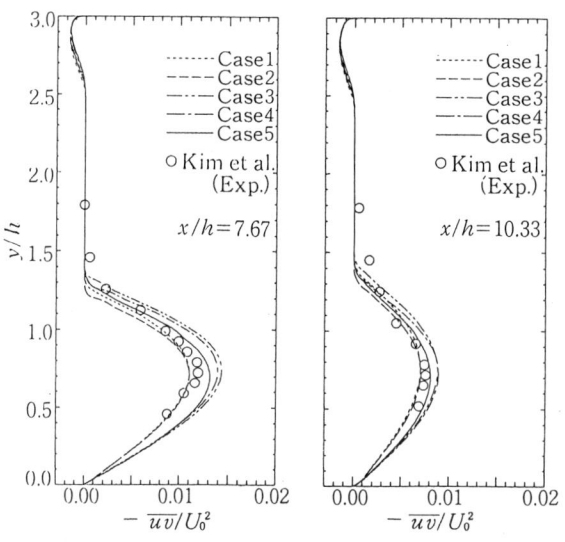

Fig.4 The distribution of the turbulent shear stress

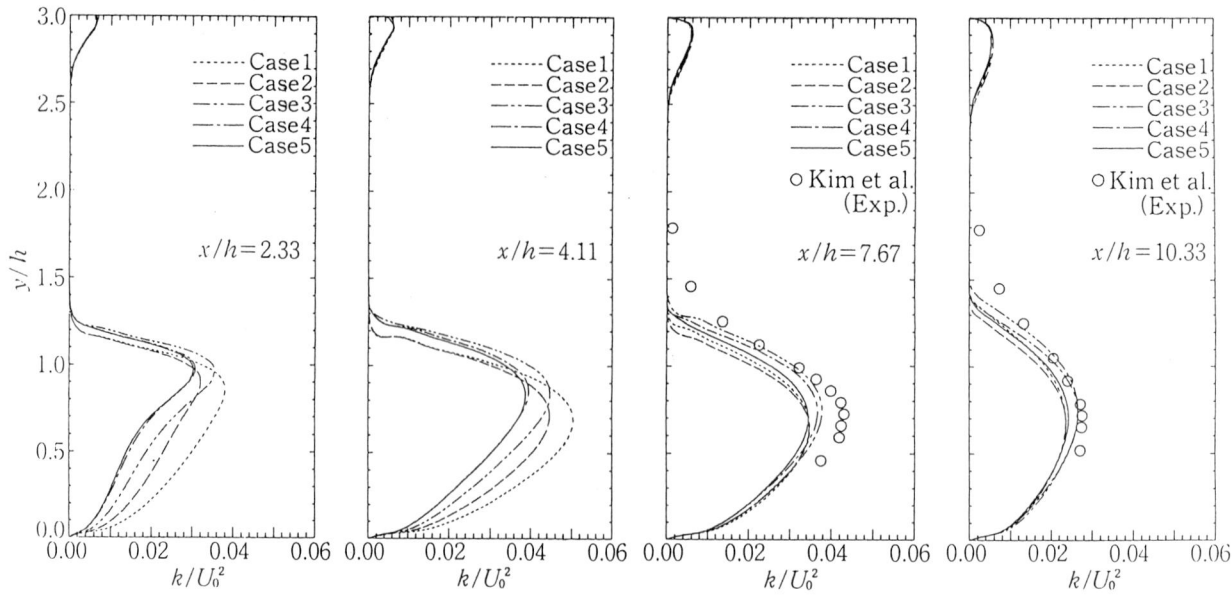

Fig.5 The distribution of the turbulence energy

Lastly, the distribution of kinetic turbulent energy k/U_0^2 is shown in Fig.5. As seen, at the point, the highest peak value of k/U_0^2 is obtained in the case 1 where the separation length of prediction is shortest, and the lowest peak value is obtained in cases 4 and 5 where the separation length is predicted longest. Although it is difficult to make strict investigation due to the lack of reliable experimental data in the recirculation zone, it can still be conjectured from the above observation that the substantial improvement in the prediction of the separation length is obtained due to the restraint of the overestimation of turbulent kinetic energy in the recirculation zone.

Figure 6 indicates the pressure distribution along the streamwize direction on the step wall side and opposite side. In the case 2 by the isotropic model, the pressure recovery zone is shifted upstream; in the contrary, the remarkable improvement can be obtained by the anisotropic mode, case 4. It is noted that the difference is not only the reflection of the movement of the reattachment point, and that the correction of the normal stress can produce the improvement of pressure distribution even when the contribution of the modification of the shear stress modeling is not considered.

4. CONCLUSION

According to the calculation of the backward facing step flow, it is shown that the anisotropic model has some improvements of flow predictions where the anisotropy of Reynolds stress plays an important role.

The separation length predicted by the standard isotropic k-ε model is 20% shorter than the experimental value. It can be improved by the MK's model constants or anisotropic models, and in the case when they are combined, the separation length is predicted close to the

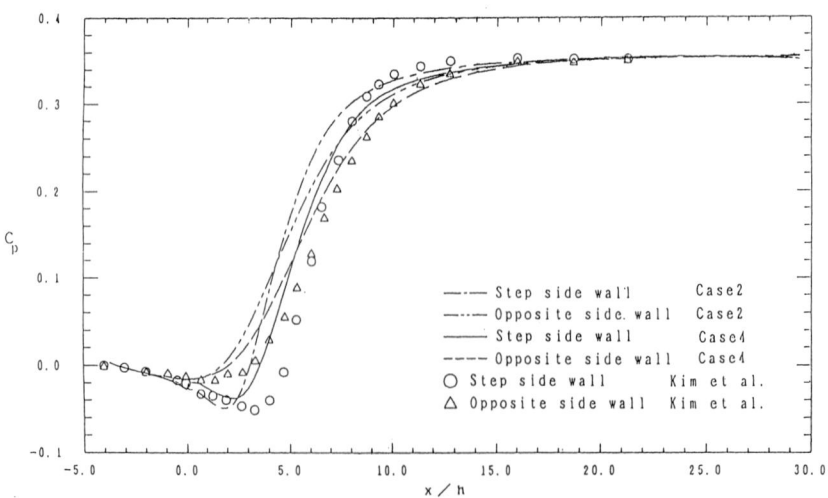

Fig.6 The pressure distribution on the step side wall and the opposite side

experimental value.

It is estimated that the substantial improvement can be obtained by the restraint of overestimation for the turbulent kinetic energy. The anisotropic model is seem to have this tendency.

Although the scheme of convection term in the momentum equation has little effect on the mean velocity distribution, for the turbulence intensity and turbulent shear stress, relatively better prediction is obtained by use of the skew upwind scheme than by use of hybrid scheme, especially in the vicinity of the reattachment point.

In this study, the total influence of calculation factors are mainly researched in the limitation of reference data. Recently, the details are investigated with the experiments[10] and the calculations[11] by the large eddy simulation, and there become appear the data even in the recirculation zone. It will be possible and necessary after this that the calculation results should be examined by more detail informations.

REFERENCE

(1) Nishizima,S. and Yoshizawa,A., AIAA J., 25, p414, 1987.

(2) Speziale,C.G.,J.Fluid Mech., 178, p459,1987.

(3) Myong,H., Dr.Eng.Thesis,Univ. of Tokyo,1988.

(4) Speziale,C.G. and Ngo,T., Int.J.Eng.Sci., 26, p1099, 1988.

(5) Kim, J., Kline, S.J. and Johnston, J.P., Technical Report MD-37, Thermosciences Div., Dept.of Mechanical Eng.,Stanford Univ.,1978.

(6) Myong,H.K. and Kasagi,N.,JSME Trans. 54-507, B, p3003, 1988.

(7) Launder,B.E. and Spalding,D.B, Comp.Math. Appl.Mech.Eng., 3, p269, 1974.

(8) Celenligil,M.C. and Mellor,G.L., J.Fluids Eng.,107, p467, 1985.

(9) Avva,R.K., Kline,S.J. and Ferziger,J.H., AIAA paper 88-0611, 1988.

(10) Itoh,N. and Kasagi,N., J.of Flow Visualization Soc.Jap.,34, p245,1989.

(11) Morinishi,Y. and Kobayashi,T., Engineering Turbulence Modeling and Experiments, (Edited by Rodi and Gaic),Elsevir,1989.

C413/077

A proposal for improving the modelling of turbulence

R K DUGGINS, BSc, PhD, CEng, MIMechE, FIEAust
Department of Mechanical Engineering, Australian Defence Force
Academy, Canberra, Australia

SYNOPSIS A modification is proposed to the k-ε turbulence model for the modelling of turbulent swirling flows, the main feature being the use of two mutually perpendicular components of the turbulent viscosity to take account of swirl-induced changes in the turbulence. When tested against experimental results for swirling flow in a pipe it has been shown to perform well. It has also been used to calculate the swirling flow through an abrupt contraction and the results have provided an improved insight into the nature of this type of flow. On the other hand, the dearth of experimental data available for this configuration has limited the extent of the comparisons.

NOTATION

k	kinetic energy of the turbulence
ε	dissipation rate
μ	turbulent (or eddy) viscosity
r, z, θ	radial, axial and angular coordinates
μ_{rz}	component of μ in the rz plane
$\mu_{r\theta}$ (=μ_{tan})	component of μ in the $r\theta$ plane
ρ	density
$-\rho u'v'$	Reynolds shear stress
u, v, w	components of velocity
$\partial u/\partial y$	transverse gradient of x-direction mean velocity
x/R	distance from pipe entrance, non dimensional with respect to radius.
s	degree of swirl
c_w	empirical constant in the equation defining μ_{tan}

1 INTRODUCTION

The most widely used model for describing the turbulence in a fluid is the k-ε model which defines the spacial distribution of turbulent viscosity μ appropriate to a particular case in the course of representing realistically both the creation and the subsequent transport and dissipation of the turbulence through the flow. It has generally worked well for many simple flows but there appears to be an in- creasing belief that it is unsatisfactory when one has the task of describing the turbulence in more complicated flows, for example ones with considerable swirl like those being considered in the present investigation.

In swirling flows the swirl causes an increase in turbulence levels and, since there is a decrease in nett energy transfer from large eddies to small ones, the rate of dissipation is lower too. Hence, particularly along the swirl axis, the length scales are magnified.

For a turbulence model to be effective in swirling flows, these important features clearly need to be taken appropriately into account and a question which arises is whether the normally-used k-ε model has sufficient flexibility for this to be done. Can it be achieved, for example, by assigning new values to the empirical constants within the working form of the turbulent energy equation? Evidence will be presented shortly which we contend indicates that the answer is 'No', the differences between k-ε computed data and the corresponding empirical data being so great.

This therefore leads to another question - should a completely new turbulence model be sought or will a modification to the basic k-ε model be more appropriate? As indicated earlier, we have chosen the latter in the belief that there is much to be gained both from preserving the link with the treatment of non-swirling flows and from retaining the valued attributes of the basic model, notably (i) the realistic way in which it handles the production, convection, diffusion and dissipation of the turbulence kinetic energy, (ii) its well-proven status, enhanced by the many improvements that have been achieved over the years from its application to a wide range of problems, and (iii) the straight forward nature of the numerical procedures whether the equations be parabolic or elliptic.

In the next section of the paper we review experimental data obtained by other investigators which we contend indicate that in strongly swirling flows it is no longer adequate to regard μ as a single-valued parameter at a particular point in the flow-field; different values require to be inserted in the equations of motion for the different mutually perpendicular directions. Unfortunately the k-ε model in its basic form does not permit this to be done.

Later in the paper we propose a modification to the basic model to incorporate this increased flexibility and suggest a way of quantifying the different components of μ to be used. The revised model is then put to the test by applying it to two flows involving swirl, that in a pipe and that in an abrupt contraction.

2 REVIEW OF SOME PREVIOUS STUDIES OF SWIRLING TURBULENT FLOWS

For the sake of brevity, only three investigations are reviewed here; the reader is referred to the monograph on Swirl Flows by Gupta et al (1) for a more comprehensive consideration.

The first set of data relates to the radial distributions of axial and tangential velocity obtained by Weber et al (2) for a typical transverse plane in an expanded flow. The investigators themselves also computed the corresponding theoretical profiles using the k-ε model, and both the theoretical and the experimental curves are shown in Fig 1. The disagreement is seen to be very considerable for both the axial and the tangential velocities, so much so that, if one were to attempt to retrieve the situation merely by replacing the empirical constants in the working form of the energy equation, the prospects for success would be poor.

The second set of data relates to the fully-developed region of a jet for which the investigators, Lilley and Chigier (3), obtained measurements of both mean velocities and turbulent fluctuations, again for a typical transverse plane. In turn the two main components of μ were calculated from these velocity measurements using expressions like

$$\mu = \overline{-\rho u'v'} \Big/ \frac{\partial \bar{u}}{\partial y}$$

enabling radial distributions of μ_{rz} and $\mu_{r\theta}$ to be plotted, as shown in Fig 2.

It is clear that the two components are very different from each other notably on the centre-line of the jet where μ_{rz} attains its maximum value but where $\mu_{r\theta}$ falls to zero. Similarly, when the dimensionless radius is about 0.15, it is now $\mu_{r\theta}$ which is at its peak while μ_{rz} is falling rapidly with increasing radius. Clearly, to have to choose a single value for μ at either of these locations, as is required in the k-ε treatment, is invidious and it is not surprising that the model often yields results like Weber's which are in poor agreement with experiment. The two components of μ are very different from each other but, as intimated earlier, the basic k-ε model does not have the capacity to take the difference into account.

The final study to be reviewed here is by Baker (4) of swirling flow in a circular pipe. Because of the wide usage of pipes, there is a wealth of experimental data available for the pipe configuration, Baker's being typical. In addition, the simplicity of the shape means that theoretical analysis can be readily undertaken, for example by using the k-ε model.

Fig 3 shows radial profiles of dimensionless tangential velocity for two transverse planes along the pipe, the left-hand side of the figure being at a dimensionless distance of $x/R = 30$ from the entrance and the right-hand side at $x/R = 70$. In this instance the k-ε computation was carried out by us and our calculated results are plotted together with Baker's corresponding experimental one. [The further sets of data shown in the figure, labelled 'k-ε modified for anisotropy', will be considered shortly].

It is clear from Fig 3 that the basic k-ε model again yields results which are in poor agreement with experiment. The calculated velocities are everywhere much smaller than the corresponding measured ones, and whereas the computed velocities are at a maximum near the wall Baker found the location of this maximum to be distant from it.

Earlier we posed the rhetorical question 'Can the basic k-ε model be modified appropriately (for swirling flows like the ones we have just described) merely by assigning new values to the empirical constants within the working form of the turbulent energy equation?' We believe that the above comparisons are further evidence that the answer is 'No'.

3 PROPOSED IMPROVEMENT TO THE k-ε MODEL

The idea of using two different viscosity components is not ours but was advanced by Lilley (5) several years ago. It appears to have attracted very little attention however in the intervening period and indeed our search of the literature suggests that the only significant attempt to employ the concept was by Boysan et al (6). The basis of their calculations was the assumption that

$\mu_{r\theta}$ and μ_{rz} were in a constant ratio

to each other but it transpired that their calculated results were not encouraging. The authors did not offer an explanation for this outcome but, with the benefit of hindsight, one could speculate that the assumed constancy of the viscosity ratio continued to deprive the k-ε model of adequate sophistication.

In our case, constancy has not been assumed. Instead we have calculated $\mu_{r\theta}$

and μ_{rz} separately and will now describe how it has been done.

The radial profile of μ which was shown in Fig 3 was a by-product of our k-ε computation for pipe-flow and is clearly not unlike the approximately parabolic radial profile for μ_{rz} obtained by Lilley and Chigier and shown in Fig 2. For this and other reasons, we decided that there was no need to change the k-ε treatment of this particular component of viscosity or to depart from employing it in the equations for axial and radial (but not tangential) momentum.

A different approach was needed, however, in relation to $\mu_{r\theta}$ (corresponding to

μ_{tan} in the equation for tangential momentum) since the radial profile of $\mu_{r\theta}$ also shown in Fig 2, bears no such similarity to our computed distribution of μ given in Fig 3. Inspiration for the way ahead was provided by the work of Rochino and Laven (7) who had studied swirling flows in ducts and had found in their case that μ_{tan} could be represented by

$$\mu_{tan} = \rho C_w^2 r^2 \left| \frac{\partial w}{\partial r} - \frac{w}{r} \right|$$

in which C_w was an empirical constant having the value of 0.034.

Incidentally, this defining equation is reminiscent of the one used for mixing length and it might be thought that, if such a simple treatment is indeed acceptable for the tangential direction, it should be used for the axial and radial directions too. The fault in the logic of this argument may be found in the final paragraph of Section 2; it would be a case of 'throwing the baby away with the bath water'.

We have adopted the above expression for μ_{tan} for our own use except that the value we have used for C_w has been 0.028 rather than 0.034. This is because Kinney (8) in subsequent work on a variety of other swirling flows found the former value to be more generally applicable.

The computer program we have used is similar to the PHOENICS program marketed by CHAM Ltd and of which descriptions are available elsewhere. The main differences relate to modifications developed by Lilley (9) for the addition of swirl components to the equations of motion and to the turbulent energy production term in the k-ε model.

4 RESULTS AND CONCLUSIONS

An assessment of our proposed modification to the k-ε model has been made by using it in relation to the same flow problem described at the end of section 2. The modified model has been used to calculate the swirling flow in a circular pipe and the computed results have been compared with those described earlier, ie. with both the corresponding computed data obtained using the basic model, and Baker's experimental results. All of this information is shown in Fig 3, the results for the modified model being labelled 'k-ε modified for anistropy'.

Although the need for conciseness dictates that only tangential velocity results be given here in detail, the calculations clearly yield a very comprehensive set of data, for example radial distributions at all sections of the pipe of the three components of mean velocity and the two components of the turbulent viscosity. In passing and with an eye on the results in their entirety we make the observation that in our opinion they are realistic with respect to both the mean flow data and the superimposed turbulence.

Turning to the profiles shown in Fig 3, it is obvious that the new ones are of the right shape and for the most part the calculated tangential velocities have magnitudes which are consistent with experiment. In addition, consideration of the two transverse planes ($x/R = 30$ and 70) together indicates that the previous over-prediction of decay rate is no longer present. Only near the axis of the pipe, at planes well downstream, does the comparison with experiment continue to be a little disappointing. With regard to the two sets of computed data (one for the basic model and the other for the modified version), here the comparison reveals how the modifications (and to some extent the turbulence anisotropy itself) are manifest. They have clearly been effective in the present task of describing the turbulent swirling flow in a pipe.

We have also used the modified k-ε model to investigate the nature and intensity of the turbulence for swirling flow through an abrupt contraction. Due to the angular momentum being largely conserved as the fluid passes through such a configuration, the swirl intensifies and secondary currents often develop and superimpose themselves on the basic axial flow. Our results also predict, not surprisingly, that as the tangential velocities themselves become larger so does the anisotropy of the turbulence associated with them.

Our calculated results are given in Fig 4. On the left hand side are shown radial profiles of dimensionless tangential velocity and pressure, and the curves on the right hand side indicate how the head loss across the contraction is affected by radius ratio and swirl number.

The calculated head loss data turned out to be broadly in line with our expectations but provided a useful quantification of the relative effects of the contraction ratio and the initial levels of swirl. The calculated velocities and pressures, on the other hand, contained some surprises, notably the major degree of anisotropy evident in the pressure distribution in the core region but very little near the wall. The profiles of tangential velocity bear a resemblance to those described earlier for swirling flow through a pipe; taking the anisotropy appropriately into account has again had the desirable effect of shifting the velocity peak closer to the pipe's axis.

Summing up, we contend that our proposed modification to the k-ε turbulence model has been shown to yield considerable benefits for the swirling flows that have so far been considered, and we are optimistic that it will do so again when extended to other similar flows. It incurs the additional complication to the calculations which arises from the use of two viscosity components but there appears to be no way of circumventing the need to increase the sophistication of the basic k-ε model.

REFERENCES

(1) GUPTA, A.K., LILLEY, D. G. and SYRED, N. _Swirl flows_. 1984, Abacus Press, Kent.

(2) WEBER, R., BOYSAN, F., SWITHENBANK, J. and ROBERTS, P. A. Computation of near-field aerodynamics of swirling expanding flows. _Proc. 21st Int. Symp. on Combustion_, Munich, FDR, 1986.

(3) LILLEY, D. G. and CHIGIER, N. A. Non-isotropic turbulent stress distribution in swirling flows from mean value distributions. _Int. J. Heat and Mass Trans_, 1971, 14, 573-585.

(4) BAKER, D.W. Decay of swirling turbulent flow of incompressible fluids in long pipes. PhD thesis, 1967, University of Maryland, USA.

(5) LILLEY, D. G. Prediction of inert turbulent swirl flows. _AIAA J._, 1973, 11, 7, 955-960.

(6) BOYSAN, F., WEBER, R., SWITHENBANK, J. and LAWN, C.J. Modeling coal-fired cyclone combustors. _Combustion and flame_, 1986, 63, 73-86.

(7) ROCHINO, A. P. and LAVEN, Z. Analytical investigations of incompressible turbulent swirling flow in stationary ducts. _Trans ASME, J. App. Mech._, 1969, 36, 151-158.

(8) KINNEY, R. B. Universal velocity similarity in fully developed rotating flows. _Trans ASME, J. App. Mech._, 1967, 34, 437-442.

(9) LILLEY, D. G. Primitive pressure-velocity code for the computation of strongly swirling flows, _AIAA J._, 1976, 14, 6, 749-756

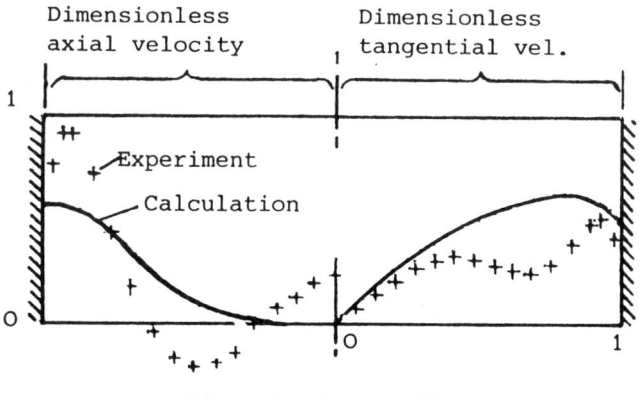

Fig 1 Radial distributions of dimensionless axial and tangential velocity for an expanded turbulent swirling flow. Major differences are apparent between the experimental results and those calculated using the basic $k-\varepsilon$ model

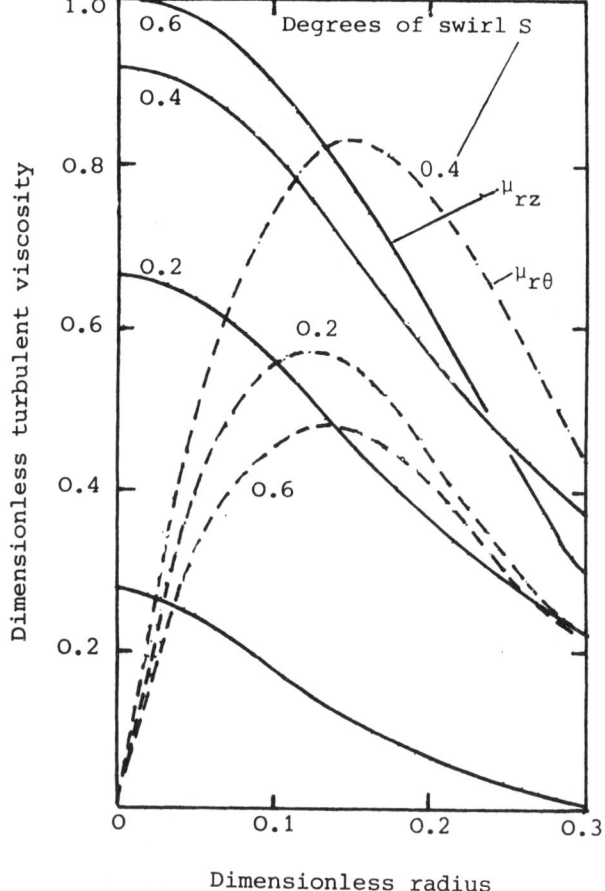

Fig 2 Radial distributions of components of dimensionless turbulent viscosity in swirling jets. The family of profiles for μ_{rz} are shown to have a completely different shape from those for $\mu_{r\theta}$.

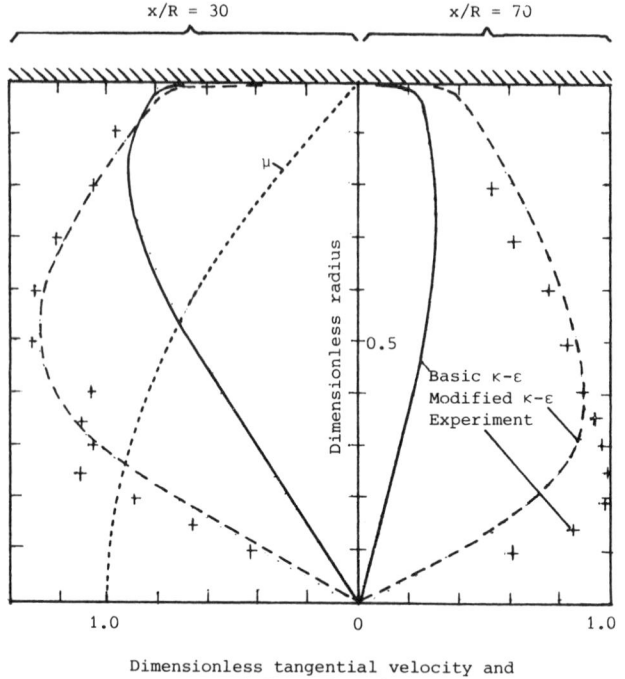

Fig 3 Radial distributions of dimensionless tangential velocity and turbulent viscosity μ for swirling flow in a pipe at two locations along it, x/R = 30 and 70. Major differences are again apparent between the experimental results and those calculated using the basic k-ε model but the differences are greatly reduced when the modified model is used instead

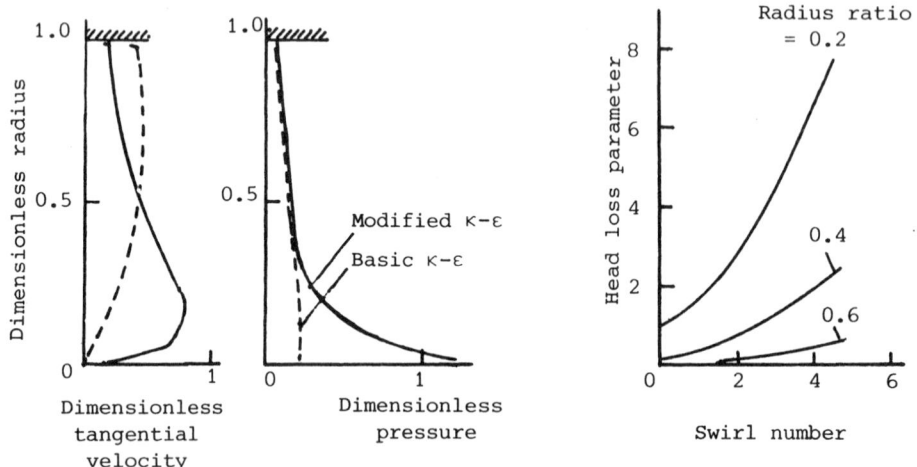

Fig 4 Data for an abrupt contraction calculated using the basic and modified k-ε models. The left-hand figures show the differences in velocity and pressure predicted by the two models and the right-hand one indicates the effects on head loss of radius ratio and swirl number

C413/022

Computational study on three-dimensional flow around a heavy-duty truck body

S NISHIKAWA, BE, MJSAE and Y WATANABE, BE, MJSAE
Nissan Diesel Motor Company Limited, Japan
K FUJITANI, ME, MJSAE, R HIMENO, PhD, MJSAE, MJSME and
M TAKAGI, PhD, MJSAE, MJSME
Nissan Motor Company Limited, Japan

SYNOPSIS The flow around a heavy-duty truck is simulated by third-order upwind-difference scheme the Navier-Stokes equations, approximated by finite-difference equations. To evaluate the effectiveness of the numerical simulation, a 1/6 scale truck model is tested in a wind tunnel on moving and fixed ground. The calculated flow structures around the vehicle agree well with the test results. The influence of the tyre rotation on the flow around the wheel arch is simulated well qualitatively. The simulated and experimental drag coefficients values also agree well quantitatively.

1 INTRODUCTION

With recent rapid advances in computational fluid dynamics (CFD) coming from the progress of supercomputers and computational methods, various three-dimensional analyses of the air flow around automobile bodies and through engine compartments have been carried out, especially for passenger cars. However, there have been few applications of CFD to heavy-duty truck bodies so far, because their bluff body shapes are too complex to map into computational grids (1). We had first analysed the air flow around the cab using a half-cab model with mudflap and tyre but without the rear body (2). This showed the applicability of CFD to the aerodynamics of a heavy-duty truck body. The main purpose of the investigation was to explain the mechanism of dirt build-up on vehicle surfaces. The calculated results were encouraging, and the CFD analysis gave useful information on the flow around the cab side and wheel arch, and also on the dirt build-up under wet conditions. This paper deals with the second phase, the extension of the model to include a van body. To evaluate the effectiveness of the computational analysis, a wind tunnel test is carried out on a 1/6 scale model truck on moving and fixed ground.

This paper compares the calculated and measured results and discusses the influence of tyre rotation.

2 BASIC EQUATIONS

The air flow around vehicle bodies is incompressible. The basic equations governing viscous incompressible flow are the well-known continuity and Navier-Stokes equations. The air flow is simulated by third-order upwind-difference scheme of these equations without turbulence models.

These equations can be described as follows with a coordinate system fixed on the vehicle moving with velocity in a stationary fluid :

$$\nabla \cdot V = 0 \qquad (1)$$

$$\frac{\partial V}{\partial t} + \{(V - V_0) \cdot \nabla\}V = -\nabla P + \frac{1}{Re}\Delta V \qquad (2)$$

where
- V : nondimensional velocity vector
- t : nondimensional time
- V_0 : nondimensional vehicle velocity vector
- P : nondimensional pressure (local static pressure)
- Re : Reynolds number

Equation (2) is approximated by finite-difference equations and solved by third-order upwind-difference scheme. For pressure calculation, Poisson's equations with respect to pressure of high Reynolds number flow are derived from the divergence of both sides of Eq.(2) :

$$\nabla P = -\nabla \cdot \{(V - V_0) \cdot \nabla\}V - \frac{\partial D}{\partial t} \qquad (3)$$

where D means the divergence of the velocity vector, i.e. $D = \text{div } V$. The velocity obtained by Eq.(2) does not normally satisfy Eq.(1), so the velocity field is corrected using the Marker and Cell method. Hence, the unsatisfied part in Eq.(1) will be built to Eq.(3) as a corrective term to prevent the accumulation of numerical errors. Eq.(3) is also approximated by finite-difference equations, and the second term on the right side is transformed as follows :

$$\nabla P = -\nabla \cdot \{(V - V_0) \cdot \nabla\}V - \frac{D^{n+1} - D^n}{\Delta t} \qquad (4)$$

If D^n becomes non-zero in a certain time interval due to a calculation error, then D^{n+1} will be replaced by zero at the next interval, learning a corrective term. The third term of Eq.(3) can be neglected due to the high Reynolds number, and the final Poisson's equation is reduced to :

$$\nabla P = -\nabla \cdot \{(V - V_0) \cdot \nabla\}V + \frac{D^n}{\Delta t} \qquad (5)$$

Velocity and pressure are obtained by solving Eq.(2) and (5) simultaneously. Drag and lift forces acting on the vehicle can be obtained by integrating the pressure distribution over the vehicle body surface. Equations (2) and (5)

(a) Calculation model

(b) 1/6 scale model

Fig.1 Calculation model and testing model

are solved as finite-difference equations of partial derivatives with respect to time and position. If the physical coordinate system were used, the approximation accuracy would worsen near solid body surfaces. Therefore, the equations are transformed into those in a generalized coordinate system and then approximated by finite-difference equations.

The second-order Euler implicit scheme is used for the time integration. All spatial derivatives, except that of the convective term, are approximated by the third-order upwind scheme. Poisson's equations are also approximated by second-order central differences. The resulting difference equations are solved by the point SOR (Successive Over-Relaxation) method.

(a) Body surface grid

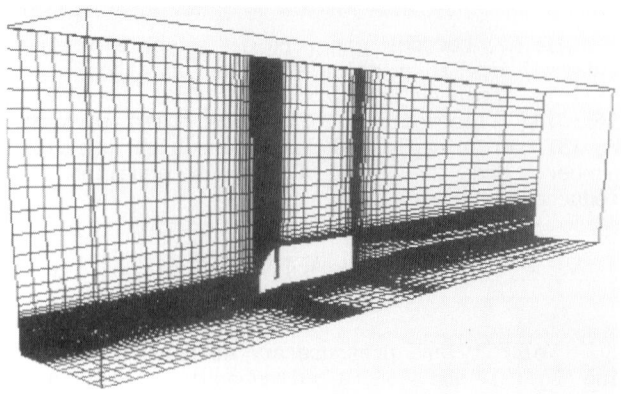

(b) Spatial grid

Fig.2 Grid system

3 MODEL AND GRID SYSTEM

Based on the model illustrated in Fig.1 left, a grid system is generated algebraically from the body shape CAD data.

The multi-block transformation technique is used for grid generation as it is suitable for the analysis of the flow around objects with complicated shapes. The three-dimensional grid over the body surface is generated by zoning the vehicle body surface into certain blocks for easier separation of the grid. The total grid system is obtained by combining the grids for each block. In the first phase, the body surface grid has been generated by dividing the cab body into six blocks : the cabin, wing, bumper, frame, tyre and mudflap (2). The vehicle body surface grid is generated by adding to these a block for the rear body. The transfinite interpolation method (3) is used for generating the inner grid of each block.

Figure 2 shows the vehicle surface and space grid system. The total number of grid nodes was 725,040 (144x53x95). Assuming that the effect of the rear wheels on the flow is small, they are ignored in the grid generation.

4 COMPUTER SYSTEM

Figure 3 shows the computer network used for the CFD analysis. It integrates the preprocessing for the grid generation and the postprocessing for displaying the calculation results. A mini-supercomputer Convex C240 is used for testing the grid systems and computing the particle traces for visualising the flow. Graphics work stations IRIS 4D series generate the grids and display the calculation results. Videotape recorders are connected to the IRIS through a controller which takes pictures frame by frame, to make animated films. The computation of the flow itself is carried out on either a supercomputer CRAY X-MP432EA (at Nissan) or a supercomputer HITAC S810/5 (at Nissan Diesel).

The computed results are displayed by the following process :

(1) Calculated series data of unsteady flow-velocity and pressure fields of all the grid points are recorded on magnetic tape.
(2) The data on the tapes are transferred to the C240.

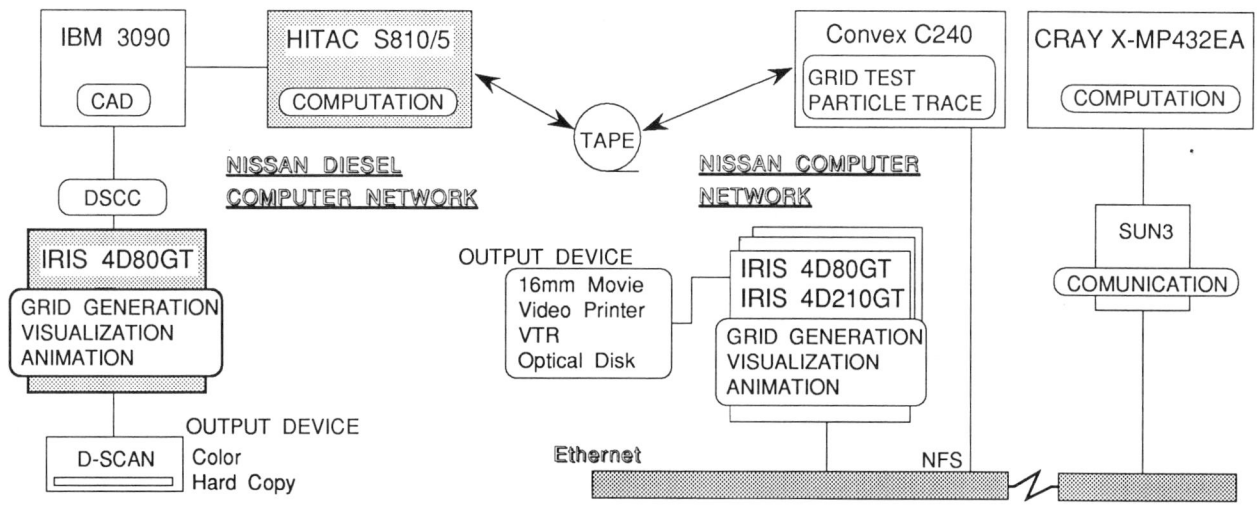

Fig.3 Computer network

(3) Start points of virtual tracers are set by IRIS.
(4) Paths of virtual tracers are calculated on the C240 from the velocity data at each grid point.
(5) Calculated results are displayed on the CRT of the IRIS.

5 WIND TUNNEL TEST

Figure 1 right shows the 1/6-scale model truck used in the wind tunnel test. It was created by a CAM system from the same body shape CAD data as for the computational model. Due to the size limitation, and the fact that only one rear axle with single tyres was attached to it, its total length was limited to 1.2 meters. The flow was observed by smoke and tufts on the surface in a semi-open scale wind tunnel with a belt moving at the wind speed, about 6 m/s. Drag and lift coefficients were measured at a wind speed of 25 m/s. Due to the excessively long computation time (in spite of the constantly busy computer operation), only one case of tyre rotation was studied. To evaluate the effect of the tyre rotation on the flow, the wind tunnel test was carried out with and without the tyre rotation.

6 RESULTS AND DISCUSSION

Calculation was carried out at Reynolds Number Re = 500,000, a representative length of 1 meter, and a non-dimensional time increment 0.001. The data averaged from time 30 to 50 are used for the following results.

6.1 Flow around the cab side

Figure 4(a) shows the calculated velocity vectors near the cab side. As indicated in the figure, the flow goes against the direction of vehicle motion over the whole observation area. Figure 4(b) shows the flow observed using tufts on the surface, indicating that the flow runs upstream.

Figure 5 compares calculated stream lines and the flow observed by smoke in the wind tunnel. In both figures, the flow runs downstream from the cab corner and separates at the cab side. Thus, both calculated velocity vectors and stream lines agree qualitatively with the test results.

6.2 Flow around the wheel arch

Figure 6 shows the calculated stream lines around the wheel arch. A part of the flow running downstream under the body comes up from the fender space and swirls out. Compared to the model without the rear body (2), the position where the flow emerges is further upstream.

Figure 7 shows the flows observed using smoke on moving and fixed ground. The tyre rotation influences the flow around the wheel arch. When the belt is not moving, as shown in the right figure, the flow emerges from a much wider area of the wing area, and the smoke can't be seen easily due to the flow diffusion. Comparing Figs. 6 and 7, the calculated flow structure agrees well with the results with the tyre rotation. Therefore, the tyre rotation must be considered in an analysis of the flow around the wheel arch.

6.3 Drag and lift coefficients

Table 1 compares the calculated and measured values. The drag coefficients agree well, but the lift coefficients do not. This discrepancy may be mainly caused by the different lower vehicle structures between the calculation and experimental models. Furthermore, ignoring the rear axle and wheels in the computation alter this discrepancy. Besides, no influence of the moving belt on the drag and lift coefficients was observed at this test in this case.

Table 1 Comparison between calculation and test results

	Drag (Cd)	Lift (Cl)
Calculation results	0.486	-0.091
Test results	0.466	-0.165

7 CONCLUSIONS

The results are summarized as follows ;

(1) The calculated results agree qualitatively with the experimental results, and so the simulation is effective in analysing the flow around a heavy-duty truck body.
(2) To simulate more exactly local flows such as the flow around the wheel arch, it is necessary to model conditions such as the existence of the rear body, mudflaps, and the tyre rotation.
(3) The simulated and experimental drag coefficients agree well quantitatively.
(4) The predicted lift coefficient does not agree with the measured value. The calculation model must be improved.

REFERENCES

(1) Drollinger, R.A. Heavy Duty Truck Aerodynamics. SP-688, Feb. 1988.
(2) Nishikawa, S., Watanabe, Y., Fujitani, K., Himeno, R., Takagi, M. Numerical Simulation of Flow Around a Heavy-Duty Truck Body. SAE Paper 890599, 1989.
(3) Himeno, R., Takagi, M., Fujitani, K., Tanaka, H. Numerical Analysis of the Airflow around Automobiles Using Multi-block Structured Grids. SAE Paper 900319, 1990.

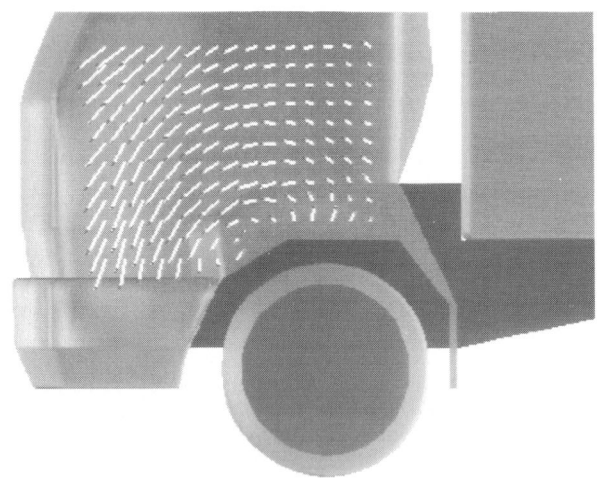

(a) Calculation results

(b) Test results

Fig.4 Velocity vectors near cab side

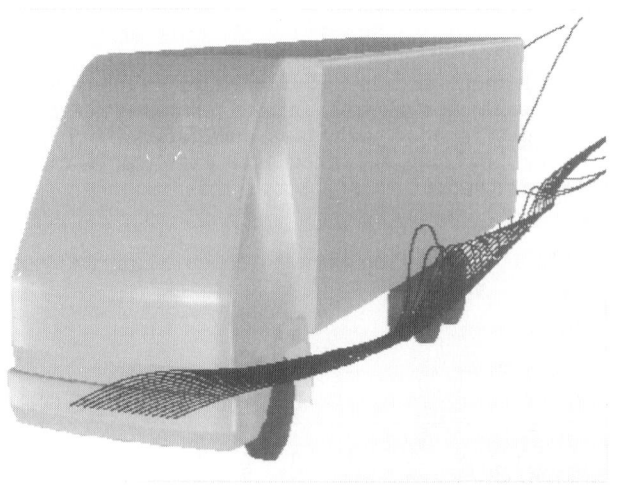

(a) Calculation results
(Stream lines)

(b) Test results
(by smoke)

Fig.5 Flow around cab side

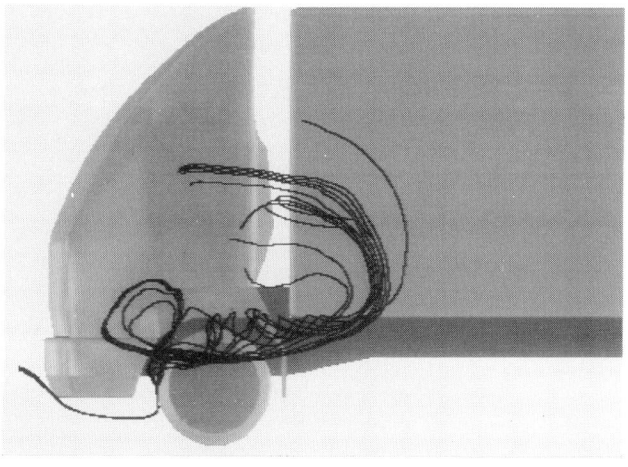

Fig.6 Stream lines around wheel arch

(a) with tyre rotation

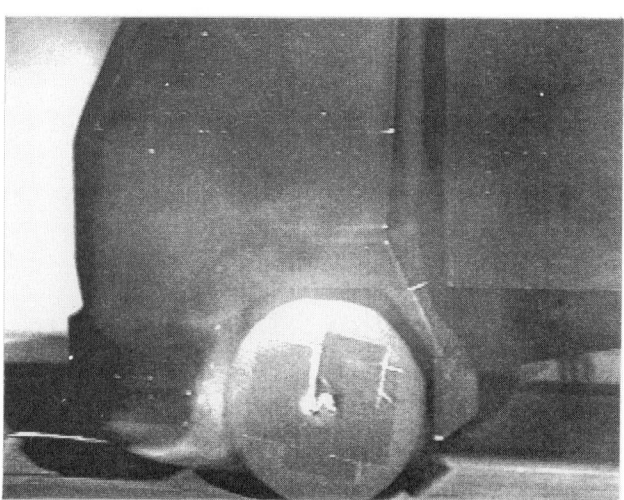

(b) without tyre rotation

Fig.7 Flow around wheel arch

C413/041
Turbulent flow and heat transfer predictions for cross-corrugated rotary regenerators

M CIOFALO, Dr-Ing
Department of Nuclear Engineering, University of Palermo, Italy
M W COLLINS, MA, PhD, DSc, MIEE, MINucE
Thermo-Fluids Engineering Research Centre, City University, London
G PERRONE, Department of Nuclear Engineering, University of Palermo, Italy

SYNOPSIS Flow and heat transfer predictions are presented for a corrugated geometry representative of power station rotary regenerators. The choice and the geometrical characteristics of the computational domain are discussed and the influence of Reynolds number, boundary conditions and corrugation geometry on overall performances is analysed. Turbulent flow conditions are considered in this study, and the k-ε turbulence model is used. The work is part of a comprehensive theoretical and experimental research programme, aimed at optimizing the geometry and the operating conditions of these heat exchangers.

1 INTRODUCTION

Air heaters are essential components of fossil-fuelled power plants (1). Typically, they cool the flue gases leaving the final water-heating stage (economizer) from 300 to 100 °C, and deliver warm air at 250 °C to the furnace. In a 500 MWe unit the air heaters may recover some 100 MW of low grade heat thus increasing substantially the plant efficiency. Moreover, preheating the combustion air makes the use of lower grade fuels possible and, in coal-fired plants, provides the means of drying the fuel.

Air heaters can be classified as direct transfer-type exchangers, or recuperators, and storage-type exchangers, or regenerators. The latter are cheap and more compact and are generally preferred in large modern power plants. Regenerators rely on the heat storage capacity of a matrix of closely packed corrugated steel plates which are exposed alternately to the hot flue gases and to the cool combustion air. In the Ljungström design the matrix rotates between stationary ducts, while in the Rothemuhle design the matrix is stationary and rotating hoods distribute the two streams through it. The two designs are essentially equivalent; the heat storage matrix is assembled in the form of a squat cylinder, typically having a vertical axis, a depth of two metres and a diameter of ten. Two such units serve a 500-MWe boiler; each of them is crossed by flue gas and air flow rates of the order 200 m^3/s, with speeds of 3-6 m/s. The surface to volume ratio can be as high as 500 m^2/m^3.

The performances required for an air heater are high heat transfer rates, low head losses and low fouling. These depend mainly on the geometrical design of the heat transfer elements. Several arrangements have been used or proposed, differing mainly in the shape and size of the corrugations formed by the steel plates; some are described, for example, in ref.2. The design on which this study is focussed is the so called 'cross-corrugated', sketched in Fig.1.

Global design correlations for rotary regenerators can be found in handbooks (3). However, the various corrugation designs have been developed mainly on an empirical basis. Although both operating and laboratory data are available (in the form of overall heat transfer and head loss coefficients) the phenomena which cause them are not well understood. Thus it would be of considerable value to supplement available bulk data by local flow and heat transfer measurements and three-dimensional numerical simulations. Such studies should identify the dependence of overall performances on fluid flow patterns and regimes (laminar - transitional - turbulent), thus leading to optimized designs which could well result in significant financial advantages.

On the above basis, a comprehensive experimental and predictive research program has been undertaken. Experimental work involves wind tunnel measurements using thermocromic liquid crystals for the local wall temperatures (4) and particle-image velocimetry for the local flow field (5). Also, bulk performance data are available from Marchwood Engineering Laboratories (2). Predictive work involves three-dimensional numerical simulations using mainly the Harwell-FLOW3D code (6). Preliminary results are presented in ref.7.

Under the operating conditions of rotary regenerators (Reynolds numbers of 1500-3000) transitional flow is expected. As is well known, this is very difficult to simulate numerically by existing models. Thus, it was decided to obtain laminar solutions up to Re = 1250 and turbulent predictions at higher Reynolds numbers, with some overlapping of the two ranges. This should give upper and lower bounds to the heat transfer and head loss performances in the transitional range and, by comparison with experimental data, should help to characterize the transition to turbulence and its effects.

The present paper is focussed on turbulent flow and heat transfer predictions. Laminar results have been presented in a companion paper (8), while preliminary turbulent predictions have been reported in ref.9.

2 CORRUGATION GEOMETRY, COMPUTATIONAL DOMAIN AND BODY-FITTED GRID

Although the exchangers themselves are substantial pieces of engineering equipment, they are composed of a large number ($\sim 10^8$!) of nominally identical, small geometrical elements. For the cross-corrugated design, a unitary cell can be identified as sketched in Fig.2. The orthogonal frame Oxyz used in the simulations is also shown. Each unitary, diamond-shaped cell has two inlets (W,D) and two outlets (E,U) and can be chosen as the computational domain. Its geometry is completely specified by the parameters P, H, s and ϑ as indicated in Fig.2. All relevant geometrical parameters can be derived from these as shown in Table 1. The reference geometry for which most results are presented here has $\vartheta = 45°$, P/H = 1.59 and s/H = 1/15.

Table 1 - Geometrical quantities characterizing the unitary cell (computational domain)

Quantity	Symbol	Expression
Corrugation pitch	P	-
Corrugation height	H	-
Wall thickness	s	-
Corrugation angle	ϑ	-
Internal height	H_f	$H - s$
Internal volume	V	$P^2 H_f / \sin\vartheta$
Lateral surface	S	$2PL/\sin\vartheta$
Hydraulic diameter	D_{eq}	$4V/S$
Flow cross-section	A_c	$2PH_f \sin(\vartheta/2)/\sin\vartheta$
Inlet-outlet length	a	$P\cos(\vartheta/2)/\sin\vartheta$

The geometric correlations are based on the assumption of perfectly sinusoidal corrugations. The length L in Fig.2 and Table 1 is then given by:

$$L = \frac{2 E(\alpha, \pi/2)}{\pi \cos\alpha} P \quad (1)$$

in which E is the elliptic integral of the second kind (10):

$$E(\alpha, \phi) = \int_0^\phi (1-\sin^2\alpha \sin^2\phi')^{\frac{1}{2}} d\phi' \quad (2)$$

and the angle α is given by:

$$\alpha = \arcsin\{ (\pi H_f/P)/[1+(\pi H_f/P)^2]^{\frac{1}{2}} \} \quad (3)$$

As pointed out in ref.9, the three-dimensional computational domain of Fig.2 can be mapped into a cube having two inlet and two outlet half-faces, and is symmetric for a 180° rotation around the main flow direction. Taking advantage of the above features, a body-fitted grid was generated by using the method described in ref.8. For the present turbulent flow predictions, however, care was taken that the near-wall grid points (control volume centres) lay in the fully turbulent region outside of the viscous sublayer $y^+ < 11$, as required by the k-ϵ wall-function model used here. Thus, the wall-adjacent layer of control volumes was imposed an approximately uniform thickness δ, specified as a fraction of H in the grid-generation phase.

An example is reported in Fig.3. Fig.3(a) shows the basic two-dimensional auxiliary grid extending over one half of the fluid core region on a face (say, face D of Fig.2). It is generated by deforming the rectangular region ABCD into the sinusoid-shaped region AB'CD (8). Algebraic formulae are then used to generate an additional layer of near-wall fluid cells, further layers in the wall thickness, and an outer layer of dummy control volumes (used to impose boundary conditions). The grid is then reflexed and translated in order to span the three-dimensional computational domain; the result is shown in Fig.3(b). It has 24^3 control volumes and $\delta/H = 0.1$. The wall thickness is resolved by a single layer of control volumes, which is sufficient for the present simulations as thermal boundary conditions are explicitly specified at the wall-fluid interface; a finer discretization of the wall could be used in studying conjugate conduction-convection problems. Further details of the grid generation are given in ref.11.

If G is the mass flow rate crossing the unitary cell, the mean velocity can be defined as:

$$\bar{v} = G/(\rho A_c) \quad (4)$$

and the Reynolds number can be based on this and on the hydraulic diameter, D_{eq} in Table 1.

The most relevant performance parameters of the exchanger are the head loss per unitary cell and the mean heat transfer coefficient. From a balance of mechanical energy in the unitary cell one has for the head loss, expressed in pressure units (i.e., as energy dissipated per unit volume):

$$|\Delta p|_L = -\frac{1}{G} \int_{D,U,W,E} (p+\rho u^2/2)\rho \underline{u} \cdot d\underline{A} \quad (5)$$

in which the integral is extended to the four faces D,U,W,E of Fig.2 and $d\underline{A}$ is the elementary area vector pointing outwards. In fully developed flow $|\Delta p|_L$ is simply the mean pressure drop between the inlet faces D,W and the outlet faces U,E. This, however, is not true in developing flow (entrance region of the exchanger).

The mean heat transfer coefficient can be defined as:

$$\bar{h} = \overline{q''_w} / (\overline{T_w} - \hat{T}_f) \quad (6)$$

in which $\overline{q''_w}$ and $\overline{T_w}$ are the wall heat flux and the wall temperature, surface-averaged over S, while $\hat{T}_f$ is the fluid temperature, volume-averaged over V.

Both $|\Delta p|_L$ and $\bar{h}$ can be made nondimensional as:

$$C_p = |\Delta p|_L / (\tfrac{1}{2} \rho \bar{v}^2) \quad (7)$$

$$Nu = \bar{h} D_{eq} / \lambda \quad (8)$$

λ being the thermal conductivity of the fluid.

3 GOVERNING EQUATIONS, NUMERICAL METHODS AND BOUNDARY CONDITIONS

The equations to be solved are the continuity, momentum (Navier-Stokes) and energy equations which, for an incompressible, constant-property fluid in steady-state turbulent flow can be written as:

$$\frac{\partial u_j}{\partial x_j} = 0 \qquad (9)$$

$$\frac{\partial u_j u_i}{\partial x_j} = -\frac{1}{\rho}\frac{\partial p}{\partial x_i} + \frac{1}{\rho}\frac{\partial}{\partial x_j}(\mu+\mu_t)\frac{\partial u_i}{x_j} \qquad (10)$$

$$\frac{\partial u_j T}{\partial x_j} = \frac{1}{\rho}\frac{\partial}{\partial x_j}(\frac{\mu}{\sigma}+\frac{\mu_t}{\sigma_t})\frac{\partial T}{\partial x_j} \qquad (11)$$

These are written in 'conservation' form, and with summation over repeated indices implied; x_i are cartesian coordinates (i.e., x,y,z in Fig.2) and u_i are the corresponding cartesian velocity components (the discretized form of the same equations on body-fitted grids is discussed in ref.12). The k-ε model was used here in its standard form (13) to express the turbulent viscosity μ_t, and the turbulent Prandtl number σ_t in eq.11 was assumed to be 0.9. Linear-logarithmic 'wall functions' were used in the form proposed in (13) to formulate wall boundary conditions, and the value 11.225 was used for the (nondimensional) thickness of the viscous sublayer. The Prandtl number σ was assumed to be 0.72 (air).

The finite-difference method implemented in Harwell-FLOW3D, Release 2 is described in ref.12. All quantities are defined at the centroids of hexahedral control volumes; the Rhie-Chow method (14) is used to prevent 'chequerboard' oscillations. The algorithm SIMPLEC (15) was used to couple the pressure and velocity fields; the convergence criterion adopted in most runs was that the mass-source residual (expressing the amount by which the continuity equation is not satisfied by the solution) fell below 10^{-6} times the cell mass flow rate, G. The hybrid-upwind discretization scheme was used for the convective terms.

Both uniform-temperature and uniform-heat flux thermal boundary conditions were tested. Results are presented here for the latter condition only. The relative performances of different geometries depend only to a minor extent on the particular condition chosen.

Two different sets of boundary conditions can be imposed on inlet-outlet faces D,U,W,E:

a) These faces can be defined as periodicity surfaces, i.e. surfaces on which the flow variables repeat themselves periodically from face D to U and from W to E. In discrete form, this is equivalent to imposing, for each flow variable X:

(D-U periodicity):
$$X_{1,j,k}=X_{NI-1,j,k} \text{ and } X_{2,j,k}=X_{NI,j,k} \qquad (12)$$

(W-E periodicity):
$$X_{i,j,1}=X_{i,j,NK-1} \text{ and } X_{i,j,2}=X_{i,j,NK} \qquad (13)$$

in which the three indices i,j,k identify a control volume in the grid; i increases from 1 to NI going from face D to U, while k increases from 1 to NK going from face W to E. Because of symmetry, NI=NK for the system considered.

If periodic boundary conditions are imposed the intrinsically non periodic quantities p and T are replaced by their periodic components p^*, T^* (i.e., 'true' p and T plus a term varying linearly along the main flow direction). Pressure losses have to be balanced by adding to the RHS of the momentum equation (10), written for the x- and z-directions, the source terms (driving pressure gradients):

$$F_x = F_z = |\Delta p|_L / (\sqrt{2}\, a) \qquad (14)$$

Similarly, heat input into the fluid has to be balanced by adding to the RHS of the energy equation (11) the source (or rather, sink) term:

$$S_T = -\frac{\rho\,(u+w)}{\sqrt{2}\,aGc_p}\int_S q_w'' \, dS \qquad (15)$$

in which u and w are the cartesian velocity components along x and z respectively, and c_p is the specific heat at constant pressure of the fluid.

The above method is appropriate to simulate the flow and temperature fields in the generic cells of the exchanger, away from intakes (fully-developed conditions).

b) as an alternative, faces D and W can be defined as inlets, i.e. surfaces on which the values of all flow variables (except p) have to be specified (Dirichlet boundary conditions), while the opposite faces U and E are treated as outlet surfaces, on which zero-normal derivative (Neumann) boundary conditions are imposed to the same quantities. The heat exchanger matrix can be assumed to be composed of identical 'paths' of consecutive cells like that shown in Fig.2. For the first (entrance) cell of any path, uniform inlet velocity profiles, having the value $2\bar{v}$ and aligned with the main flow direction, can be reasonably assumed. Sensible values of k and ε, representative of the inlet-stream conditions, and any uniform reference temperature, have also to be imposed. Now, the outlet distributions computed for this cell can be used as inlet distributions for the next cell, and so on. Within the approximation implied by assuming zero-normal derivative conditions at the outlet faces, this method will give the distribution of head losses, heat transfer coefficients etc. along the first consecutive cells from the entrance, thus allowing the assessment of entrance effects. Also, this method is the only viable way to simulate wind tunnel experiments, which are inevitably far from fully developed conditions. As more and more cells are considered, results are expected to approach those obtained under fully developed flow (periodicity) assumptions.

4. RESULTS AND DISCUSSION (PERIODIC FLOW)

Periodic flow simulations were run by imposing the Reynolds number, Re^o, and adjusting the driving pressure gradient $|\Delta p|_L$ in eq.(14) at each SIMPLEC iterations (after a 'settling' phase of about one hundred iterations) according to the actual Reynolds number Re as follows:

$$|\Delta p|_L^{(new)} = (1-\beta)\,|\Delta p|_L^{(old)} + \beta\,|\Delta p|_L^{(old)}\,Re^o/Re \quad (16)$$

in which β is an underrelaxation factor (6). It was set at 0.9 in most cases.

The typical behaviour of $|\Delta p|_L$ and Re as functions of the number of iterations is shown in Fig.4 for Re^o = 2500 and a 24^3-volumes grid. Up to 2000 iterations were needed for a satisfactory convergence to be attained on $|\Delta p|_L$, Re and the mass-source residual.

The sensitivity of computed C_p and Nu to the number of control volumes in the grid was studied for Re^o = 2500 and for a fixed thickness, δ = 0.2·H, of the wall-adjacent layer of control volumes. A 32^3-volumes grid was the finest that could be treated on the IBM-3090 computer used (16-Mbytes RAM). Results from a 24^3- and a 32^3-volumes grid still differed by about 1 percent, so that complete grid-independence could not be demonstrated. As CPU times were about 2.5 times higher for the 32^3- than for the 24^3-grid, the latter was used in most runs. Considering the comparative and preliminary nature of the present study, some residual amount of grid-dependence was judged to be tolerable.

At Re^o = 2500, the choice δ = 0.2 H gave, for the nondimensional distance of the near-wall grid points (control volume centres) from the wall, an average value of about 15. At lower Reynolds numbers, this value became less than ~11 (the dimensionless thickness of the viscous sublayer) at some wall locations. As discussed in ref.16, this condition violates the assumptions underlying the use of 'wall functions' and leads to overpredicting heat transfer rates and, to a lesser extent, wall shear stresses. On the other hand, a further increase in the thickness of the near-wall layer of control volumes would impair the resolution of the flow field and introduce other (truncation) errors. Thus, a compromise was adopted here: the thickness δ was kept = 0.2·H at all Reynolds numbers in the range 1250 to 5000, but results for the lower Reynolds numbers were corrected for sublayer effects by assuming the error to be a function of δ^+ only, and studying the dependence of C_p and Nu upon δ^+ for Re^o = 2500 and varying δ. Details are given in ref.11. Significant corrections were found to be necessary only on Nu and for Re^o < ~2000. Only minor corrections were required on the wall shear stress and C_p.

The dependence of C_p and Nu on the Reynolds number is shown in Fig.5 for the 'reference' geometry. For comparison purposes, the corresponding quantities for a straight circular duct having the same hydraulic diameter D_{eq} and equivalent length a are also reported (17):

$$C_p = 4fa/D_{eq} \quad (17)$$

(with the friction coefficient f given by the Moody chart for smooth pipes), and:

$$Nu = 0.116 \cdot (Re^{2/3} - 125) \cdot Pr^{1/3} \quad (18)$$

(simplified form of the Hauser correlation for $L/D \to \infty$ and constant-property fluids).

The head loss coefficient computed for the corrugated geometry is about twice that for a smooth pipe, and decreases as $Re^{-1/2}$. The Nusselt number is significantly higher than in a smooth pipe at low Reynolds numbers, but increases only as $Re^{1/2}$ so that the advantage decreases with increasing Re.

The flow field computed for Re = 2500 is shown in Fig.8 (periodic solution) in the form of the in-plane velocity vector plots on the midplane y=0 (left) and on the plane midway between faces D and U (right). Such plots will be compared directly with experimental double-exposure, light-sheet images obtained by particle-image velocimetry (5). The predicted flow field scales almost exactly with Re in the range investigated.

The dependence of C_p and Nu on the corrugation angle ϑ (for a given P/H) and on the pitch to height ratio P/H (for a given ϑ) was investigated and results are shown in Fig.6 for Re = 2500. As ϑ increases from 30° to 60°, the Nusselt number increases almost linearly while the head loss coefficient exhibits a minimum at $\vartheta \cong 40^o$. As P/H increases from 1.5 to 2.1, the head loss coefficient increases while the Nusselt number remains almost constant. It should be stressed that such comparisons are purely preliminary and should be interpreted with some care, because different flow rates with different corrugation geometry can give the same Reynolds number.

5 RESULTS AND DISCUSSION (DEVELOPING FLOW)

When inlet-outlet (developing flow) conditions were used, SIMPLEC convergence was faster than under periodicity assumptions because the flow rate was imposed and not allowed to vary during successive iterations. About 500 iterations were required for convergence. The sensitivity of the results to the number of control volumes (grid-dependence) was comparable with that found for periodic flow. Results are presented here for the 24^3-volumes grid.

The behaviour of C_p and Nu in the first few consecutive cells is shown in Fig.7 for Re^o = 2500 and periodic-flow results are also reported for comparison purposes. For the same grid and Reynolds number the flow fields in the first three cells (in-plane velocity vector plots on the same two 'slices' described above) are shown in Fig.8, and can be compared with the periodic solution.

These results suggest that the flow and temperature fields evolve rapidly towards fully-developed conditions. The entrance region, in which C_p and Nu differ significantly from their periodic values, extends over only a few cells, i.e. a negligible fraction of the overall flow 'path'. It is noteworthy that in laminar flow (8) the flow and temperature fields were found to develop much more slowly, the entrance region extending to as many as 10-15 consecutive cells.

6 COMPUTATIONAL REMARKS

All simulations were run on the IBM 3090 (without Vector Facility) of the University of

Palermo Computing Centre. Release 2.1 of the code Harwell-FLOW3D was used with minor 'ad hoc' modifications.

Typical CPU times were about 3-4 seconds per iteration for the 24^3-volumes grid. A fully converged solution (mass source residual below 10^{-6} G) required about 2000 iterations when periodicity was imposed, and about 500 when inlet-outlet conditions were used. Thus overall CPU times were about 2 hours and 1/2 hour, respectively. These figures increased about 2.5 times if a finer 32 -volumes grid was used. CPU times should be reduced about two-three times by the use of an IBM 3090 VF (Vector Facility) or of a CRAY machine.

Storage requirements were close to 120 real (4-bytes) locations per grid point, i.e. about 8 Mbytes for the 24^3-volumes grid and about 16 Mbytes for the 32^3-volumes one (which was the finest affordable).

7 CONCLUSIONS AND FUTURE WORK

Flow and temperature fields, and bulk head loss and heat transfer coefficients, were computed for Reynolds numbers in the range 1250 to 5000 under turbulent flow assumptions for the unitary cell of a cross-corrugated rotary regenerator. The Harwell-FLOW3D computer code, the k-ε turbulence model with 'wall functions', and the SIMPLEC pressure-velocity coupling algorithm, were used on a co-located, body-fitted grid.

Results based on inlet-outlet boundary conditions developed rapidly towards the corresponding periodic values, showing that entrance effects extend over only a minor fraction of the exchanger. Under periodic (fully developed) flow assumptions, the head loss coefficient was predicted to decrease roughly as $Re^{-1/2}$, and to be about twice that of a straight circular duct having the same hydraulic diameter and equivalent length. The Nusselt number was predicted to increase roughly as $Re^{1/2}$, and to be significantly higher than in a circular duct only at Re < 3000.

The head loss coefficient was found to increase with the pitch to height ratio and to have a minimum for a corrugation angle of about 40°. The Nusselt number was found to be almost insensitive to the pitch to height ratio and to increase with the corrugation angle ϑ. However, only the ranges P/H = 1.5 to 2.1, ϑ = 30° to 60° were investigated here.

The above turbulent flow predictions, together with laminar predictions such as those described in ref.8 and with wind tunnel flow and heat transfer measurements, will form the basis for a better understanding of the phenomena occurring in power station rotary regenerators and of the dependence of bulk performances upon local flow and temperature fields at the transitional Reynolds numbers typical of their operating conditions. Future work includes the extension of the parametric study outlined in this paper to a wider range of Reynolds numbers, pitch to height ratios and corrugation angles, as well as to different and more general shapes. The use of a low-Reynolds number turbulence model and of Direct and Large-Eddy Simulation are also planned as is the cross-comparison of wind tunnel results, bulk performance data and numerical predictions.

AKNOWLEDGEMENTS

This work was sponsored by PowerGen, U.K. The authors wish to thank Mr. P.E.Chew and Mr. A. Marshall of Marchwood Engineering Laboratories for their helpful suggestions and for permission to publish the present results.

REFERENCES

(1) CHOJNOWSKI, B. and CHEW, P.E. Getting the Best Out of Rotary Air Heaters. CEGB Research Journal, May 1978, pp.14-21

(2) CHEW, P.E. Rotary Air Preheaters on Power Station Boilers. Proc. Institute of Energy Symposium 'Waste Heat Recovery and Utilisation', Portsmouth, UK, September 1985

(3) ROHSENOW, W.M. et al. (eds.) Handbook of Heat Transfer Applications, Chapter 4 (Compact Heat Exchangers). McGraw-Hill, 2nd ed., 1986

(4) STASIEK, J. and COLLINS, M.W. Local Heat Transfer and Fluid Flow Fields in Crossed Corrugated Geometrical Elements for Rotary Heat Exchangers - Report No.1: Design of Wind Tunnel and Review of Liquid Crystal Thermography. T.F.E.R.C. Report, The City University, London, October 1989

(5) STASIEK, J., SHAND, A., CIOFALO, M. and COLLINS, M.W. Local Heat Transfer and Fluid Flow Fields in Crossed Corrugated Geometrical Elements for Rotary Heat Exchangers - Report No.2: Interim Report of Work Done to End of January,1990. T.F.E.R.C. Report, The City University, London, February 1990

(6) BURNS, A.D., JONES, I.P., KIGHTLEY, J.R. and WILKES, N.S. Harwell-FLOW3D, Release 2: User Manual. Harwell Report AERE-R (Draft), July 1990

(7) CIOFALO, M., PERRONE, G. and COLLINS, M.W. Predictions for Laminar and Turbulent Flow and Heat Transfer for a Cross Corrugated Geometry - Report No.4: Final Report on Work Done to End of March 1990. T.F.E.R.C. Report, The City University, London, June 1990

(8) CIOFALO, M., COLLINS, M.W. and PERRONE, G. Laminar Flow and Heat Transfer Predictions in Cross-Corrugated Rotary Regenerators. Proc. 8th Nat. Conf. of UIT (Unione Italiana di Termofluidodinamica), Ancona, Italy, 28-30 June, 1990

(9) FODEMSKI, T.R. and COLLINS, M.W. Computer Simulation of Flow and Heat Transfer in a Corrugated Geometry Using the Code FLOW3D, Release 2 (Final Report). T.F.E.R.C. Report, The City University, London, November 1988

(10) BEYER, W.H. (ed.) Handbook of Mathematical Science. 6th ed., CRC Press, Boca Raton, Florida, 1987

(11) PERRONE, G. Simulazione numerica del moto e dello scambio termico in rigeneratori rotanti. Graduation thesis in Nuclear Engineering, Univeristy of Palermo, Italy, 1990 (in Italian)

(12) BURNS, A.D. and WILKES, N.S. A Finite Difference Method for the Computation of Fluid Flows in Complex Three-Dimensional Geometries. Harwell Report AERE-R 12342, 1987

(13) LAUNDER, B.E. and SPALDING, D.B. The Numerical Computation of Turbulent Flows. Comp. Meth. Appl. Mech. Eng. 1974, $\underline{3}$, pp.269-289

(14) RHIE, C.M. and CHOW, W.L. Numerical Study of the Turbulent Flow Past an Airfoil with Trailing Edge Separation. AIAA Journal 1983, $\underline{21}$, pp.1527-1532

(15) VAN DOORMAL, J.P. and RAITHBY, G.D. Enhancements of the SIMPLE Method for Predicting Incompressible Fluid Flows. Numer. Heat Transfer 1984, $\underline{7}$, pp.147-163

(16) CIOFALO, M. and COLLINS, M.W. k-ε Predictions of Heat Transfer in Turbulent Recirculating Flows Using an Improved Wall Treatment. Numer. Heat Transfer 1989, part B, $\underline{15}$, pp.21-47

(17) PERRY, R.H. and CHILTON, C.H. (eds.) Chemical Engineers' Handbook. McGraw-Hill, 5th ed., 1973

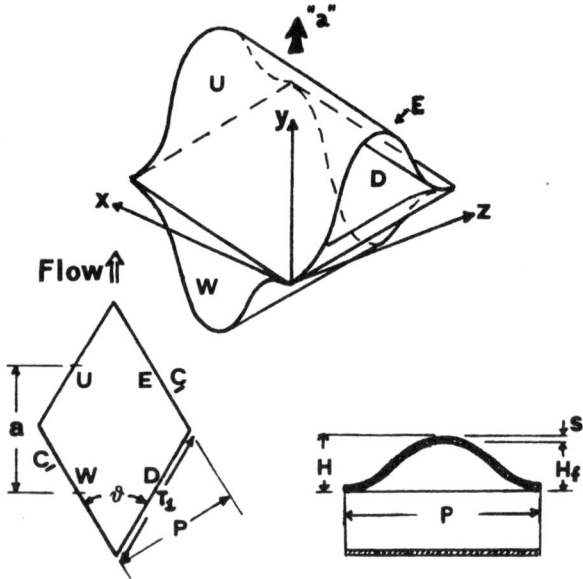

Fig. 2: Unitary cell (computational domain). Top: perspective view; bottom left: section y=0; bottom right: section C-C normal to the corrugations.

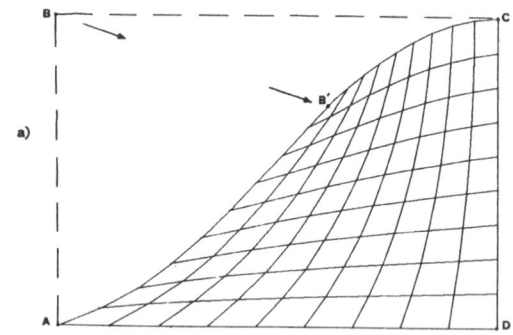

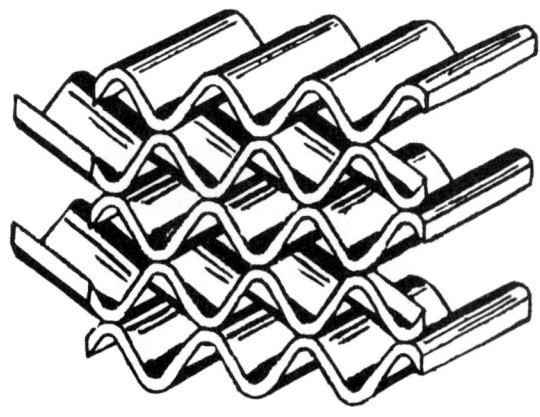

Fig. 1: Cross-corrugated heat transfer elements.

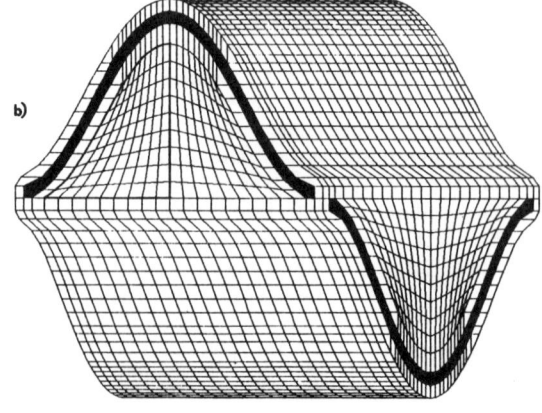

Fig. 3
Computational grid having 24^3 control volumes
a) auxiliary two-dimensional grid on a face
b) complete three-dimensional grid solid wall

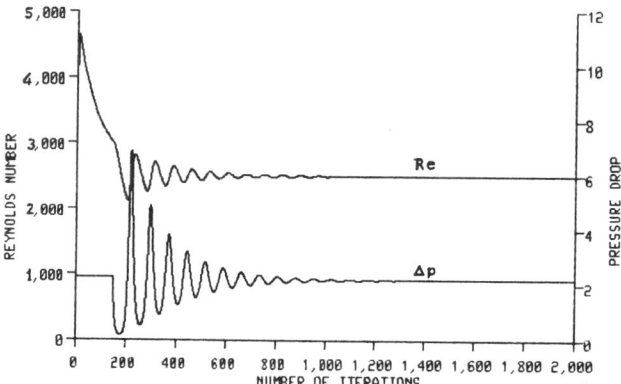

Fig. 4: Behaviour of Re and Δp in a periodic flow simulation (Re°=2500, 24^3 grid).

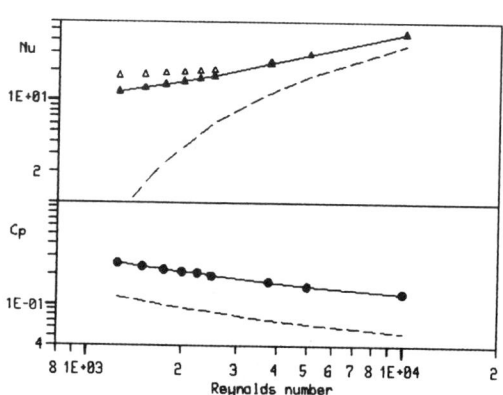

Fig. 5:
Dependence of Nu and C_p on Re.
-Top: Nusselt number. △: raw predictions;
-▲-: predictions corrected for sublayer effects; ---: Hauser correlation /17/.
-Bottom: pressure drop coefficient.
-●-: predictions; --- : Moody chart.

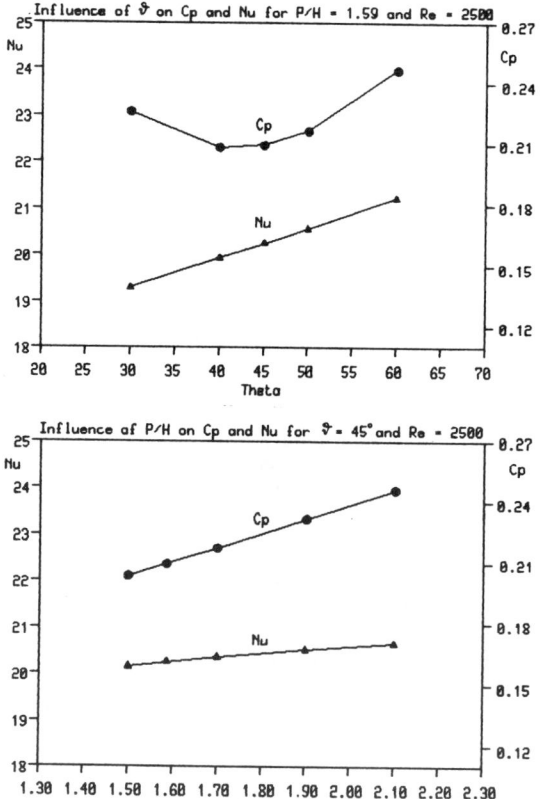

Fig. 6: Dependence of C_p and Nu on pitch/height ratio (bottom) and corrugation angle (top). Re=2500, 24^3 grid, periodic flow.

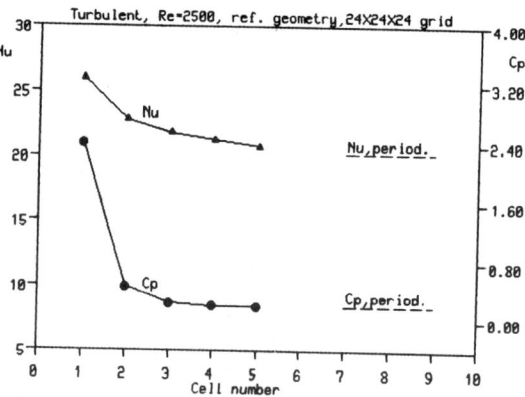

Fig. 7:
Behaviour of C_p and Nu in consecutive cells from intake (Re=2500, 24^3 grid).

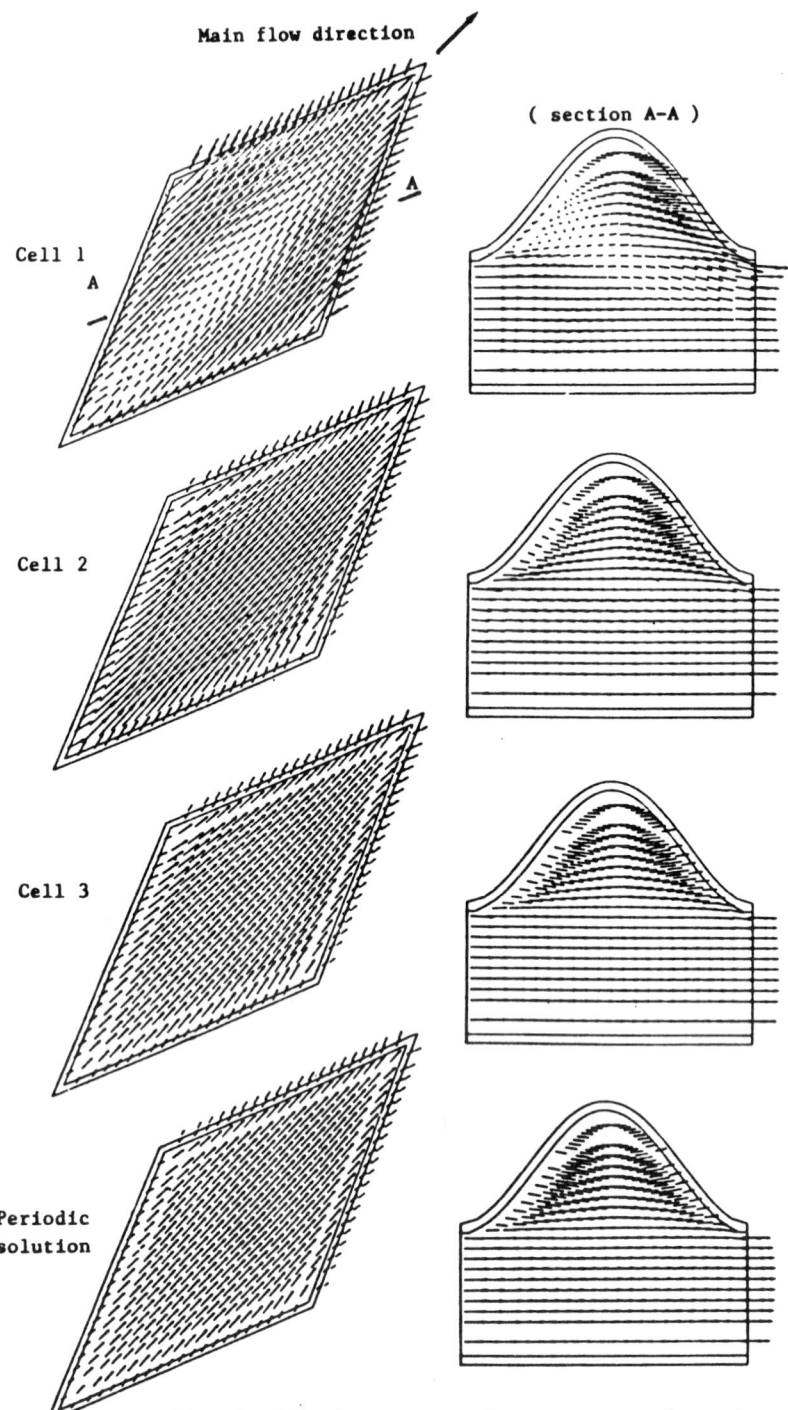

Fig. 8: Velocity vector plots on two selected planes for the first three consecutive cells and for periodic flow (Re=2500, 24^3 grid). Left: midplane y=0; right: midplane parallel to faces U,D.

C413/046

Numerical simulation of transitions in axial-rotary flows in an annulus

T J LOCKETT, MEng
AEA Petroleum Services, United Kingdom Atomic Energy Authority
Harwell, Oxfordshire
S M RICHARDSON, BSc(Eng), PhD CEng, MIChemE
Department of Chemical Engineering, Imperial College of Science
Technology and Medicine, London
W J WORRAKER, BSc, PhD,
AEA Petroleum Services, United Kingdom Atomic Energy Authority
Harwell, Oxfordshire

Numerical techniques have been used to predict the value of the axial flow parameter (Re) at which Taylor vortices appear for a given value of rotational parameter (Ta) in a concentric cylinder geometry. Agreement with literature values is excellent for the ranges studied. Preliminary quantitative results for the effect of non-Newtonian fluid behaviour on the transition values are also presented.

NOMENCLATURE

b	annulus gap width (R_2-R_1)
D_H	hydraulic diameter (2b)
i, j, k	indices to directions in the three dimensional computational grid
K	consistency in power law or Herschel-Bulkley fluid model
L	axial length of annulus
n	index in power law or Herschel-Bulkley fluid model
$NIJK$	number of nodes in the three dimensional computational grid
r, θ, z	radial, azimuthal, axial coordinate system (see Figure 1)
R	Radius of cylinder wall
Re	Reynolds number ($2U_m b\rho/\mu$)
Re'	Generalised Reynolds number ($U_m^{2-n} D_H^n \rho/K$)
Ta	Taylor number ($b^3 R_m \Omega^2 \rho^2/\mu^2$)
Ta'	Generalised Taylor number ($b^{2n+1} R_m^{3-2n} \Omega^{2(2-n)} \rho^2/K^2$)
U, V, W	velocity components in z, r, θ directions respectively
γ	rate of strain (s^{-1})
τ_y	yield stress in Herschel-Bulkley fluid model
Ω	rotation rate (rad/s)

Subscripts:

m	mean value
$1, 2$	value evaluated at the inner and outer cylinder walls respectively
c	value at bifurcation, critical value

1. INTRODUCTION

The removal of rock cuttings and their subsequent transportation to the surface is crucial to the performance of a bore-hole drill, the difficulties of suspension being made all the more acute by the modern practice of cutting highly deviated wells. A fluid, drilling mud, is circulated down through the drill centre and returns, with the cuttings, by way of the annular gap surrounding the drill.

The components of such a flow are combined rotary and axial motion. Initial suspension of the solids is brought about by the highly turbulent flow in the bit region, but thereafter the non-Newtonian properties of the mud in laminar flow maintain the transport. It is postulated that significant advantages would be offered by exploiting a more complex flow pattern generated by a naturally occurring instability, and it is to test this hypothesis that the current numerical simulation work has been undertaken.

Experimental works relating to the issue of bifurcation in fluid flows between rotating concentric annuli began with Taylor[1] who observed a transition from a purely laminar (azimuthal, Couette) flow to a flow containing recirculating vortices in the other two spatial directions. Figure 1 shows the geometry upon which the work was based, and forms the basis for the computational simulation. Similar more extensive work was reported by Donnelly and Simon[2] with a greater emphasis being placed on the measurement of the torque transmitted to the outer cylinder.

The situation was extended by Kaye and Elgar[3] and Snyder[4] who superimposed a flow in the axial direction onto the rotary motion. The resulting laminar—turbulent and laminar—laminar+vortices transitions form a flow map with the Reynolds number (Re) and Taylor number (Ta) describing the axial and rotary flows respectively. Figure 2 gives a schematic representation of the four regions reported. More modern measurement techniques have been applied by Coney et al[5,6,7,8] allowing velocity profiles in three dimensions and wall shear stress to be determined. All of these studies used either air or water as the fluid, both of which show a Newtonian relationship between stress and rate of strain.

Recently, Naimi et al[9] have detailed the results from combined axial and rotary flow experiments using non-Newtonian fluids. Taylor vortices were observed for both a Power law fluid and a fluid exhibiting a yield stress. However, details of the non-Newtonian properties were not given. A wide annular gap was used (i.e. $b/R_m = 0.5$), for which the Newtonian transition curve has not been reported, and the experiments have not extended beyond slow axial flow ($Re'_c \leq 3$).

Engineering experiments on large laboratory rigs capable of simulating solids conveying in deviated wells with real drilling muds have been reported by Azar et al[10,11]. None of the results indicated the presence of vortices since the experiments were conducted in the laminar regime (see Figure 2). It appears however, that similarity with typical drilling practice might not have been imposed simultaneously upon geometric, axial flow and rotary flow parameters.

In engineering applications such as electric motors or drilling, heat transfer across the annular gap or heat removal by the fluid is involved. Becker and Kaye[12] and Bjorklund & Kays[13] carried out experimental studies concerned with the effects of heat transfer. Temperature profiles within the fluid, and a qualitative regime map were given. Naimi et al[9] included a study of the effect of heat transfer on their non-Newtonian fluids.

A complementary, theoretical, linear perturbation analysis by Taylor[1], restricted to thin annuli (i.e. $b/R_m \leq 0.1$), gave predictions of transition values which proved to be in excellent agreement with the experimental observations. Davey[14] conducted a non-linear analysis which was valid for any gap width, transition point and transmitted torque agreeing well with the data of Donnelly and Simon[2]. DiPrima[15] and Chandrasekhar[16] considered the stabilizing effect that a superimposed mean axial flow has on the transition to vortices by approximating the azimuthal velocity component by its mean value and the axial velocity component by a parabolic profile and its mean value respectively. Chandrasekhar[17] went on to develop a more refined theory for the limit as axial flow tends to zero. Both show agreement with the work of Snyder[4] to within experimental error. Again, only Newtonian fluids have been treated.

Numerical studies have been fairly limited. Coney & El-Sharrawi[18] examined the entrance region of a laminar (no vortices) combined axial and rotary flow, while many other papers have considered flows in which bouyancy driven natural convection interacts with Taylor vortices and therefore fall outside of the scope of the current project.

2. APPROACH

The simulations to date have all been carried out using the HARWELL-FLOW3D computational fluid dynamics code to simultaneously solve the mass and momentum (Navier-Stokes) equations of classical continuum mechanics. It is a three-dimensional finite difference code which is capable of computing solutions to both steady state and transient problems on general grids in non-orthogonal coordinate space. It has been widely validated for laminar fluid flows and readily permits one of a number of differencing and solver techniques to be selected. It uses a non-staggered grid mesh (i.e. values for velocity components are stored at the same positions in physical space as the pressure) and employs the Rhie-Chow algorithm to maintain coupling of odd and even grids, thus reducing the tendency of solutions to adopt 'checkerboard' oscillations.

The simulated annulus was described by solid wall boundary conditions at the inner and outer radii (no slip, no penetration) and by periodic conditions in the azimuthal direction to join up $\theta = 0$ with $\theta = 2\pi$. Conditions prevailing in a very long annulus ($L/D_H \gg 1$) were imposed by also using a periodic condition between the two ends in the axial direction. The common technique of selecting a grid distribution which has smaller distances between nodes in the areas likely to exhibit sharp velocity gradients (e.g. near walls) was not used, since the position of these areas was not known a priori. Hence, a regular grid in cylindrical polar coordinates was chosen (see Figure 1) and had typical size 20 x 20 x 12 cells in the axial, radial and azimuthal directions respectively.

Some preliminary results were obtained for the nominally two dimensional case of toroidal vortices generated by a rotary flow in the absence of a mean axial flow, before the full combined problem was tackled. While the rotary flow may be imposed by assignment of the tangential velocity components at the wall, the axial flow must be imposed by the addition of a constant pressure drop term during the periodic updating in the axial direction of the pressure values. For values of Re and Ta in the laminar regime, a flow solution almost equal to the sum of Poiseuille, $U(r)$, and Couette, $W(r)$, solutions is obtained. i.e. linear superposition holds to a good approximation.

As alluded to in several of the experimental works covering combined flows, the imposition of a mean axial flow either causes vortices to be carried as complete units in the axial direction or may cause the generation of a single helical vortex which is convected axially — the two possibilities depending upon the relative strengths of axial and rotary motions. Both of these situations are transient to a stationary observer but may be rendered steady by choosing the grid reference frame to have the same axial velocity as the vortices. Since the velocity required to bring about this 'vortex freezing' is *not* equal to the mean axial velocity (Snyder[4]), it must be determined by a predictor/corrector method within the code.

A more convenient way of implementing this technique proved to be the pre-setting of reference frame velocity coupled with an adjustment of the applied axial pressure drop. The response of helix velocity relative to the computational frame was found to be approximately first order with iteration count (decaying exponentially to a new position) following a step change of the applied axial pressure drop, although as yet this information has not been used to assist the convergence rate.

Convergence may be judged by examining the change with continued iteration count, of the mass source residual (terms needed to satisfy the continuity equation exactly), the relative velocity of the vortices, the applied axial pressure drop and the values of velocity component norms defined by equation (1). While the first two quantities tend to zero, the others should approach a constant value.

$$normx = \frac{1}{NIJK} \sum_{ijk} x_{ijk}^2 \quad ; \quad x = U, V, W \quad \ldots (1)$$

Newtonian simulations were subsequently used to initialise the solutions of non-Newtonian cases. A Herschel-Bulkley yield-power law model was used since this has been reported to describe mud behaviour in the range of shear rates $0-1000$ s^{-1} very well; see Houwen[18] and Alderman et al[19]. The yield part of the model was however modified by introducing a $tanh(\dot{\gamma})$ function multiplier which acted at very low shear rates ($\dot{\gamma} \leq \dot{\gamma}^*$) and removed the need to find yield surfaces within the flow. The relationship between stress and rate of strain was thus given by equation (2).

$$\tau = \tau_y \, tanh\left(\frac{3\dot{\gamma}}{\dot{\gamma}^*}\right) + K\dot{\gamma}^n \quad \ldots (2)$$

3. RESULTS AND DISCUSSION

Initial validation of the code was performed by comparing predicted values for the torque transmitted to the stationary outer cylinder (derived by trapezoidal integration of wall shear stress) with those reported by Donnelly & Simon[2] (see Figure 3). Clearly, the torque values in the laminar regime (Ta < 1850) are in excellent agreement with the experimental data, as is the predicted value for the bifurcation occurrence — distinguished by the change of slope. Post critical behaviour is less well predicted, a linear variation with correct initial slope being obtained instead of the non-linear response seen in practice. A sensitivity analysis has shown this response to be grid independent and unaffected by the method of discretisation (see §5). The poor representation of post-critical behaviour did not hamper progress in predicting the onset of vortices however.

The norm of the radial direction velocity component (V) was found to be linearly dependent upon the square of the axial flow Reynolds number for converged solutions in the laminar+vortices regime, for a given value of Taylor number. This behaviour was essentially as expected, since many bifurcation problems show a linear relationship between some measure of the amplitude of the disturbance (in this case normv) and a measure of the displacement from the critical condition (Re^2). At present, the reason for this quantity in particular being the relevant measure of displacement from the critical condition is not totally understood. Furthermore, since this norm approximates to zero for solutions in the laminar regime, the value of Re at which the transition occurs may be found by extrapolation. The extended linearity characteristics, observed previously in the torque graph for purely rotary flows, improves the reliability of this extrapolation, and thus only relatively few points are required for the prediction of each transition value.

The predicted transition values of Re at a fixed Ta are shown in Figure 4 along with the experimental data of Snyder[4] and the theoretical values of DiPrima[15] and Chandrasekhar[17]. The graph clearly shows that the introduction of an axial motion progressively increases the stability of the laminar regime, such that a higher Taylor number is required to sustain vortices in the flow. In common with Snyder's observations, the numerical solution also revealed the phenomenon whereby the forward velocity of the vortices was found to be 10–15% in excess of the mean axial velocity.

The agreement with experimental literature values for Re < 100 is quite remarkable, and these results have formed the Newtonian base from which the non-Newtonian transition may be explored. For Re > 100, the numerical results start to deviate from the experimental values, with the transition Re being under-predicted. Grid refinement has enabled the difference to be reduced; however, some discrepancy still remains.

4. PRELIMINARY NON-NEWTONIAN RESULTS

By setting τ_y equal to zero and n = 1.0 in equation (2), the Herschel-Bulkley non-Newtonian model reduces to the well known Power law and Bingham formulations, respectively. Preliminary simulations have now been carried out for each of these simpler fluid models. However, only a limited range of parameters have so far been considered. These are viewed as intermediary steps, useful for characterising the solution behaviour, before extension to the complete generalised Newtonian model is attempted.

The introduction of a shear rate dependent viscosity calls for different dimensionless parameters to be used; Re', Ta' replace Re and Ta for the power law case, and are defined in the nomenclature. Since Ta' and Re' are not independent of n, primitive variables have been used when comparing the flow map for non-Newtonian fluids with the results shown in Figure (4) for Newtonian fluids.

Solutions for Re > 0 were again found by the method of matching the reference frame velocity to the forward axial velocity of the vortices. The introduction of the secondary motion at the onset of vortices greatly increases the shear rate near the outer wall of the annulus, and the most viscous part of the fluid moves from there to become located at the two 'eyes' of the vortices.

4.1 Power law fluids

The critical values of Re' were determined for n = 0.8 by a method exactly analogous to that described for the Newtonian case, except that the radial direction norm appeared to be linearly dependent upon Re'$^{1.67}$ in place of the squared relation mentioned earlier. This might be indicative of a general '2/(2–n)' power relationship, an expression which is drawn from a desire to maintain linearity between normv and the square of the axial velocity. Results for n = 0.6 seem to confirm this, however, other values of 'n' have yet to be computed. Figure 5 summarises the results for power law fluids, and includes the numerical results for the Newtonian case, n = 1.0.

The shear thinning behaviour of the fluid has caused the laminar+vortices regime to advance into areas previously characterised by purely laminar flow. This observation is entirely to be expected, since, for shear rates greater than $1s^{-1}$, the effective viscosity is lower than the Newtonian equivalent case, and the effective Taylor number correspondingly higher. Thus, flows previously in the laminar regime become able to sustain vortices.

4.2 Bingham Model Fluids

Taylor vortices have been predicted for flows of a modified Bingham fluid, of plastic viscosity 0.01 Pas and yield point of up to 5 Nm^{-2}, in the absence of a superimposed axial pressure drop. Since a linear relationship between the norm and Taylor number has yet to be found, only upper and lower bounds for the critical values, as a function of yield stress, may be presented (see Table 1). Even though the results are of a preliminary nature, it is clear from the table that the introduction of even a small yield stress profoundly stabilizes the purely laminar flow solution with respect to the solution containing vortices. A greatly increased speed of rotation is therefore required before vortices are generated, and this result is in qualitative agreement with the experimental work of Naimi et al[9].

Table 1: Influence of yield stress on bifurcation

τ_y	Ta_c
0	1858
1.5	4660 - 4900
3.0	7050 - 7250
5.0	10060 - 10300

5. SENSITIVITY

In common with true experimental work, this kind of numerical experimentation must be subjected to appraisal, especially in those areas for which the predictions show poor agreement with literature values. The basic areas which need to be examined are those of solution independence with regard to computational grid, method of discretisation and 'arbitrary' quantities.

All of the Newtonian calculations have been performed using a Sun 4–330 workstation holding 8 M-bytes of RAM. While the memory limitation restricted the grid mesh size to a maximum of 144000 nodes, the speed of calculation reduced the available grid size still further, a 20 x 20 x 12 grid being used for most of the work. A 40 x 20 x 12 grid was used as the next stage in grid refinement while even larger grids (60 x 30 x 22) were implemented on a Cray2 super-computer. The solution was shown to be insensitive to spacing in the azimuthal direction. While solution changes were observed between the 20 and 40 axial cell grids (e.g. Ta = 9600, Re_c = 128.8, 135.0 respectively), significant changes were not revealed by further refinement.

Little grid refinement of the non-Newtonian fluid solutions has been completed. The extra results which are available have been shown alongside the results from the 20 x 20 x 12 grid on Figure 5. Although the exact values of the critical parameters are expected to change slightly as more refined grids are used, the initial indications are that this movement will be small (<5%) and that the general form of the curves will be unchanged.

For reasons of speed and robustness of operation, the 'HYBRID' first order accurate differencing scheme was used for all of the calculations. The 'QUICK' second order scheme has subsequently been used to check some of the solutions. The torque values shown in Figure 3 were not affected by this alteration, and the solutions to the combined flow problem for small axial flows have also been found to be insensitive to this change. At high values of Ta and Re, convergence difficulties were encountered with the scheme, and these have yet to be surmounted.

The intention of this work was to simulate a long annulus by way of periodic boundary conditions, thereby reducing the calculation to that of the flow in a single pair of vortices. The length of annulus required to hold such a pair was however, not known (the "wavenumber selection" problem). The value of this quantity was therefore arbitrarily set in all of the simulations, with the only justification coming from the excellent agreement with experimental data. Inspiration for the choice of length was drawn from Taylor's original observations that each vortex occupies a length of annulus approximately equal to the gap width between the cylinders, however, sensitivity to such a parameter must be included in the study.

To a limited extent, the vortices appeared capable of expanding or contracting to fit into the available space, ±15% being typical limits before convergence difficulties were encountered. The grid length was found to significantly affect the flow field at solution, both for the pure rotary flow and the combined flow cases. Consequently, the radial velocity component norm, used for finding the critical value of Re by extrapolation, was also significantly affected. Extrapolating normv data generated by grids of differing physical length however, yielded one value of the critical Reynolds number, since all of the lines converged to a single point.

6. CONCLUSIONS AND FUTURE WORK

The transition from laminar flow to laminar flow containing Taylor vortices has successfully been calculated for a Newtonian fluid in a thin annulus, both in the absence and in the presence of a superimposed axial pressure drop. The method is based on the use of steady state simulations, made possible by the careful choice of a moving reference frame, to generate relatively few data points from which the critical parameter may be predicted by linear extrapolation. The results are in qualitative and quantitative agreement with literature experimental and theoretical values up to the highest so far predicted at $Re_c = 135$.

A modified Herschel-Bulkley model has been implemented to represent non-Newtonian fluid behaviour, and exercised for the two restricted cases of Power law and Bingham fluids. Shear thinning fluids, characterised by $n < 1$ in the power law model, have been shown to increase the range of primitive variables for which vortices are sustained in the flow relative to the Newtonian equivalent case. The introduction of a yield stress causes a marked increase in the stability of the laminar regime, and consequently, vortices are not generated until much higher rotation speeds are attained. Both of these non-Newtonian trends are in qualitative agreement with the work of Naimi et al[9]. More extensive comparison was not possible due to incompatibility of geometry and a lack of information regarding the fluid properties.

For the Power law model, early indications have led the authors to suspect that the quantity normv may be linearly related to U_m^2 and hence $Re^{2/(2-n)}$ for all post-critical solutions, regardless of the value of index n. Sensitivity with regard to computational grid has yet to be completely explored. However, while small adjustment to the position of the transition boundary is probable, major reshaping seems unlikely.

The main aims of planned future work are as follows:

i) To pursue further non-Newtonian simulations, both in terms of the extent of deviation from Newtonian behaviour, and by combining the simple Power law and Bingham models into the complete Herschel-Bulkley equation.

ii) Extend the regime boundary to higher values of axial and rotary flow parameters.

iii) Examine the effect of non-isothermal operation, both with and without temperature-dependent viscosity parameters.

iv) By making use of body fitted coordinate systems, it should be possible to investigate the effect of allowing the location of the inner cylinder within the outer to become eccentric.

v) Consider the implications that flow regime has on the ability of a fluid flow to transport solid particles, taking into account the reverse effect, namely the influence of the particles on the flow.

ACKNOWLEDGMENTS

The work has been funded under the UKAEA corporate research programme, and copyright on both written material and figures has been retained by the UKAEA. The results form part of the Ph.D. research of T.J. Lockett, and the project is overseen in collaboration with the Department of Chemical Engineering at Imperial College, London.

© UKAEA 1990.

REFERENCES

1. Taylor G.I., "Fluid friction between rotating cylinders. Part I: Torque measurements", Proc. Roy. Soc. A, Vol 157, 546–564, (1936).

2. Donnelly R.J. and Simon N.J., "An empirical torque relation for supercritical flow between rotating cylinders", J. Fluid Mech., Vol 7, 401–418, (1960).

3. Kaye J. and Elgar E.C., "Modes of Adiabatic and Diabatic fluid flow in an annulus with an inner rotating cylinder", Trans. A.S.M.E., 753–765, April 1958.

4. Snyder H.A., "Experiments on the stability of spiral flow at low axial Reynolds numbers", Proc. Roy. Soc. A, Vol 265, 198–214, (1962).

5. Coney J.E.R. and Simmers D.A., "The determination of shear stress in fully developed laminar axial flow and Taylor vortex flow using a flush-mounted hot film probe", DISA information Vol 24, 9–14, May 1979.

6. Abdullah Y.A.G. and Coney J.E.R., "Adiabatic and Diabatic flow studies by shear stress measurements in annuli with inner cylinder rotation", J. Fluids Engng., Vol 110, 4, 399–404, 1988.

7. Simmers D.A. and Coney J.E.R., "The experimental determination of velocity distribution in an annular flow", Int. J. Heat and Fluid Flow, Vol 1, 4, 177–184, 1979.

8. Simmers D.A. and Coney J.E.R., "A Reynolds Analogy solution for the heat transfer characteristics of combined Taylor vortex and axial flows", Int. J. Heat and Mass Transfer, Vol 22, 5, 679–689, 1979.

9. Naimi M., Devienne R. & LeBouche M., "Etude dynamique et thermique de l'ecoulement de Couette-Taylor-Poiseuille; cas d'un fluide presentant un seuil d'ecoulement", Int. J. Heat & Mass Trans., Vol 33, 2, 381–391, (1990).

10. Okrajni S.S. and Azar J.J., "The effects of mud rheology on annular hole cleaning in directional wells", SPE Drilling engineering, 297–308, (August 1986).

11. Tomren P.H., Iyoho A.W. and Azar J.J., "Experimental study of cuttings transport in directional wells", SPE Drilling Engineering, 43–56, (Feb 1986).

12. Becker K.M. and Kaye J., "Measurements of Diabatic flow in an annulus with an inner rotating cylinder", Trans. A.S.M.E., J. Heat Trans., 97–105, May 1962.

13. Bjorklund I.S. and Kays W.M., "Heat transfer between concentric rotating cylinders", Trans. A.S.M.E., J. Heat Trans., 175–186, August 1959.

14. Davey A., "The growth of Taylor vortices in flow between rotating cylinders", J. Fluid. Mechs., Vol 14, 336–368, 1962.

15. Di Prima R.C., "The stability of a viscous fluid between rotating cylinders with an axial flow", J. Fluid. Mechs., Vol 9, 621–631, (1960).

16. Chandrasekhar S., "The hydrodynamic stability of viscous flow between coaxial cylinders", Proc. Nat. Acad. sci., wash., Vol 46, 141–143, 1960.

17. Chandrasekhar S., "The stability of spiral flow between rotating cylinders", Proc. Roy. Soc. A, Vol 265, 188–197, (1962).

18. Coney J.E.R. and El-Shaarawi M.A.I., "Laminar heat transfer in the entrance region of concentric annuli with rotating inner walls.", J. Heat Transfer, Vol 96, 4, 560–562, Nov 1974.

19. Houwen O.H. and Geehan T., "Rheology of Oil-based muds", 61st Annual Technical Conference and Exhibition, Soc. Pet. Engrs., New Orleans, LA, 5–8th October 1986. (SPE paper No: 15416)

20. Alderman N.J., Gavignet A., Guillot D. and Maitland G.C., "High-temperature, high-pressure rheology of water-based muds", 63rd Annual Technical Conference and Exhibition, Soc. Pet. Engrs., Houston, TX, 2–5th October 1988. (SPE paper No: 18035)

© UKAEA 1990

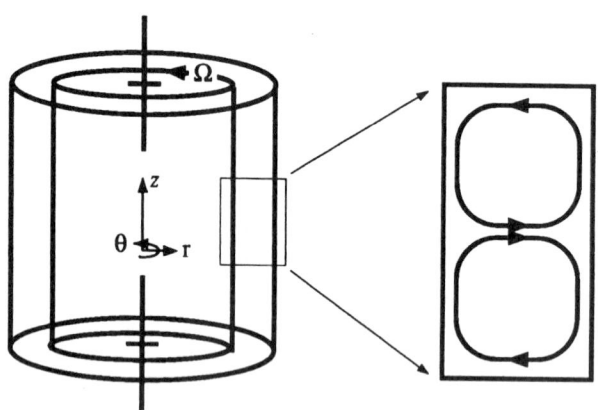

Fig 1 Concentric cylinder geometry with exploded view of Taylor vortices.

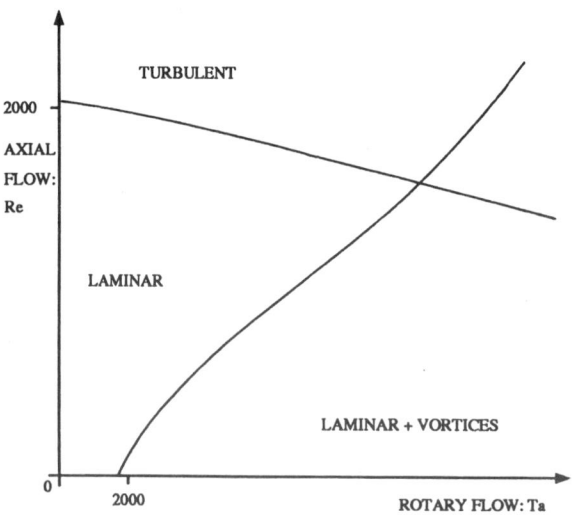

Fig 2 Schematic diagram of axial-rotary flow regime map.

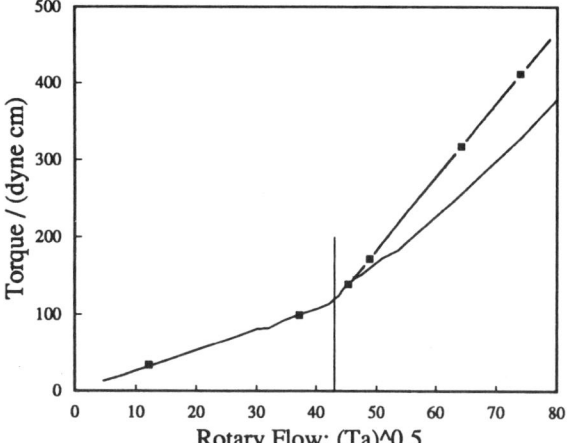

Fig 3 Comparison of torque values for rotary flow. — Experimental data of Donnelly & Simon[2], ■ Numerical predictions of present study.

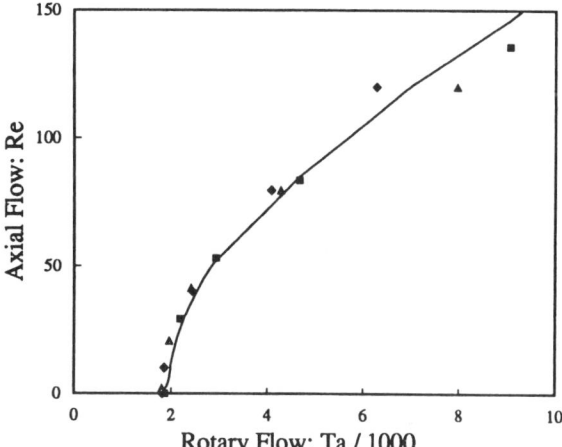

Fig 4 Comparison of predicted and observed transition between Laminar (top left) and Laminar + Vortices (bottom right) flow regimes. — Experimental data of Snyder[4], ■ Numerical predictions of present study, ▲ Theoretical predictions of DiPrima[15], ♦ Theoretical predictions of Chandrasekhar[16].

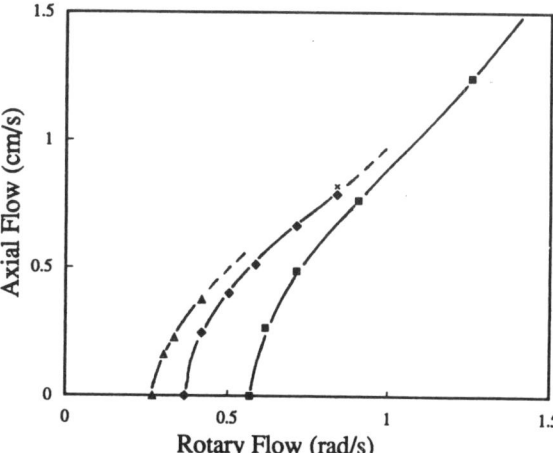

Fig 5 Numerical predictions for Newtonian and Power law fluid transitions between Laminar (top left) and Laminar+Vortices (bottom right) flow regimes. $\rho = 1585$ Kgm^{-3}, $K = 9.2$E-04 Pas^n, $R_2 = 6.28$E-02 m, $R_1 = 5.97$E-02 m. ■ Power law index, n, = 1.0, ♦, x n = 0.8, ▲ n = 0.6.

C413/043

The rotating matrix swirling gas burner: performance and computational design

W A ABD AL-MASSEEH, BSc, MSc, D BRADLEY, BSc, PhD, CEng, FIMechE, FRS, FInstP, P H GASKELL, BSc, PhD, A ISHIKAWA, BSc and A K C LAU, BSc, PhD
Department of Mechanical Engineering, University of Leeds

SYNOPSIS The characteristics of a rotating matrix burner are examined experimentally and also through a laminar flamelet mathematical model. The model embraces the important effects of flame strain, which can be high enough to quench a flame. This model is coupled with a direct stress hydrodynamic turbulence model, which consists of sixteen differential equations. Numerical solutions are compared with observations of combustion in a novel rotating matrix swirl burner and there is good agreement between prediction and observation. Blow-off velocities are also predicted and are in good agreement with those measured.

1 INTRODUCTION

The introduction of swirl to the incoming axial flow of fuel and air to a burner can have an appreciable influence on flame structure, and angled swirlers for the primary air have been used in gas turbines since the early Whittle engine (1). For predominantly axial flow, the circumferential velocity of swirl is an extra source of momentum and kinetic energy that can modify the mean flow pattern and enhance the transport processes. The swirling component of momentum can induce a flow divergence that can lead also to recirculation. The dissipation of the swirl kinetic energy enhances the mixing processes. Flow circulation can have a large influence on the flame shape and stability. The present paper will show by both mathematical modelling and experiment how, for a premixed flame, an increase in swirl can considerably reduce the flame length. A shorter and more intense flame, with a higher volumetric heat release rate, is favoured for the more compact power units. On the other hand, for some furnaces a longer flame is to be preferred. It is highly desirable to develop appropriate design procedures for the particular requirements.

For diffusion flames swirl can enhance and closely control the mixing of fuel and air, particularly when secondary, and even tertiary air streams are employed. This creates the possibility of controlled staged combustion, such as might reduce those NO_x emissions that originate from fuel nitrogen in coal combustion (2). Angled swirlers, variable vane swirlers, tangential entry flows, rotating pipes and rotating matrices have all been used to generate swirl. Reference (3) summarizes many of these development. The present paper presents results for swirl created by a novel rotating matrix of hypodermic tubes and a gaseous premixture. This configuration can give infinitely variable control of the swirl for any axial velocity. A strained laminar flamelet mathematical model is employed for the combustion (4,5), of the form used in an earlier analysis of combustion, in which swirl was induced by vanes in an annulus (6). With appropriate modelling of the mixing process it can be applied also to diffusion flames (7).

From the rotating matrix a methane-air premixture discharges into a cylindrical glass tube, and the flame is photographed over a range of axial velocities and swirl numbers. The blue flame radiation is principally due to chemiluminescence from the C_2 and CH radicals in the reaction zone, the intensity of which correlates well with the volumetric chemical heat release rate (8). Consequently, the photographed intensity is compared with the computed volumetric heat release rate. The agreement between the two is qualitatively good, but this is not intended in this paper as a rigorous quantitative validation of the model. Rather it is a demonstration that the spatial distribution of the reaction zone can be well predicted over a wide range of flow conditions. Another observation of considerable practical importance is the flow rate at flame blow-off and this is well predicted over a range of conditions. The model is thus shown to reveal the principal characteristics of the various flame forms that can be generated by the rotating matrix burner. The infinitely variable speed control of the matrix suggests a useful technique for the control of combustion processes.

2 ROTATING MATRIX SWIRL BURNER

Figure 1 is a schematic drawing of the rotary matrix swirler and burner which were built for the present study. The swirler steel tube is 140mm long with inner and outer diameters of 12 and 16 mm. Swirl is generated by rotating this tube with a variable speed 1.5 kW d.c. motor through a cascade of belt driven gears. The overall gear ratio is 18 and the motor maximum speed is 3,000 rpm imparting a maximum speed of 54,000 rpm to the rotary tube. This presents some problems of sealing around the swirler tube and a high flow rate of recirculating lubricating oil is necessary to cool the tube and the seals. Premixed methane-air, at atmospheric pressure and 290K, enters the tube and discharges into a silica glass burner tube, in which combustion occurs. This is of 39.2 mm diameter and 0.5 m length. A tube length of one meter also was used and this had no effect on the flame. Twenty five hypodermic thin-walled tubes of 1 mm diameter and 12 mm length are packed into the end of the swirler tube. This creates solid body rotation of the gas at the tube exit. The swirl number, S, defined as the ratio of the axial fluxes of angular and axial momenta divided by the tube radius, in such a case gives $W = 2US(r/R)$, where W is the tangential velocity at radius r, U the mean axial velocity and R the radius of the tube.

Since both the axial and angular momenta of the premixture are independently variable by means of a valve and a motor speed controller, the present swirler can conveniently provide a wide range of conditions from which to assess both the burner performance and predictive power of the mathematical model.

3 STRAINED LAMINAR FLAMELET MODEL

For subsonic compressible flows in industrial burners and furnaces, combustion affects the hydrodynamic field through the density changes across the flame brush. If the ideal gas law is invoked, these can be obtained from the temperature field which is determined from the energy equation. Accurate modelling of the volumetric heat release rate, the source term in this equation, is not easy because of the complex interaction of turbulence and chemistry. The detailed chemical kinetics of many higher hydrocarbon fuels are uncertain. Even where they are known, current limitations in computing power make it impractical in turbulent flames to solve the conservation equations for all the chemical species. In principle, a composite probability density function (pdf) of all relevant hydro-thermochemical variables is required to describe the nonlinear interactions between fluid mechanics and chemistry (9). Again, the computational effort can be enormous.

The laminar flamelet concept considerably simplifies the modelling of the heat release rate. This concept rests upon the predominating role of molecular processes in flame propagation, with the consequence that a turbulent flame might be regarded instantaneously as an array of laminar flames. The chemical processes of premixed flames are strongly dependent upon temperature and it is convenient to express them in terms of the dimensionless temperature rise, which acts as a single reaction progress variable, $\theta = (T - T_u)/(T_b - T_u)$. Here T is the gaseous temperature and subscripts u and b denote unburnt and burnt adiabatic states. The relationship between the thermochemical states and in particular, the heat release rate q, and θ is given by a mathematical model of the laminar flame.

The situation is complicated by the fact that straining of the flame by the flow field can affect the heat release rate and other parameters. Laminar flame models (10,11) are available which express the volumetric heat release rate, q_l, in such a flame that is subjected to a strain rate, s, in terms of both θ and s. In turbulent combustion at a given point, both θ and s will be fluctuating. Such fluctuations can be expressed by $p(\theta,s)$, a joint pdf of θ and s. The flamelet model expresses the mean turbulent volumetric heat release as,

$$\bar{q} = \int_{-\infty}^{\infty} \int_0^1 q_l(\theta,s) \, p(\theta,s) \, d\theta ds \qquad (1)$$

If it is assumed that θ and s are uncorrelated and that, to a justifiable first approximation, q_l is unaffected by strain until a limit quenching strain rate, s_q, is attained, above which value no heat release occurs and all strain rates are positive, then Eq. (1) becomes,

$$\bar{q} = \int_0^{s_q} p(s)ds \int_0^1 q_l(\theta)p(\theta)d\theta \qquad (2)$$

If $p(s)$, the pdf of flame straining is given by a quasi-gaussian pdf (5,12), then the first integral in Eq. (2) is,

$$\int_0^{s_q} p(s)ds = erf(\frac{s_q}{\sqrt{\pi \bar{s}}}) \qquad (3)$$

A further problem is the relationship between $\bar{s}$, the mean strain rate, and the turbulent energy dissipation rate per unit mass, ϵ. Bradley and Lau (5) tentatively have suggested an interim expression of $\bar{s} = 0.07(\epsilon/\nu)^{0.5}$

Earlier (13), it has been shown that the well known $k - \epsilon$ model of turbulence over-predicts the decay of swirl and a direct stress model is to be preferred. In ref. (6), such a model for variable density flows has been proposed, coupled to a strained flamelet model. Its predictions are in encouraging agreement with temperatures measured in annular turbulent swirling premixed methane-air combustion. This model is now used to predict not only flame position in the swirl burner, but also the blow-off limits. Strained laminar flame data, $q_l(\theta)$ and s_q for methane-air mixtures were kindly supplied by Dixon-Lewis (10,14) from a mathematical model that employed detailed chemical kinetics. A beta function is employed for $p(\theta)$ with first and second moments of θ obtained from the energy equation (4,5).

The transport equations for the first two moments of temperature are,

$$\bar{\rho}\frac{D\tilde{T}}{Dt} = -\frac{\partial \overline{\rho u_m'' T''}}{\partial x_m} + \bar{q}/C_p \qquad (4)$$

$$\bar{\rho}\frac{D\widetilde{T''^2}}{Dt} = \frac{\partial}{\partial x_m}C_{T'}\bar{\rho}\frac{k}{\epsilon}\overline{u_m'' u_n''}\frac{\partial \widetilde{T''^2}}{\partial x_m}$$

$$+ 2\overline{\rho u_m'' T''}\frac{\partial \tilde{T}}{\partial x_m} + \frac{2\overline{T''q''}}{C_p} - C_T\bar{\rho}\frac{\epsilon}{k}\widetilde{T''^2} \qquad (5)$$

where $\frac{D}{Dt}$ denotes a total derivative, ρ the density, u_m the Cartesian velocity component in the x_m coordinate direction, C_p specific heat at constant pressure, $C_{T'}$ and C_T are empirical constants given in (6), $\overline{(\ )}$ and $\widetilde{(\ )}$ denote time and Favre mean quantities and $(\ )''$ a Favre fluctuating quantity. The first two moments of θ are directly related to those of T by :

$$\tilde{\theta} = (\tilde{T} - T_u)/(T_b - T_u) \qquad (6)$$

$$\widetilde{\overline{\theta''^2}} = \widetilde{\overline{T''^2}}/(T_b - T_u)^2 \qquad (7)$$

and the former moments are used to evaluate the beta function $p(\theta)$,

$$p(\theta) = \frac{\theta^{\alpha-1}(1-\theta)^{\beta-1}}{\int_0^1 \theta^{\alpha-1}(1-\theta)^{\beta-1}\,d\theta} \qquad (8)$$

where α and β are the parameters of the function,

$$\alpha = \tilde{\theta}[\tilde{\theta}(1-\tilde{\theta})/\widetilde{\overline{\theta''^2}} - 1] \qquad (9)$$

$$\beta = (1-\tilde{\theta})[\tilde{\theta}(1-\tilde{\theta})/\widetilde{\overline{\theta''^2}} - 1] \qquad (10)$$

The overall model comprises one continuity equation and fifteen transport equations for the sixteen dependent variables, namely, three velocity components, one pressure, six Reynolds stresses, one temperature, three turbulent heat fluxes, one temperature fluctuation and one turbulent dissipation rate. Full details are given in ref (6). For the present two-dimensional, swirling condition, the sixteen elliptic transport equations were written in a general form for a cylindrical coordinate system and solved by a pressure correction method (15). A higher order bounded scheme was used for the convection terms in the general transport equation to ensure good numerical accuracy and prevent nonphysical oscillations in the numerical solutions (16). A non-uniform grid system of either 35x32 or 32x32 was employed, which covered the interior domain of the glass burner. The assumed boundary conditions were : (i) zero derivatives for all variables at the burner exit, (ii) standard wall function treatment at solid boundaries, (iii) zero derivatives, or zero values for all variables at the axisymmetric centre line, depending on whether they are odd or even functions of the radial distance; (iv) value of variables at the burner entry are those for a fully developed pipe flow and (v) adiabatic conditions. A 35 x 32 grid system was used for high swirl and the grid lines were concentrated in the region of rapid changes at the burner entry.

4 PERFORMANCE OF ROTATING MATRIX BURNER

4.1 Flame Size and Shape

The flame photographs, for an equivalence ratio of 0.84, in Figs 2 and 3 show the effects of increasing the swirl number by increasing the speed of rotation for mean axial velocities from the matrix of 15 and 30 m/s. The exposure time was 33 ms with Kodak ISO 800 film. Increasing the swirl shortens and thickens the flame, whilst drawing it closer to the matrix. The intensities of the flame emissions from C_2 and CH bands are indicative of the mean volumetric heat release, $\bar{q}$, computed contours of which are shown underneath each photograph. Bearing in mind that the contours relate to a diametral plane, whilst the photographs are a two dimensional record of integrated emissions through the tube, the ability of the model to predict flame position is good.

Predicted velocity vectors for the axial-radial plane are shown in Fig 4 for a mean axial velocity of 15 m/s and reveal the aerodynamics of flame stabilisation. Without swirl, the flame is stabilised by an elongated recirculation zone initiated by the sudden expansion step. An increase in swirl number increases the flow divergence shortening this zone and causing the observed flame shortening and flattening. For S = 0.48 the flame is in contact with the wall and spreads along it. A further increase in swirl (S = 0.84) creates a second, central, zone of recirculation. This forces the heat release zone upstream towards the central region of the matrix. The heat release is concentrated there and towards and along the wall, with the flame stabilised by the two recirculation zones. With the higher axial mean velocity of 30 m/s, the maximum speed of rotation of the matrix tube of 54,000 rpm limited the swirl number to a value of about 0.42, too low to create central recirculation.

4.2 Flame Blow-off

For practical operation, in addition to knowing where the heat release is occurring during normal running, it is also necessary to know the flow rate at which the flame will blow off the burner. This was measured by gradually increasing air and methane flows at constant equivalence ratio, ϕ, until blow-off occurred. The mean axial velocities from the matrix for this are shown by the circled points in Fig 5. These data points are for no swirl, because to obtain a significant swirl number at such high axial velocities would necessitate much higher speeds of rotation than were possible. This is different from fixed vane type burners (6) which inherently maintain the same swirl number for all axial velocities.

Clearly, with no matrix rotation there is no central recirculation to affect the mechanism of blow-off. This must be explained in terms of the structure of the step recirculation zone and again we look to the mathematical model for explanation. Figure 6(a) shows the velocity vectors for this zone, together with the isotherms of mean temperature. The flame, of equivalence ratio 0.73, is stable at the mean axial velocity of 100 m/s. Hot gases are recirculated to the incoming premixture. Flame stability is possible because of the relative positions of the recirculation zone and the flame.

As the axial velocity increases, so also does the axial length of the recirculation zone, but to a decreasing degree, until further increase in velocity has little effect. In the central region the flame moves further downstream as the axial velocity increases. To some extent, this is countered by the increase in turbulent burning velocity consequent upon the increase in r.m.s turbulent velocity, but this effect is of diminishing influence as the velocity increases (17). At the higher axial velocities as the central flame moves away from the recirculation zone with further increase in velocity, the recirculation of hot gases diminishes and, with it, their stabilising influence. Eventually the flame blows off.

Figure 6(b) shows a split flame at the limit of stability. The part of the flame that has moved downstream is inherently unstable and as it blows off, so also must the other part that is associated with some degree of recirculation. A further increase in velocity creates a turbulent lifted flame located downstream and this also is revealed by the computed solutions. However, once again, in practice this could not be stable. Computations revealed these three regimes and the blow-off velocity was defined from interpolation of results at different axial velocities and equivalence ratios. The associated matrix exit mean axial velocities are shown by the full line curve in Fig 5. This is close to the experimentally measured blow-off velocities, suggesting the model to be valid over the full operating regime up to blow-off.

Although the phenomenon of flame straining has been known for many years (18) it is only recently that the reduction of heat release rate due to it has been incorporated in mathematical models (5). Because of the high flame straining that exists just prior to blow-off, it is of interest to quantify the effect of the allowance for this phenomenon. Accordingly, flame straining was neglected by putting $s_q = \infty$ in Eq. (2), to make the first integral unity, and blow-off flow rates were computed. These were found to be about four times higher than those measured experimentally, a clear indication of the necessity of incorporating the effects of flame straining into mathematical models. To this might be added the necessity of direct stress turbulence modelling because of the inadequacies of the $k - \epsilon$ model in swirling flows.

5 CONCLUSIONS

1. The characteristics of a novel, rotating matrix swirl burner, with sudden expansion geometry, have been studied experimentally.

2. Photographs show how increasing the swirl shortens and thickens the flames. At sufficiently high swirl numbers a central flame stabilisation recirculation zone is created.

3. These observations are supported by a direct stress, strained laminar flamelet mathematical model. The mechanisms by which step induced recirculation and central recirculation create a stable flame are explained by the model.

4. Blow-off velocities have been measured experimentally in the absence of swirl. The model predictions are in good agreement with these.

5. The importance of incorporating the effects of flame straining in mathematical models has been demonstrated.

ACKNOWLEDGEMENT

The authors wish to thank Dr. R. M. Davis of British Gas plc, Midlands Research Station, for his continued help and encouragement.

REFERENCES

(1) WATSON, E.A. and CLARKE, J.S. Combustion and combustion equipment for aero gas engines. J. Inst. Fuel, 1947, 21, 2-34.

(2) MATTHEWS, K.J. and MOBSBY, J.A. NO_x reduction in power station system. NO_x generation and control in Boiler and Furnace. Plant Symposium, Portsmouth, 1988, 84-108, Institute of Energy (South Coast Branch).

(3) GUPTA, A.K., LILLEY, D.G. and SYRED, N. Swirl Flows, 1984 (Abacus Press).

(4) BRADLEY, D., KWA, L.K., LAU, A.K.C. and MISSAGHI, M. Laminar flamelet modeling of recirculating premixed methane and propane-air combustion. Combustion and Flame, 1988, 71, 109-122.

(5) BRADLEY, D. and LAU, A.K.C., The mathematical modelling of premixed turbulent combustion. Pure & Applied Chemistry, 1990, 62, 803-814.

(6) ABD AL-MASSEEH, W.A., BRADLEY, D., GASKELL, P.H. and LAU, A.K.C. Turbulent premixed, swirling combustion : direct stress, strained laminar flamelet modelling and experimental investigation. Twenty-third Symposium (International) on Combustion, 1990, The Combustion Institute, in press.

(7) BRADLEY, D., GASKELL, P.H. and LAU, A.K.C. A mixedness-reactedness flamelet model for turbulent diffusion flames, Twenty-third Symposium (International) on Combustion, 1990, The Combustion Institute, in press.

(8) HURLE, I.R., PRICE, R.B., SUGDEN, T.M. and THOMAS, A. Sound emission from open turbulent premixed flame. Proceedings of the Royal Society of London, 1968, A303, 409-427.

(9) POPE, S.B. Pdf methods for turbulent reactive flows. Progress in Energy and Combustion Science, 1985, 11, 119-192.

(10) DIXON-LEWIS, G. Numerical modelling of strained flames with complex chemistry. Proceedings of Workshop on Gas Flame Structure, 1988, USSR Academy of Sciences, Part 2, 3-37.

(11) KEE, R.J., MILLER, J.A., EVANS, G.H. and DIXON-LEWIS, G. A computational model of the structure and extinction of strain, opposed flow premixed methane-air flames. Twenty-Second Symposium (International) on Combustion, 1989, The Combustion Institue, 1479-1494.

(12) ABDEL-GAYED, R.G., BRADLEY, D. and LAU, A.K.C. The straining of premixed turbulent flames, Twenty-Second Symposium (International) on Combustion, 1989, The Combustion Institute, 731-738.

(13) ABD AL-MASSEEH, W.A., BRADLEY, D., GASKELL, P.H. and LAU, A.K.C. The numerical predictability of strongly swirling flows, Seventh Symposium on Turbulent Shear Flows, Stanford, 1989, 21.3.1-21.3.6.

(14) DIXON-LEWIS, G. and ISLAM, S.M. Flame modeling and burning velocity measurement. Nineteenth Symposium (International) on Combustion, 1982, The Combustion Institute, 283-292.

(15) PATANKAR, S.V. Numerical Heat Transfer and Fluid Flow, 1980 (McGraw Hill).

(16) GASKELL, P.H. and LAU, A.K.C. Curvature-compensated convective transport : SMART, a new boundedness-preserving transport algorithm, International Journal for Numerical Methods in Fluids, 1988, 8, 617-641.

(17) ABDEL-GAYED, R.G., BRADLEY, D. and LAWES, M. Turbulent burning velocities : a general correlation in terms of straining rates. Proceedings of the Royal Society of London, 1987, A414, 389-413.

(18) KLIMOV, A.M. Laminar flame in turbulent flow, Zh. Prikl. Mekh. Tekhn. Fiz., 1963, 3, 49-58.

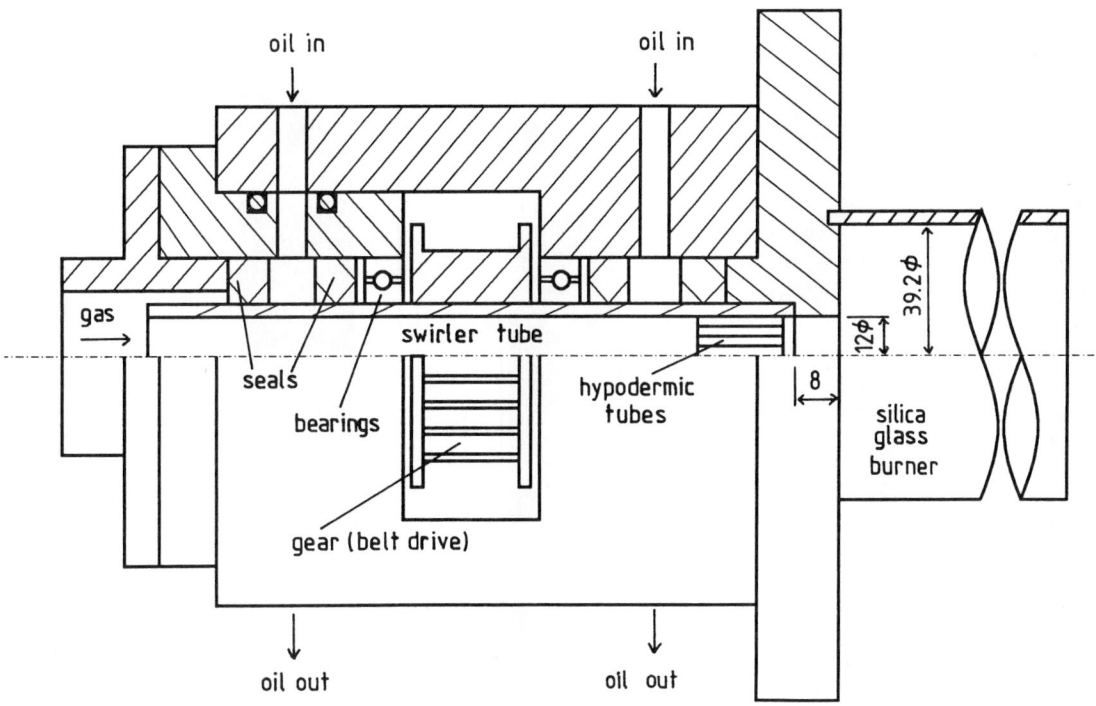

Fig 1

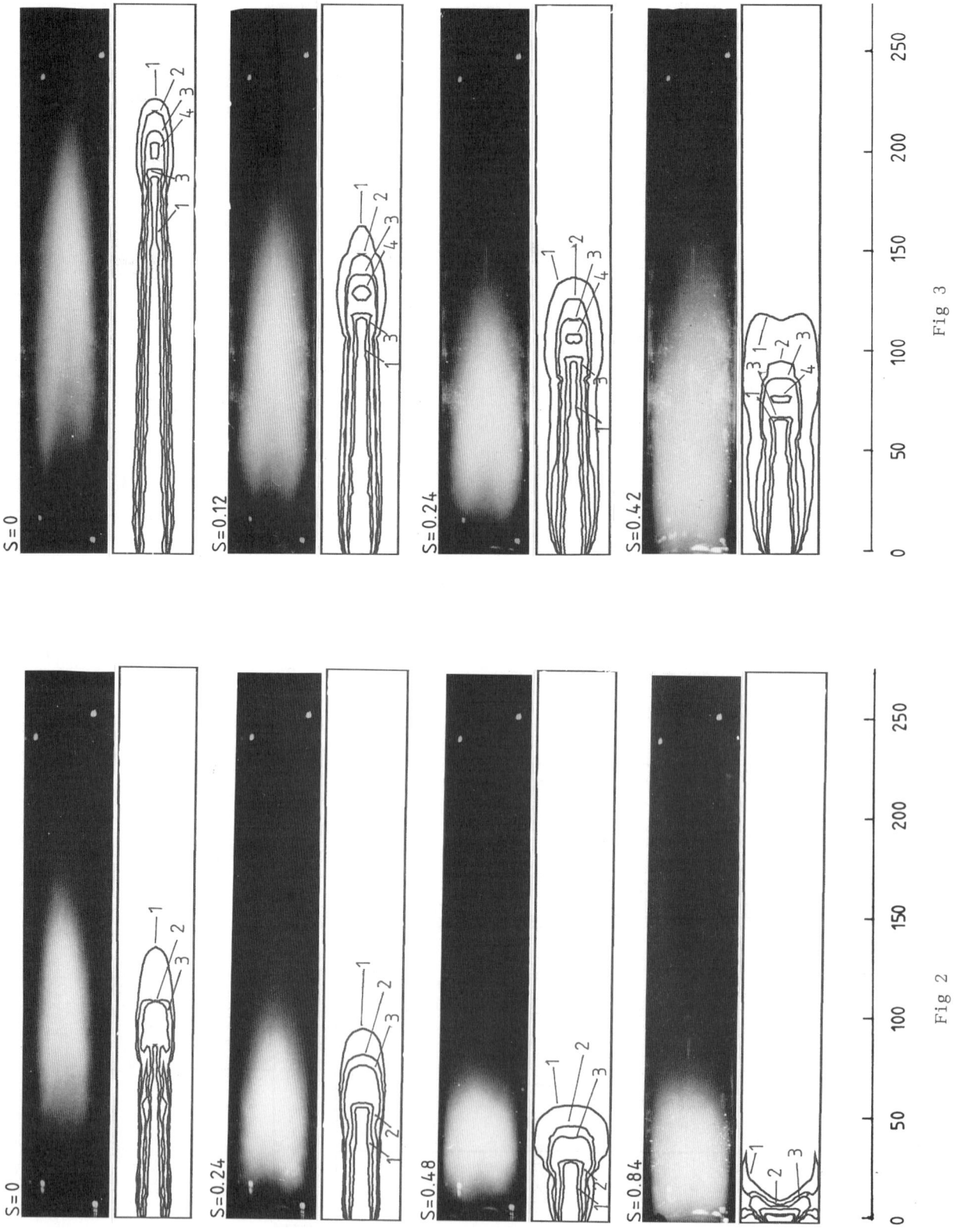

Fig 3

Fig 2

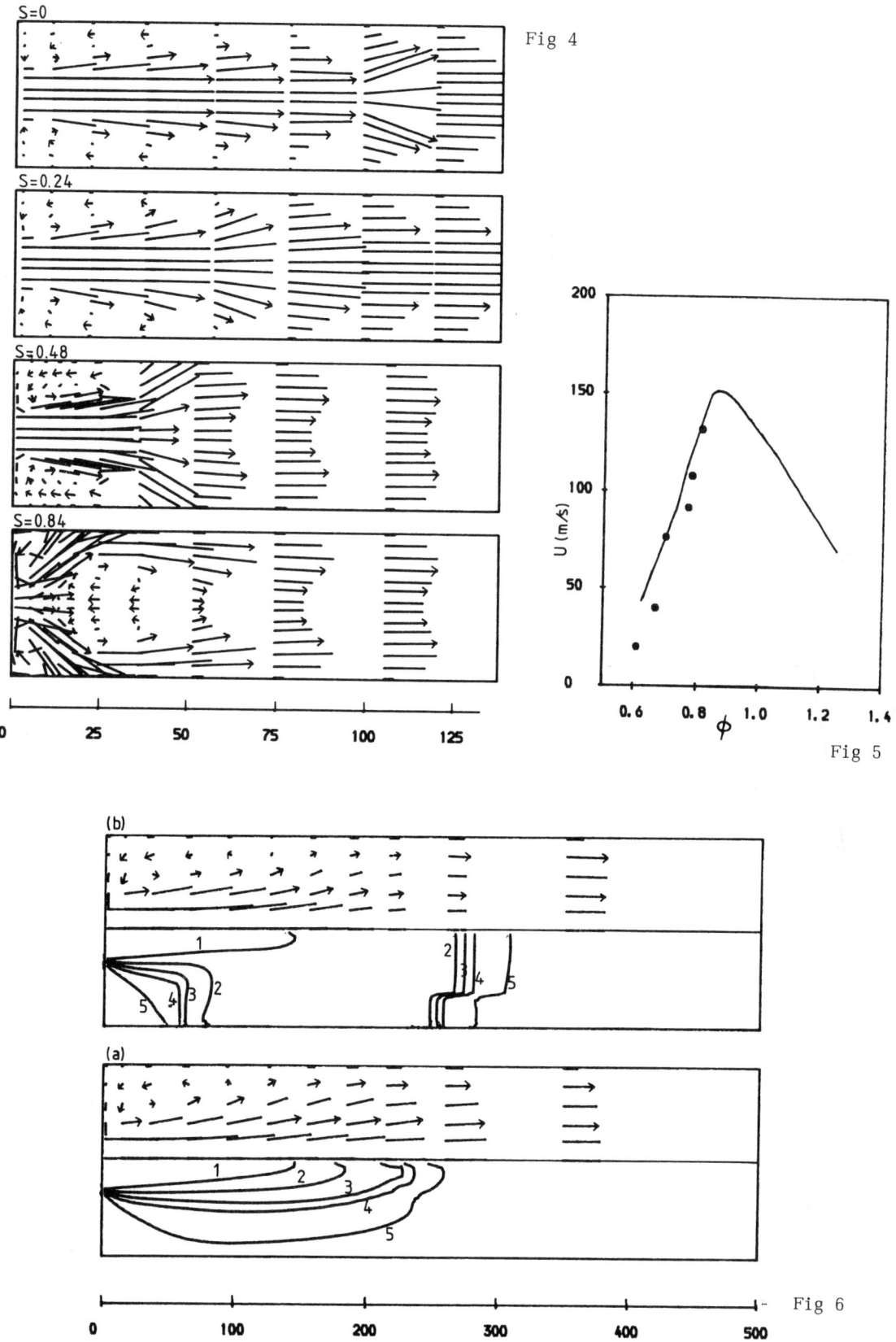

Fig 4

Fig 5

Fig 6

C413/054

Modelling gas-fired furnace flow and combustion using the PCOC code

S A BELTAGUI, BSc, PhD, MInstC, University of Glasgow
P J STOPFORD, AEA Technology, Harwell Laboratories, Didcott, Oxfordshire
R N FUGGLE, National Engineering Laboratory, East Kilbride, Glasgow
A M A KENBAR, BSc, MSc, Glasgow University
T RALSTON, BSc, National Engineering Laboratory, Glasgow
N MARRIOTT, AEA Technology, Harwell Laboratory, Didcot, Oxfordshire

SYNOPSIS This paper reports on comparisons between the predictions of the HTFS combustion modelling program, PCOC, and a range of measurements from NEL furnace, emphasising the modelling of the flow and combustion patterns.

The present burner has high-speed radial fuel-jets producing rapid turbulent pre-mixing of the reactants. Such flames have often proved difficult to predict due to limitations in the modelling of turbulence and chemical kinetics.

This paper describes calculations using the 'Eddy Break-up' combustion model. Supplementary information on inlet conditions were deduced from isothermal experiments. The predictions are compared with in-flame velocities and temperatures recently measured in the NEL furnace, which encompass a range of four burner swirl settings.

NOTATION

A	Eddy break-up constant (Model 1)	Dimensionless
B	Eddy break-up constant (Model 2)	Dimensionless
D	Damköhler number	Dimensionless
k	Kinetic energy of turbulence	m^2/s^2
M	Species mass fraction	kg/kg
R	Rate of chemical reaction	$kg/m^3 s$
r	Inlet radius	m
S	Swirl number	Dimensionless
s	Stoichiometric air-fuel ratio	Dimensionless
u	Axial velocity	m/s
v	Radial velocity	m/s
w	Swirl velocity	m/s
ε	Rate of dissipation of turbulent kinetic energy	m^2/s^3
ρ	Density	kg/m^3
τ	Time	s

Subscripts

c	Critical
ch	Chemical
f	Fuel
o	Oxidant
p	Product
t	Turbulent

1 INTRODUCTION

The successful prediction of swirling gaseous diffusion flames has been demonstrated by earlier HTFS work and others, eg (1). These works covered burner geometries consisting of two concentric pipes resulting in comparatively simple model boundary conditions with no pre-mixing. In industrial applications, however, burners often consist of a central gun which introduces several high-speed jets of fuel gas radially into the swirling air flow within a surrounding pipe. This report describes an extension of the PCOC program to allow the modelling of such burners.

The immediate objective is to model a large and comprehensive set of data which has been measured on the furnace at the National Engineering Laboratory (NEL). The data comprise isothermal (2, 3) and combusting velocities, major species concentrations and temperatures as a function of the air swirl.

The NEL furnace, shown schematically in Figure 1, consists of a 1 m internal diameter vertical cylinder, 3 m high. The rig is water-cooled in separate sections to facilitate measurement of heat-transfer distribution and is provided with ports for access by a range of probes. The facility is more fully described elsewhere (2-4).

The burner, (Figure 1), has a fuel injector consisting of sixteen radial gas jets which create intense turbulence to induce rapid mixing within the burner quarl. The resulting pre-mixing of fuel and oxidant means that the combustion rate will depend on the pre-heating of reactants and chemical kinetics of reactions and not just the rate of mixing. Any model of this type of burner must successfully account for both the pre-mixing and combustion phases in order to obtain accurate predictions.

This paper uses isothermal mixing data, (2,3) to avoid the necessity for 3-D turbulence modelling of the radial-jet flow. The partially pre-mixed nature of the combustion process is then taken into account by incorporating an 'eddy break-up' model into PCOC.

Before the calculations are described, the PCOC furnace code and the 'eddy break-up' model are briefly reviewed.

2 THE PCOC CODE

PCOC is an extension of the earlier HTFS furnace codes, TUFC and NERO, to include pulverised coal combustion. For gaseous flames, PCOC is essentially equivalent to TUFC except that the burner quarl can now be represented as a series of steps in the finite difference grid. The code solves the steady-state conservation equations for mass, momentum, energy and mixture fractions in an axisymmetric furnace. The solution is found iteratively using line relaxation and the SIMPLEC velocity-pressure coupling method of Reference (5). Turbulence is represented by the standard k-ε model and radiative heat transfer by a single grey gas approximation and finite-ordinate discretisation. The 'mixed-is-burnt' model, used by TUFC and PCOC for gaseous diffusion flame calculations, is described briefly in the next section.

3 REVIEW OF TURBULENT DIFFUSION FLAME MODELS

Early studies of turbulent jet flames lead to the development of two different classes of model, both of which were based on the assumption that the rate of combustion was controlled by the mixing of fuel and oxidant rather than by the chemical kinetics. The 'mixed-is-burnt' model (6) assumes that the chemical processes are so fast that fuel and oxidant do not co-exist at the same time and place. As a result, concentration fields can be described by a single conserved scalar, the 'mixture fraction'.

Turbulent fluctuations are included by assuming a shape for the pdf describing the fluctuations in mixture fraction. The model has been validated for a wide range of furnace sizes by various workers in HTFS and others eg (1).

The alternative approach is the 'eddy break-up' or 'eddy dissipation' model of Spalding (7) and Magnussen and Hjertager (8) which assumes that the reactants are homogeneously mixed in the fine-scale dissipative eddies of the turbulence. In the limit of fast chemistry, the reaction rate of the fuel is dependent on the concentration of the limiting reactant:

$$R_f = A\rho \frac{\varepsilon}{k} \mathrm{Min}\left(M_f, \frac{M_o}{s}\right) \quad \text{Model 1}$$

where M_f and M_o are the mean mass fractions of fuel and oxidant respectively, s is the stoichiometric ratio and A is a dimensionless constant which is usually obtained by fitting the predictions to the data. The choice of A is discussed below. Comparisons between the 'mixed-is-burnt' and 'eddy break-up' model, for example (9), generally indicate close agreement between the two models for near-stoichiometric conditions.

4 PARTIALLY-PREMIXED COMBUSTION MODELLING

When the fuel and oxidant co-exist in the same turbulent eddies, the reaction rate is controlled by chemical kinetics rather than by turbulent mixing. The 'mixed-is-burnt' assumed

pdf shape approach depends crucially on the impossibility of the co-existence of reactants in order to uniquely define a mixture fraction. Consequently, an extension to partially-premixed conditions is difficult if not impossible. On the other hand, the 'eddy break-up' model is easily extended.

The approach adopted by Bakke and Hjertager (10) is to quench the reaction at a cut-off point determined by the value of the Damköhler number (D), defined as the ratio of the turbulent and chemical time-scales (τ_t/τ_{ch}), ie

$$R_f = \frac{A\rho}{\tau_t} \mathrm{Min}\left(M_f, \frac{M_o}{s}\right) \quad D > D_c$$
$$= 0 \quad D < D_c \quad \text{Model 2}$$

where $\tau_t = k/\varepsilon$ and $D_c = 10^{-3}$. For natural gas combustion in air, a modified Arrhenius expression for τ_{ch} is given by Beltagui and Maccallum (11). This model has the advantage that τ_{ch} and D_c have direct physical relevance so they can be easily adjusted to apply to different reactants. Ideally, τ_{ch} should be measured in a turbulent flame as the temperature in eddies containing reacting gas may be significantly higher than the bulk temperature.

One way of avoiding explicit chemical modelling is to postulate that the reaction rate in eddies containing a cold unburnt mixture of reactants is proportional to the mixing rate with the surrounding hot combustion products. This phenomenological approach lead Magnussen and Hjertager (8) to a simple extension of the 'eddy break-up' model:

$$R_f = \frac{\rho}{\tau_t} \mathrm{Min}\left(AM_f, \frac{AM_o}{s}, \frac{B M_p}{1+s}\right) \quad \text{Model 3}$$

where M_p is the mass fraction of products and B is another constant. This model has been widely used for partially-premixed gaseous combustion, including coal volatiles combustion (12), with the constants A = 4 and B = 2 chosen empirically.

In Section 6 all three models are compared with the data from the NEL furnace.

5 COMPUTATIONAL DETAILS

The NEL furnace geometry was discretised by an axisymmetric grid of 54 axial and 42 radial intervals with 588 nodes in the quarl region. The QUICK higher-order differencing scheme was used to avoid the possibility of smearing the near-burner flow details by numerical diffusion.

The measured profiles of axial and swirl velocity at the inlet plane (2) were used as input to the program boundary conditions. The radial velocity was fixed by the requirement that the radial momentum of the fuel jet is conserved. The mixture fraction at the inlet was measured by helium tracer experiments performed earlier (2).

Values of k and ε were assumed proportional to the mean velocity values from the measurements reported in Reference 13. The effect of varying these will be discussed in the next section.

The radiation model assumes an emissivity of 0.2 over the quarl and 0.85 on the furnace wall. The mean absorption coefficient for the furnace gases was taken to be 0.355 m^{-1}.

Calculations were performed with under-relaxation fractions of 0.05 for density and reaction rate and 0.3-0.5 for the other variables.

About 2000-3000 iterations were required to reduce the dimensionless mass and enthalpy residuals to 3×10^{-4} and 10^{-2} respectively, at which point the calculations had effectively converged. Each calculation took about 8 minutes of CPU time on the Harwell CRAY-2 supercomputer.

6 RESULTS

The first calculations using the diffusion flame models, 'mixed-is-burnt' and the 'eddy break-up' (EBU) Model 1, greatly over-predicted the reaction rate as expected. As a result, the central recirculation zone (CRZ) was predicted to be over 1 m long compared with the measured 0.25-0.3 m.

A comparison of the centre-line axial velocity and temperature predicted by the three models and the measured values is presented in Figure 2 for swirl number of 0.45. The results for the other swirl intensities are similar.

The agreement with the data was considerably improved by switching to Model 2 or 3.

For model 3, Figure 3 shows the comparison of the predicted and measured flow reversal boundaries for three different swirl numbers. The swirl number is defined by:

$$S = \frac{\int_0^r \rho u w r^2 \, dr}{r_b \int_0^r \rho u^2 r \, dr}$$

where u is the axial velocity, w the tangential velocity and r_b is the burner radius. Examples of the detailed comparisons for temperature and velocity can be seen in Figures 4-7.

The results obtained with the explicit chemical kinetic model (EBU Model 2) were almost identical with temperatures, for example, agreeing to within 20°C with those of Model 3, see Figure 2.

6.1 Flow Pattern

The length of the CRZ was found to be dependent on the assumed inlet values of k and ε. The calculations shown in Figures 4-7 assume that $k \alpha (u^2 + v^2 + w^2)$ so that k = 10, 13 and 24 m^2/s^2 for swirl numbers 0, 0.45 and 0.9 respectively. An energy dissipation rate of 3000 m^2/s^3, independent of swirl, gave the observed variation of the CRZ length with swirl. The predictions were relatively insensitive to the choice of k and ε at the inlet (both to the overall magnitude and to any cross-stream variation) since the inlet turbulence level was always small compared to that generated by the combustion process itself, the most sensitive feature was the length of the CRZ. The effect of varying k and ε on the flow boundaries predicted by Model 3 for S = 0.45 is shown in Figure 8. In general, the length of the CRZ varies approximately as $\varepsilon r^2/k^{3/2}$ where r is the inlet radius.

It is particularly encouraging that the code successfully predicts a CRZ even in the absence of swirl, a feature associated with the radial fuel injection of this burner geometry.

6.2 Temperature Distribution

Temperature distributions are illustrated in Figure 4, where 4a, 4b and 4c show results for S = 0, 0.45 and 0.9 respectively. The upper half of each figure presents the predictions of Model 3. In all three cases the predictions yield higher temperatures in restricted areas than are evident in experimental measurements. Figure 4b is typical with a predicted torroidal region, close to the burner exit, where temperatures exceed those measured by 200°C. It is also notable that the predictions suggest lower exit temperatures than recorded in the measurements. Contour shapes in predictions and measurements are very similar with large zones of the flow being reasonably predicted.

Clearly, the local gas temperatures are highly dependent on the precision of the combustion model and, to a lesser extent, on the transport of mass and radiative heat transfer in the near-burner region. Future comparisons will address the heat-transfer behaviour of the predicted system.

6.3 Axial Velocity Profiles

Radial profiles of axial velocity are presented in Figures 5a and 5b. Since the region of most interest is that closest to the burner, results are confined to the first 1050 mm of the chamber. Although 3 components of velocity were measured at 4 swirl intensities, results are presented for only 2 swirl levels in this paper. For all axial locations, measurements from near-side and farside of the furnace centre line are discriminated by the use of cross and box symbols. Whilst not always coincident, a high degree of symmetry is evident in the measured data.

In general, axial velocity profiles are well predicted. In a partially-premixed flame, such as that examined here, the gas expansion associated with combustion has a major influence on the near-flame aerodynamics.

Figure 5a is typical of the set. Close to the quarl exit, very reasonable quantitative agreement in local velocity predictions is evident. PCOC predictions agree very closely with the experimental CRZ boundary and the boundary of the outer recirculation zone. Both forward and reversed peak velocities are well predicted up to 110 mm downstream of quarl exit. However, as the flow progresses, the predictions are seen to exaggerate peak velocity values and the associated gradients. PCOC predicts a maximum velocity off the centre line in all of the profiles illustrated, whereas the measurements indicate that this feature disappears between 300 and 545 mm.

6.4 Radial Velocities

Figure 6 is one example to illustrate the comparison of measured and predicted radial profiles of radial velocity for S = 0.9.

As previously reported, (2, 3), the least confidence in aerodynamic measurements is ascribed to the radial components of velocity. Here, as in isothermal comparisons, the general shape of radial profiles is well represented. It is worth noting that the position of turning points in the experimental profiles are closely represented by the predictions. In general, however, local measured radial velocity values exceed those predicted. Results at higher swirl exhibit marginally better agreement than those at low swirl.

6.5 Swirl Velocity

Figure 7 illustrates the radial profiles of swirl velocity for S = 0.9.

It is clear that, close to the quarl exit, predictions exceed measurements by as much as a factor of two for those areas of the flow where the experimental swirl velocity is well defined. As the flow progresses downstream, the predictions and measurements do exhibit improved agreement.

7 DISCUSSION

Given the complexity of the inlet conditions, the agreement between PCOC, with combustion Models 2 and 3, and data is most encouraging. In particular, the agreement between the chemical kinetic and product eddy mixing models indicate that either can be used depending on the type of reaction rate information available.

The length of the CRZ is found to be much smaller than in the corresponding isothermal flow (3). This is due to the volume expansion associated with combustion at about 0.25 m from the burner. The calculations suggest, however, that there is still some dependence on the burner aerodynamics through the inlet turbulence level.

Two aspects of the comparison require further investigation; the temperature gradients around the CRZ and the swirl velocity are generally overpredicted. Although the mixing produced by the radial fuel injection has been allowed for in this set of calculations, other effects such as the inevitable reduction of the mean velocities, has not.

The overprediction of the temperature gradients is probably due to the simple one-step chemistry assumed in the calculations. An extension of model 2 to multi-step chemistry is planned using measured values of intermediates, such as CO.

8 CONCLUSIONS

The 'eddy break-up' model is capable of simulating the combustion of natural gas using a central-gun-type burner. The present theoretical approach should be relevant to many industrial applications which use this burner configuration. The success of a combustion model based on chemical kinetics suggests that the model can also be applied in the case of other reactants for which kinetic data of the type provided by Reference 11 are available.

The objective of HTFS work on the NEL furnace system has been to provide detailed validation of furnace modelling on a semi-industrial scale facility. Future comparisons with PCOC will allow refinement of the predictions with access to some turbulence information from LDA measurements within the chamber. LDA work will also afford the opportunity of exploring further time-averaged velocities in the near-burner field. In addition, the heat-transfer predictions of the PCOC code will be compared with the detailed heat-flux data previously published, (13).

The local measurements of NO_x have been assembled and these also offer scope for validation of the thermal NO_x model in the PCOC program.

ACKNOWLEDGEMENTS

This work was carried out under the research programme of the Heat Transfer and Fluid Flow Service (HTFS) and was supported by the Department of Trade and Industry. This paper is Crown copyright.

A M A Kenbar wishes to acknowledge the support of the Government of Iraq.

REFERENCES

(1) KHALIL, E. E., HUTCHISON, P. and WHITELAW, J. H. The calculation of the flow and heat-transfer characteristics of gas-fired furnaces. Proc. 18th Symp. (Int.) on Comb., 1981, 1927-1938, The Combustion Institute, and Harwell report AERE-R 9591.

(2) BELTAGUI, S. A., FUGGLE, R. N. and RALSTON, T. Aerodynamics and mixing within the quarl of a variable-swirl burner. 1st European Conference on Industrial Furnaces and Boilers, Lisbon, Portugal, March 1988.

(3) BELTAGUI, S. A., FUGGLE, R. N. and RALSTON, T. An isothermal study of the aerodynamics of the flow issuing from a variable-swirl burner. 1st World Conference on Experimental Heat Transfer, Fluid Mechanics and Thermodynamics, Dubrovnik, Yugoslavia, Sept 1988, pp 1548-1555. Edited by R. K. Shah et al. Elsevier Applied Science Publishers, London.

(4) BELTAGUI, S. A., FUGGLE, R. N. and RALSTON, T. Measurement and prediction of heat transfer in the NEL furnace. I.Mech.E./I.Chem.E./Heat Transfer Society, 2nd UK National Heat Transfer Conference, Glasgow, 14-16 Sept, 1988, 2, Paper C163/88, pp 1219-1232, Mechanical Engineering Publications.

(5) VAN DOORMAAL, J. P. and RAITHBY, G. D., Enhancements of the SIMPLE method for predicting incompressible fluid flows. Numer. Heat Transfer, 1984, 7, 147-163.

(6) WILLIAMS, F. A. Combustion theory. 1965, Addison-Wesley, Reading, Mass.

(7) SPALDING, D. B. Mixing and chemical reaction in steady confined turbulent flames. Proc. 13th Symp. (Int.) on Comb., 1970, 649-658, The Combustion Institute.

(8) MAGNUSSEN, B. F. and HJERTAGER, B. H. On mathematical modelling of turbulent combustion with special emphasis on soot formation and combustion. Proc. 16th Symp. (Int.) on Comb., 1976, 719-730, The Combustion Institute.

(9) WILKES, N. S., GUILBERT, P. W., SHEPHERD, C. M. and SIMCOX, S. The application of Harwell-FLOW3D to combustion problems. Harwell report AERE-R 13508, 1989.

(10) BAKKE, J. R. and HJERTAGER, B. H. The effect of explosion venting in empty vessel. Int. J. Num. Methods Engng., 1987, 24, 129-140.

(11) BELTAGUI, S. A. and MACCALLUM, N. R. L. Stability limits of free swirling premixed flames. Part II. Theoretical prediction. J. Inst. Energy, 1986, 59, 165-167.

(12) LOCKWOOD, F. C., SALOOJA, A. P. and SYED, S. A. A prediction method for coal-fired furnaces. Comb. Flame, 1980, 38, 1-15.

(13) SCHMID, C. and DUGUE, J. Movable block swirler calibration. International Flame Research Foundation Doc. No F59/y/8, April 1990.

© Crown Copyright

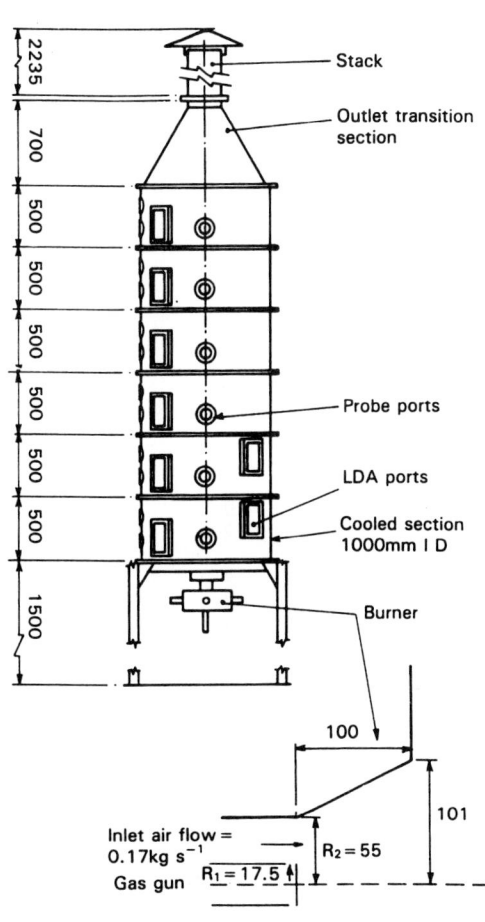

Fig 1 Schematic of furnace and burner arrangements

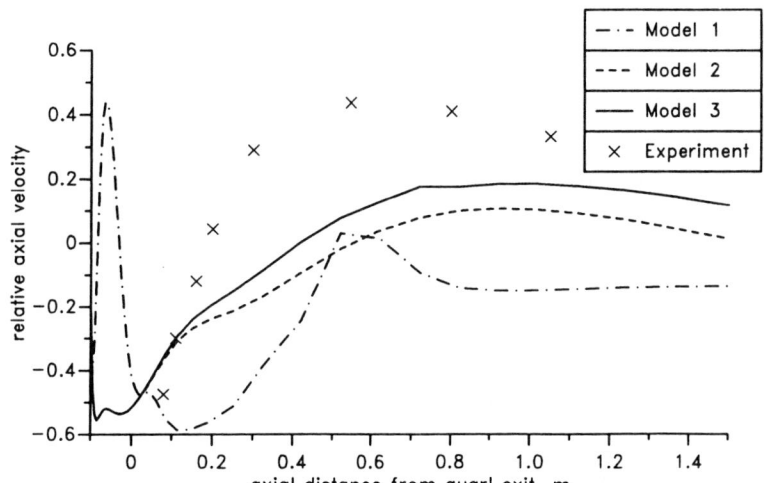

Fig 2a Predicted centre-line axial velocity compared with experiment for S = 0.45

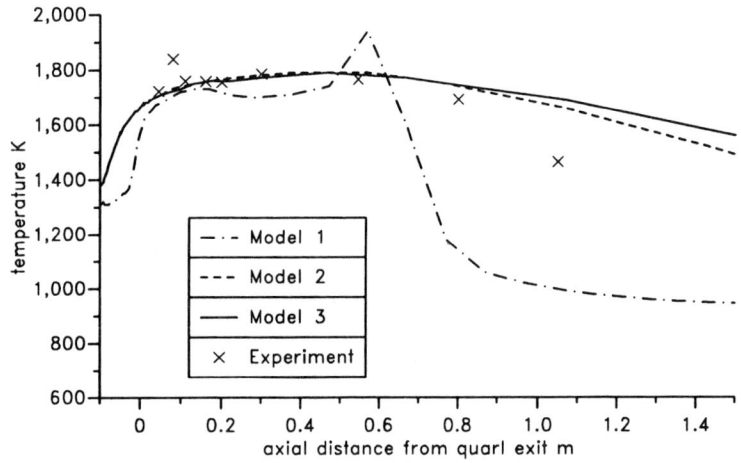

Fig 2b Predicted centre-line temperature compared with experiment for S = 0.45

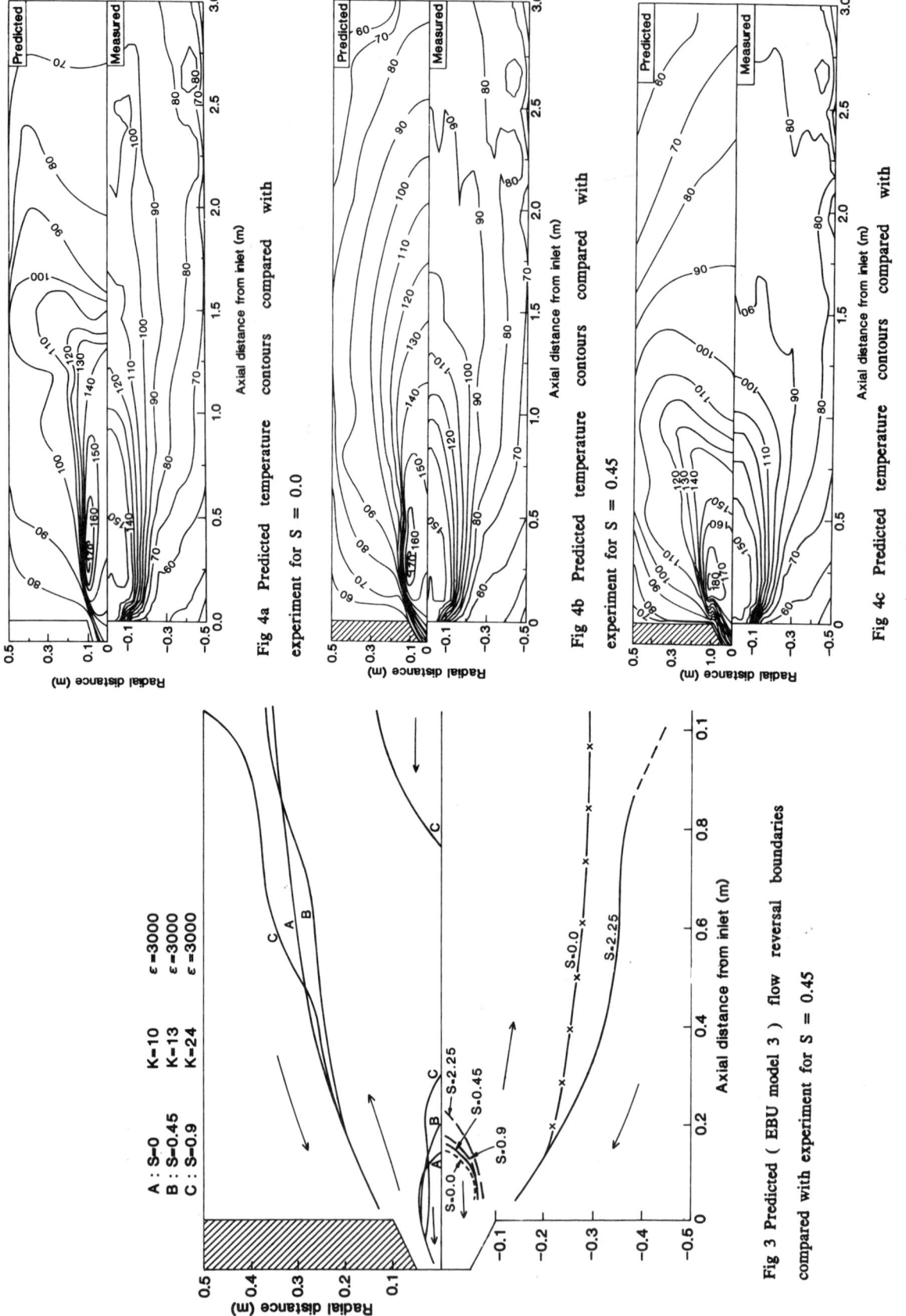

Fig 4a Predicted temperature contours compared with experiment for S = 0.0

Fig 4b Predicted temperature contours compared with experiment for S = 0.45

Fig 4c Predicted temperature contours compared with experiment for S = 0.90

Fig 3 Predicted (EBU model 3) flow reversal boundaries compared with experiment for S = 0.45

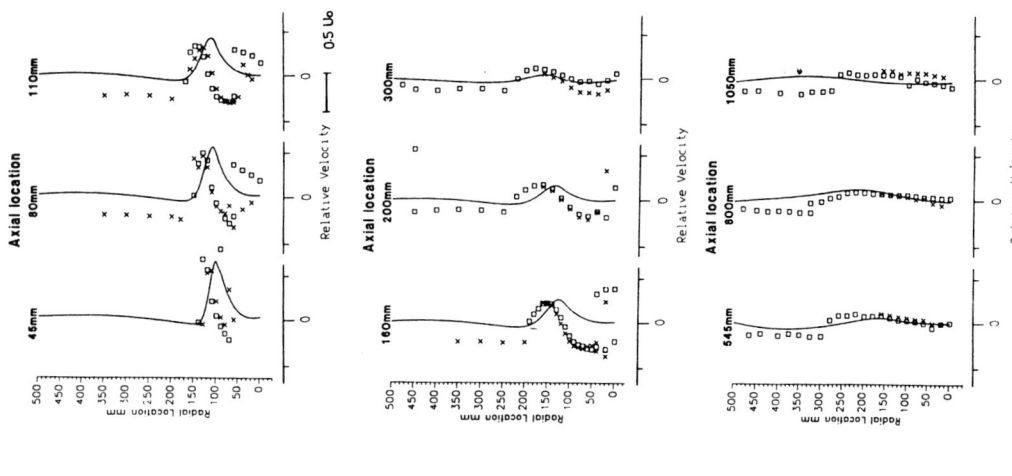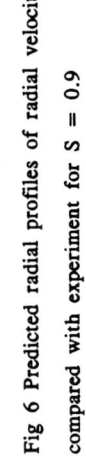

Fig 6 Predicted radial profiles of radial velocity compared with experiment for S = 0.9

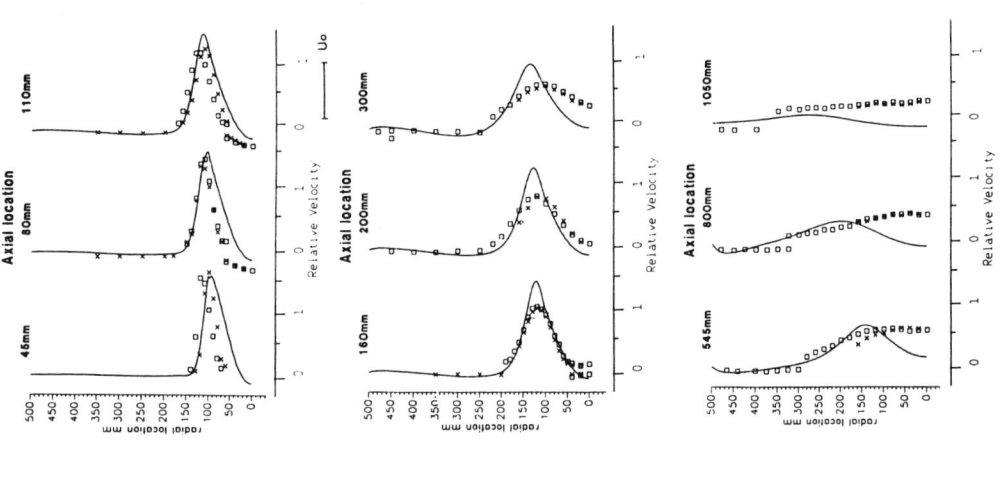

Fig 5b Predicted radial profiles of axial velocity compared with experiment for S = 0.90

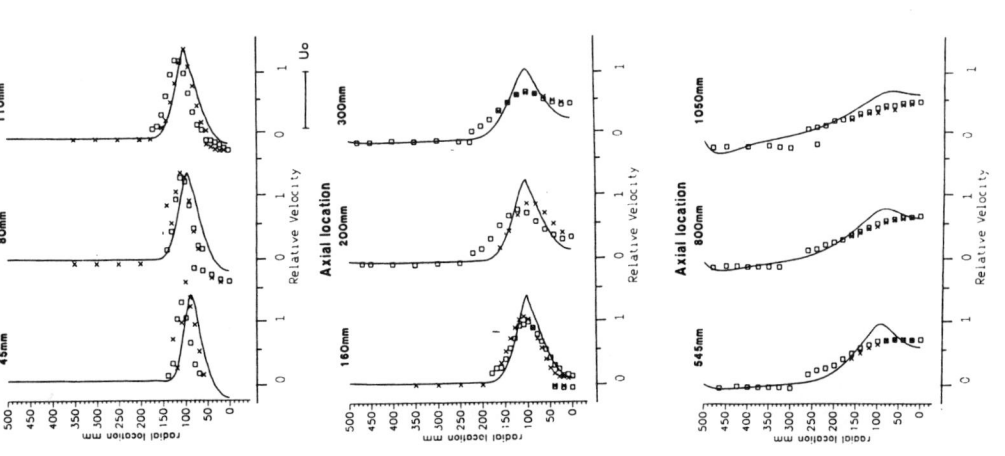

Fig 5a Predicted radial profiles of axial velocity compared with experiment for S = 0

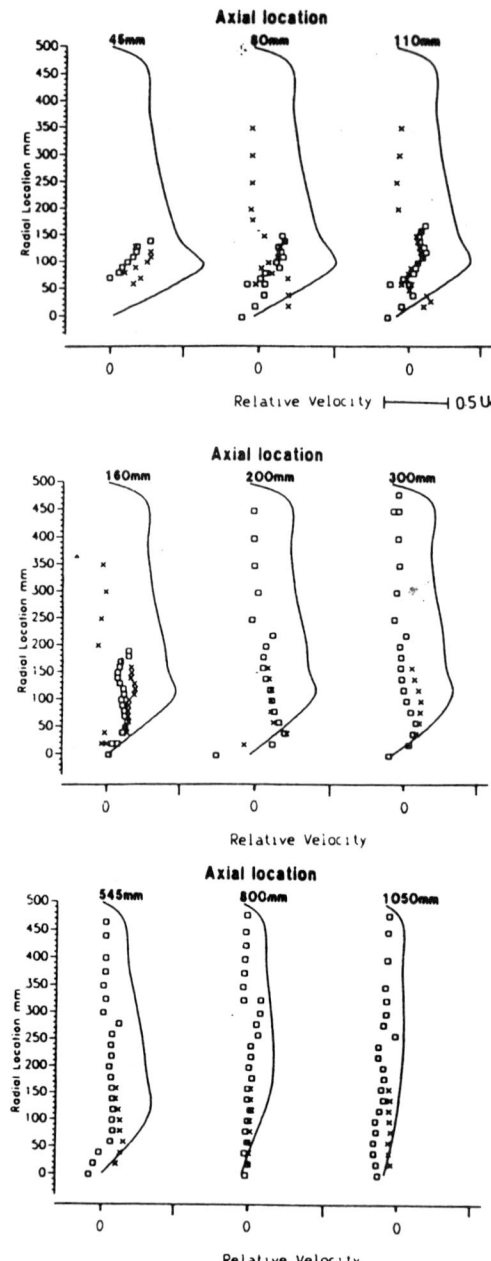

Fig 7 Predicted radial profiles of swirl velocity compared with experiment for $S = 0.90$

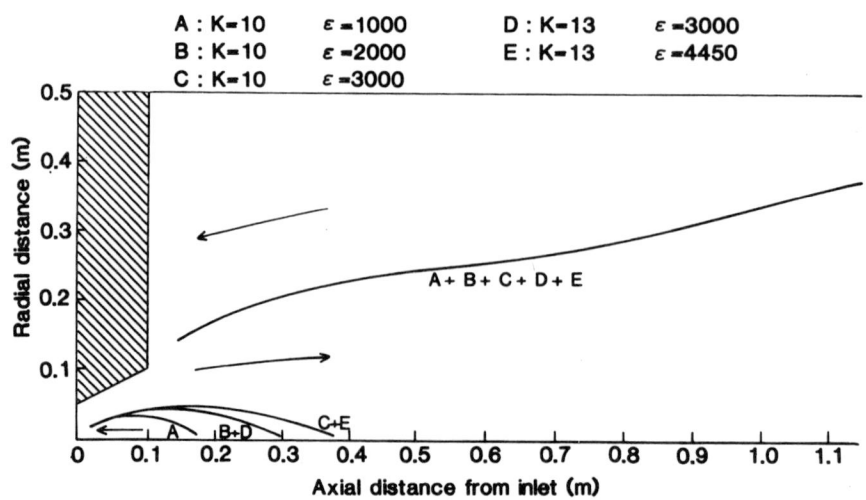

Fig 8 Flow boundaries predicted by model 3 for various choices of k and ε at inlet ($S = 0.45$)

C413/076

Exploitation of modern flow computation techniques in the design of turbine blading

C T J SCRIVENER, BSc, MSc
Rolls Royce plc, Derby
J GRANT, BSc, PhD
NEI Parsons Limited, Heaton, Newcastle upon Tyne

Synopsis

Two examples are presented illustrating the application of modern flow computation methods to facilitate the design of efficient turbine blading. The first example involves the use of viscous 2D and 3D CFD methods in the aerodynamic design of the High Pressure turbines of modern aero engine gas turbines. Understanding and minimisation of viscous losses in these low aspect ratio turbines is an invaluable aid in generating high efficiency turbine designs. Major loss sources include trailing edge flows, tip leakage flows and passage secondary flows.

The second example is that of a three-dimensional inviscid computation for the flow through the complete final stage of a large low pressure steam turbine. The method enables the twist of the fixed and moving blades to be optimised to reduce both flow incidence effects and the leaving energy from the stage. The blade geometries can be adapted to alleviate shock formation and avoid regions of potential flow separation.

1 INTRODUCTION

Modern Computational Fluid Dynamics (CFD) methods are having a very significant influence on the aerodynamic design of turbine blading for both aero engine gas turbines and steam turbines. There are two important aspects to this work. Firstly the development of larger and faster computers has permitted flow modelling of greater detail through the use of much finer calculation grids. Secondly their speed has provided the capability to do such detailed calculations at the design phase, to positively influence the blading geometry as it is designed. The ability to model greater detail can again be exploited in two ways. Firstly it can be employed to analyse geometrical complexity in the flow. Typical examples are a full blade to blade calculation for a stage (rotor and stator pair), or unsteady aerofoil flow, or the interaction of rotor and stator flows as the rotor passes through the stator exit flow.

Secondly, the flow can be modelled in much greater detail, for example the treatment of viscous flows, with the associated complexity and detail in the computational grids. This permits the understanding and optimisation of the flow to minimize the loss generation and dissipation to provide more competitive turbines with progressively increased efficiency.

The first part of this paper illustrates the application of viscous CFD methods in the design of HP turbine designs of high efficiency for aero engine application. The second part describes the application of three dimensional inviscid calculation for final stage stator and rotor, of a modern low pressure steam turbine. These methods have been particularly useful in these applications because of the combination of high Mach numbers, strong radial flows and very twisted and tapered blading. Both viscous and inviscid CFD methods have been very effective and significant in improving modern turbine designs.

2 APPLICATION OF VISCOUS CFD METHODS TO AERO ENGINE HP TURBINES

Aero engine gas turbines are constrained in the number of turbine stages employed because of weight and length considerations. These are obviously very important in such applications. Moreover because of these considerations and those of fuel consumption they tend to operate at high gas temperatures. The latter factor both improves cycle efficiency and reduces gas generator core mass flow and size. Thus aero engine high pressure turbines tend to be relatively highly loaded, high hub/tip ratio, with air cooled blading. Consequently the aerofoils are high camber with high exit Mach numbers, and total coolant flows for the stage around 10 to 12% of core mass flow. Important loss mechanisms which require optimisation and reduction in the design phase thus include trailing edge losses because of the high loading and large pressure drop, and secondary flow losses because of the high gas deflections in the blading. Viscous CFD methods have now been developed to the stage where they can and have been used to understand and optimise these important features of aero engine HP turbine design, and this is illustrated below.

2.1 Secondary Flows in Rotor Passages

Secondary flows occur when the inlet flow, with a non uniform total pressure distribution is turned in a blade row through a significant deflection. In classical terms the severity of the secondary flow is proportional to the thickness of the inlet boundary layer and the angle through which the flow is turned.

In an aero engine HP turbine the situation is more complex. The injection of cooling, sealing and leakage flows into the gaspath are all potential sources of low momentum flow, or at least deviation from mainstream gas conditions which could contribute to secondary flows. It is important that rotor secondary flows are correctly evaluated for the following reasons. Firstly rotor exit conditions must be adequately matched to the downstream blade row. Secondly they must be accommodated in the satisfactory design of the rotor itself in terms of aerofoil lift distribution and minimisation of loss.

Rolls-Royce has developed and used 3-D viscous flow calculation methods for application to Turbomachinery blading, over a number of years. One of these, the Moore Elliptic Flow Program (MEFP) has proved particularly useful because reliable and accurate solutions can be obtained relatively quickly, and using a calculation grid that is more coarse than some other methods would require. The method described in References (1) and (2) is essentially an elliptic, pressure correction method and the turbulent viscosity is calculated using a well established Prandtl mixing length model.

This method has been evaluated by comparison with a wide variety of test data. Lapworth, (3) contains a comparison with experimental data from an unshrouded research HP turbine. Measured data included detailed traversing of total pressure and flow angle at stator and rotor exit planes within the high speed turbine. Total temperature was also measured at rotor exit.

The comparison between measured and calculated rotor exit relative whirl angles, Fig 1 shows very encouraging agreement. In particular it is interesting that the detailed shape of this distribution and the radial location of the positions at which minimum and maximum turning occur is well predicted. Furthermore the variation between maximum and minimum flow angle is also well predicted. These calculations were achieved with approximately 110,000 grid points, although similar agreement was achieved in calculation with very much fewer grid points. One primary objective in developing viscous flow calculations is to predict and minimise the loss generated in a turbine during the aerodynamic design. This provides greater scope for optimisation of performance. A comparison of the distribution of rotor relative total pressure loss coefficient confirms that the shape of the measured distribution has been very successfully reproduced. The agreement in magnitude between the two loss profiles is encouraging. Predicted mass mean rotor total pressure loss coefficient at 0.08 is below the measured value of 0.12.

2.2 Tip Leakage Flows

Because of the high pressure drop across the rotor blade row, the tip leakage losses of an unshrouded HP turbine can amount to 2.5% of stage efficiency. Consequently it is very important both to understand the flow in the tip clearance region, and to minimise the consequent losses. The viscous calculation described above has been developed to calculate the flow through the tip gap as an integral element of the 3-D passage flow calculation. Examination of the velocity vectors in a transverse plane at an axial position 10% axial chord upstream of the trailing edge, Fig 2, shows that the significant features have been reproduced. These include the very high velocities in the pressure surface entry region, the separation in the gap just downstream of the pressure surface entry corner, the diffusion of the leakage flow to fill the tip gap, a recirculation in the mainstream driven by the exit leakage jet, and the entrainment of flow radially upwards on the blade aerofoil suction

surface. Velocity vectors in the pitchwise blade to blade plane, Fig 3 again show clearly the interaction between the leakage and mainstream flows. Mixing between the two flows is responsible for additional dissipation losses. The flow within the tip gap near the leading edge is dominated by the direction of the inlet flow. Moving rearwards, the pressure difference between pressure and suction surfaces increases significantly, and this causes the flow to be driven abruptly across the tip gap in a direction normal to the aerofoil. This form of leakage flow then persists along the leakage gap to the trailing edge.

2.3 Trailing Edge Flows

Modern high temperature aero gas turbine HP turbine blades need relatively thick trailing edges because of cooling requirements. Passage exit Mach numbers are also approximately sonic because of the high duty. Consequently the mixing and dissipation losses at the trailing edge can thus be equivalent to 3.5% efficiency loss.

Viscous flow calculations now permit detailed calculation of trailing edge flows and losses with high accuracy. Consequently they can and have been employed in recent HP turbine blade design to achieve a modification in trailing edge geometry providing a significant improvement in rotor blade performance. This problem has been studied by a programme of complimentary experimental and theoretical (CFD) work. For computational efficiency and economy this was a 2-D study.

The CFD method used in this case was the Rolls-Royce ANSI-2D code Ref (4) this being predominantly of the time marching type. One of the features of this code is that various turbulence models can be incorporated, evaluated and thus developed.

The initial work was carried out on a flat plate at low Mach numbers, to obtain a good understanding of the flow and to establish the validity of the CFD method. For this work, to remove uncertainty and variability in calculated boundary layer development, a trip was employed at the leading edge. Consequently it was appropriate and logical to use a Cebeci-Smith, mixing length type of turbulence model. In the flat plate experiment the profiled trailing edge showed an 18.2% reduction in measured loss when tested at an exit Mach number of 0.24. In comparison the loss reduction predicted by viscous CFD was 19.6%. This is obviously very encouraging agreement between measurement and prediction. In order to check that the improvement given by the modified trailing edge shape was equally applicable at higher Mach numbers the CFD calculation was then run at an exit Mach number of 0.58. For this case the predicted loss reduction was again 19.6%.

Based on these encouraging results, the CFD code was then used to derive reduced trailing edge loss for an HP turbine rotor aerofoil operating close to the transonic flow regime. In this case there was likely to be more transitional flow around the aerofoil, since it was inappropriate to artificially trip the boundary layer at any location. Hence it was important that the CFD calculation should be able to handle transitional flow. For this reason a one equation k-l turbulence model - described in Ref (5), was used in the ANSI-2D calculations.

The objective of the exercise was to improve the performance of the rotor profile in two phases. The first was by the adoption of a thinner trailing edge (from 1.13mm to 0.75mm) whilst retaining the semicircular trailing edge shape. The second step involved the incorporation, at the same reduced trailing edge thickness, of the profiled trailing edge shape. The results, Fig 4, show very encouraging

agreement between measured and calculated values. The effect of simply thinning is underpredicted and that of the modified shape somewhat over predicted, as was the case with the earlier examples.

Comparison of the predicted trailing edge flowfield, fig 5, shows both the detail that is necessary to achieve an accurate prediction, and the significant reduction in the recirculation region that provides the mechanism for the trailing edge loss reduction. Comparison of the total pressure contours in the viscous flow region, Fig 6, shows the thinner and shallower wake, achieved with the thinner trailing edge. It is also clear that the total pressure deficit in the base, and initial mixing region has been substantially reduced.

3 APPLICATION OF INVISCID CFD METHOD TO THE DESIGN OF THE FINAL STAGE OF A LARGE STEAM TURBINE

The steam which exhausts from the final stage of a low pressure steam turbine, after its expansion through the turbine, is condensed; its remaining energy, including the kinetic energy, is lost. In order to reduce this leaving energy or so called leaving loss and maximise turbine efficiency it is advantageous to provide as large an exhaust area as possible, within the limits which mechanical design aspects will allow.

Fig. 7 shows a longitudinal section through the final stage of a large low pressure steam turbine illustrating typical stage geometry. Maximum exhaust area is provided by achieving the largest possible rotor blade tip diameter in conjunction with the lowest possible ratio of hub diameter to tip diameter.

Low values of hub to tip ratio necessitate highly twisted rotor blading in order to match blade inlet angle with the radial variation of steam inflow angle relative to the blade. Also blade centrifugal stress considerations dictate that the rotor blade sections are highly tapered. A stacked view of the blade sections for both the fixed stator blade and the rotor blade are also shown in Fig. 7.

The large expansion ratio of the steam through the final stage also means that steam velocities are high and indeed the flow regime is transonic in nature (i.e. mixed subsonic/supersonic flow). Finally there is a requirement, in order to limit costs, to keep turbine axial length as short as possible. The flare angle of the outer flow annulus boundary may be substantial (up to $35°$) and radial components of steam flow velocity through the stage are significant. Clearly this represents a complex and highly three-dimensional flow situation.

Fortunately CFD techniques are available which enable inviscid flow computations to be performed for a complete turbine stage, including the important interaction of the stator and rotor flows.

At NEI Parsons this method, which was developed by Denton and is fully described in reference (6), has been applied to aid the development of a final stage rotor blade of length 1070mm for operation at 3000 rpm on 50 Hz grid systems. This blade has a hub to tip ratio of 0.45 and offers an exhaust area of 9.54 m^2. By contrast the largest last stage blades currently operating in UK Power Stations have an exhaust area of $7.8m^2$ and a hub to tip ratio of 0.47. Potentially a 1070mm blade can achieve an improvement of around 1% in overall turbine heat consumption, and hence in annual fuel costs, for certain highly rated turbine applications, due to the reduction in leaving loss. For example, based on current evaluation rates this would represent a 'worth', relative to initial capital cost, of around £5.6 million for a 500MW coal fired turbine generator set with one low pressure turbine, or £1.8 million for a 450MW combined cycle plant operating on natural gas. However, the decrease in hub to tip ratio further exacerbates the already severe difficulties in achieving efficient aerodynamic blading performance.

The subsections below describe briefly:- the application of the computation procedure to the design of the stage, the important design considerations in achieving reasonable efficiency, the design variations which can be considered using the predictive CFD method and some computation results for the design ultimately adopted. Finally some comparisons are presented with experimental measurements from a one quarter scale research turbine, which was tested to demonstrate the performance of this design.

3.1 Computation Method

The computation grid which was used for this problem is illustrated in Figs. 8, 9 and 10. Fig. 8 shows the grid in the meridional plane (axial-radial) while figs. 9 and 10 show the grid in the transverse circumferential plane (blade-to-blade), corresponding to the base and tip profiles of the rotor blade (Sections 1 and 19 in Fig. 8 respectively).

A grid size of 154 axial, 19 radial and 16 circumferential was chosen, giving a total of 46,816 grid points. The broken line in Fig. 8 between the stator blade outlet and the rotor blade inlet is chosen as a so called 'mixing plane'. It is not possible yet, for such a three-dimensional problem, to model the interaction of the circumferential variations in the stator blade outlet flow with the rotor. Instead the outlet flow from the stator is averaged circumferentially at the mixing plane and it is this flow which 'interacts' with the rotor.

It is evident from Figs. 9 and 10 that the circumferential pitches of the stator blade and rotor blade are not, in general, equal. It is not necessary that the circumferential grid is continuous at the mixing plane. Indeed the CFD method switches from flow relative to the fixed blade to flow relative to the moving blade at the mixing plane and it is advantageous for convergence reasons that the grid direction changes at the same point.

A total of 1.05×10^{10} floating point operations is required for a calculation on a grid of this size. This represents a computation time of around 7 hours on the Apollo DSN4000 workstation with floating point accelerator which is available, enabling such calculations to be performed routinely.

3.2 Design Considerations

Mechanical design considerations for long high speed rotor blades impose severe constraints to achieving efficient aerodynamic performance. Gyarmathy in Ref (7) has reviewed the limitations on rotor blade length. The 1070mm blade under discussion is near to his defined limit for which acceptable mechanical and aerodynamic performance can be achieved using conventional 12% Cr steel rotor blade material.

The design difficulties can be described with reference to Figs. 9 and 10. The rotor blade base section must be robust enough to carry the overall blade centrifugal load which exceeds 360 tons for each blade and so the solidity at the root is relatively high. In addition the base section must effect a high degree of turning of the flow, typically around 100°. As a result the rotor blade base section is highly cambered and the flow passage between adjacent blades has little or no convergence, as illustrated in Fig. 9.

The stage is designed to achieve nearly equal pressure drops across the stator blades and the rotor blades at the mid-height position. Because of the radial pressure gradient set up in the highly swirling flow at outlet from the stator, the pressure drop across the base section of the rotor blade is small (low reaction). The relative inlet Mach number of the steam flow onto the rotor blade is in the high subsonic range and, without careful design, local supersonic patches can develop terminated by shock waves or adverse pressure gradients which can cause boundary layer growth or induce flow separation leading to high profile losses Ref (8).

By contrast, at the tip, the cross sectional area of the rotor blade profiles must be minimised to reduce the total centrifugal load at the base. The relative inflow and outlet flow directions are similar and near circumferential. The profiles have little camber, or even reverse camber, and are highly staggered as evident in Fig. 10. The pressure drop across the tip is large. Relative inlet Mach number may be around 0.8 and outlet Mach numbers may reach 1.7 in the rapidly expanding flow. Once again care is required to ensure satisfactory loading on the blade and to avoid large profile losses.

3.3 Design Optimisation

Although the design difficulties are appreciable, and are made worse by the required reduction in hub to tip ratio, the predictive CFD method enables various design parameters to be explored in an effort to maximise efficiency. Calculations were performed to consider the effect of stator blade twist (variation in outlet angle), stator blade lean and stator blade circumferential pitching. Twist and lean can be adopted to improve rotor blade root reaction Ref (7). They also influence the steam inflow direction onto the rotor blade and the

distribution of mass flow across the turbine annulus.

The method thus enables rotor blade inlet angle to be chosen so as to reduce steam incidence losses. Finally the fixed and moving blade profiles can be 'sculptured' within the mechanical constraints which pertain to avoid the formation of shock waves or regions of adverse pressure gradient.

3.4 Computation Results for Adopted Design

In Figs. 11 and 12 results are shown for the design which was finally adopted. The stator blade sections are tapered so as to maintain a near uniform value of blade pitch to chord length ratio over its height. In addition the fixed blade is twisted and is leaned in the circumferential direction by several degrees. The rotor blade has a total inlet angle variation of 105 degrees, varying non-linearly with height.

Fig. 11 shows contours of equal flow Mach number over the suction surfaces of the stator and rotor blades in the meridional plane. The flow over the stator is supersonic at outlet over almost the full blade height. However the radial gradient in outlet Mach number was limited through the use of twist and lean.

A small patch of supersonic flow relative to the rotor blade is predicted to occur at the base and towards the leading edge. This region extends to approximately 10% blade height but the peak Mach number value is below 1.2 at the base section and the region is not terminated by a strong shock.

This is illustrated further in Fig. 12 which shows the variation of flow surface Mach number (relative to both fixed and moving blades) with axial distance, for the hub, mean and casing grid lines. At the hub the Mach number at outlet from the stator blade reaches 1.4: the maximum value on the suction surface of the stator blade is only slightly higher at 1.6 and there is little diffusion. The reaction across the base of the rotor blade is quite favourable and there is a net acceleration relative to the blade from a Mach number of 0.8 at inlet to 1.2 at outlet. So, although supersonic velocities develop on the suction surface towards the leading edge and there is some diffusion, the subsequent acceleration ensures that profile losses are not large.

The surface Mach number variation for the casing grid line, corresponding to the rotor blade tip section is shown as the broken line in Fig. 12. The choice of pitch to chord length ratio, stagger angle, and the profile geometry for the rotor blade tip section ensure that there is a well defined throat and the tip section is well loaded. With the high degree of reaction there is an acceleration from a relative Mach number of 0.70 at inlet to 1.6 at outlet. Once again a large over expansion on the suction surface is avoided and the peak Mach number is about 1.9. Some shock losses are inevitable but these have been controlled and minimised.

3.5 Model Turbine Test Results

To demonstrate the performance of this design the blading was incorporated as the final stage of an existing three-stage one-quarter scale research steam turbine. The turbine operates at 12000 rpm which is four times full scale running speed, so that aerodynamic similarity is maintained.

The casing of the model turbine was machined to accommodate the longer rotor blade resulting in a casing flare angle through the final stage of 31°. The model turbine is coupled to a water brake allowing overall performance to be measured. It is also fitted with access ways to allow traverse probes to be inserted downstream of the final stage rotor blade to measure steam velocities and flow direction.

Performance measurements showed that the peak blading efficiency (total-to-total excluding leaving loss) matched the previous build despite the increase in blade length, the associated reduction in hub to tip ratio and the increase in casing flare angle. Moreover the aimed for reduction in leaving loss was realised. Fig. 13 compares the measured and predicted radial variation in the swirl angle and kinetic energy of the flow leaving the turbine, for equivalent volumetric flow rates. The comparison is encouraging. With the aid of the prediction method a fairly uniform distribution of leaving loss was achieved which is confirmed by the test results.

4.0 CONCLUSION

The objective of this paper has been to provide an insight into some of the ways in which predictive CFD methods, backed by confirmatory experimental testing, are impacting beneficially on the aerodynamic design of turbine blading. Two important aspects of this CFD work have been described.

First, the use of viscous CFD methods to obtain a very comprehensive picture of certain local flow situations where viscous effects are of paramount importance. These viscous methods enable mechanism for loss generation to be studied and loss levels to be compared. On this basis specific blade features can be refined to achieve improved aerodynamic performance.

Second, the application of inviscid three-dimensional CFD codes to provide a prediction for the interactive flow behaviour through a complete stage of blading, stator and rotor, which has considerable geometric complexity. In this case the method allows various alternative blading geometries to be appraised which could not be considered experimentally because of the prohibitive cost. Testing need only be undertaken for the configuration judged finally to be the optimum aerodynamically.

The examples considered specifically in the paper pertain to small aero-engine HP turbine blading and the long final stage blading of a large steam LP turbine. The methods can be, and indeed are being, applied equally successfully to steam turbine HP blading and aero-engine LP blading.

At the present time CFD methods continue to advance rapidly, driven by the ever increasing capability of main frame computers and work stations. Rolls-Royce/NEI are actively engaged in the development and application of new and more powerful techniques. Latest developments include the calculation of unsteady aerofoil flows, interaction of rotor and stator flow fields, CFD calculation of film cooling flows, aerofoil 3D internal cooling flows, and sophisticated 3D input and output calculation and graphics display. It is anticipated that with the benefit of these more complex methods improvements in turbine efficiency will continue to be achieved.

5.0 ACKNOWLEDGEMENTS

The authors would like to thank their respective companies, Rolls-Royce Industries and NEI Parsons for permission to publish this paper. In addition the valuable help of their colleagues Dr B L Lapworth, Mr A Jefferson and Mr I J Rainbow at Rolls-Royce and Mr D Borthwick and Mr S Humphrey at NEI Parsons is gratefully acknowledged.

REFERENCES

(1) MOORE J G Calculation of 3D flow without numerical mixing AGARD LS140 - 3D Computational Techniques Applied to Internal Flows in Propulsion Systems. Cologne & Paris 1985.

(2) NORTHALL J D, MOORE J G & MOORE J. 3D Viscous Calculation for Loss Prediction in Turbine Blade Rows" International Conference on Turbomachinery, Cambridge Sept 1987 p63-71 (IMechE C267/87).

(3) LAPWORTH B L. Numerical Investigation of Shroudless HP Model Turbine Rotor using MEFP. RR Internal report. 1990.

(4) NORTON R J G, THOMPKINS W T HAIMES R. Implicit Finite Difference Scheme with non-simply connected grids - a novel approach. 22nd AIA A meeting on Aerospace Sciences. Reno January 1984.

(5) CHEW J W & BIRCH N T. Comparisons between numerical solutions and measurements for high speed turbine blades. RR Internal Report. 1987.

(6) DENTON J D, An improved time-marching method for turbomachinery flow calculations. Transactions ASME July 1983 Vol.105, 514-520.

(7) GYARMATHY G, SCHLACHTER W. On the design limits of steam turbine last stages. Technology of turbine plant operating with wet steam. Joint Conf. BNES, IMechE, ENS, London, 11-13th Oct 1988.

(8) GRANT J, BORTHWICK D. Fully three-dimensional inviscid flow calculations for the final stage of a large low-pressure steam turbine, Conference on Turbomachinery - Efficiency Prediction and Improvement, Cambridge, 1987. (IMechE C281/87).

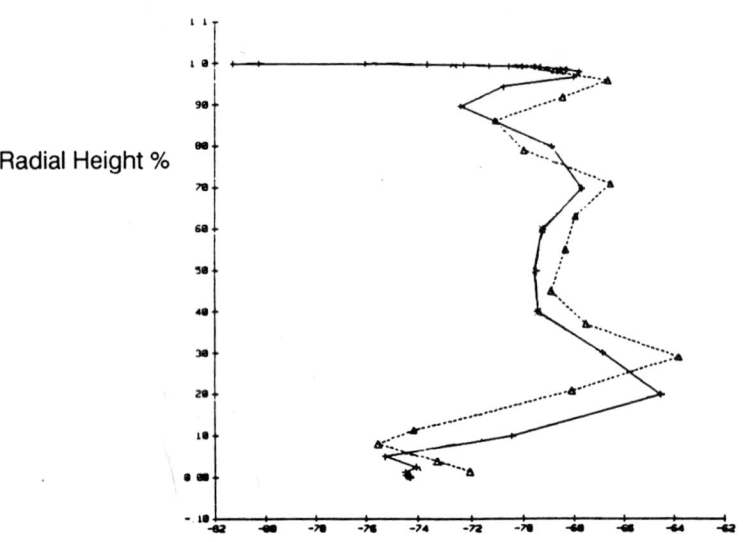

+ MEFP Prediction - Fine grid
△ Measurement

Flow angle (degrees)

1   Rotor relative exit whirl angle (radial profile of circumferential mass means)

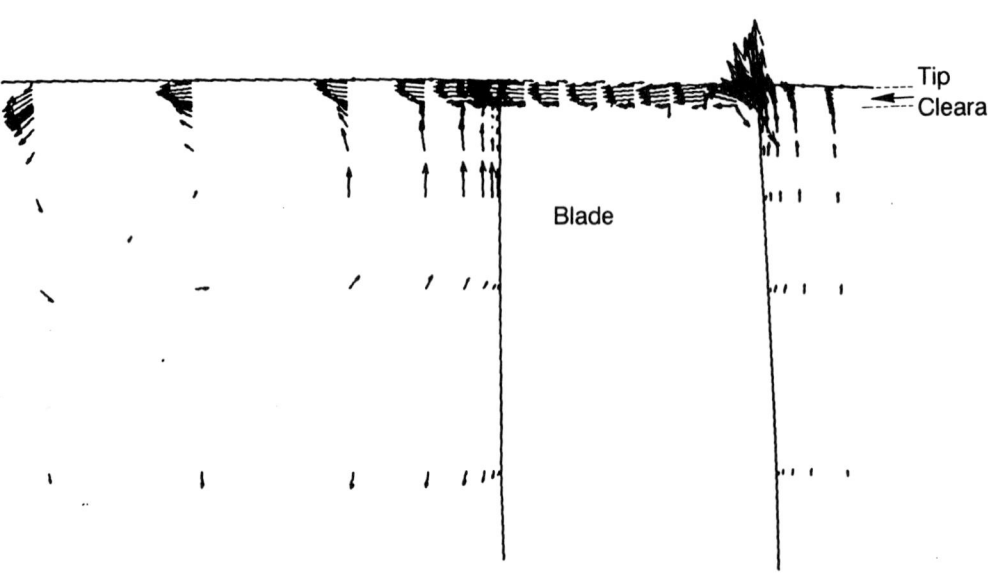

2   Secondary velocity vectors from MEFP solution in a tangential plane at 90% axial chord.

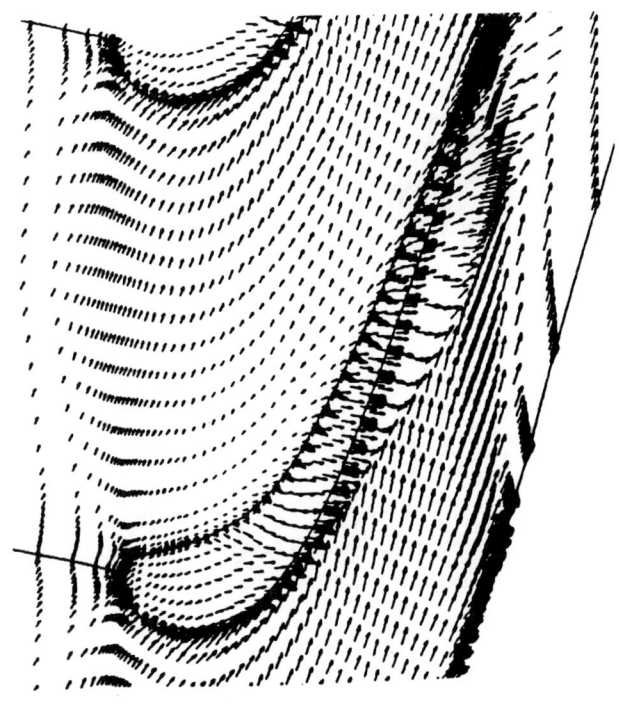

3   Velocity vectors within the rotor blade tip clearance from MEFP solution.

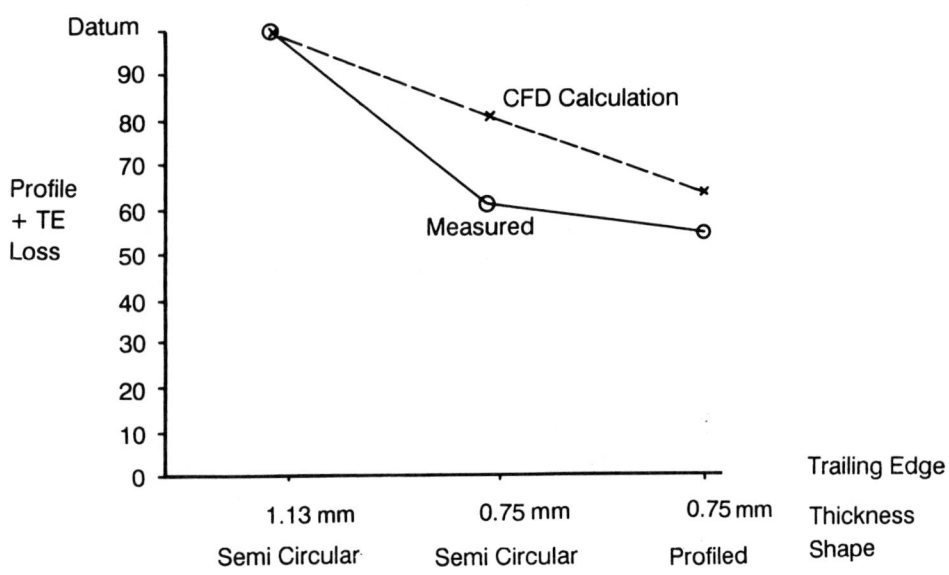

4   Rotor loss reduction.

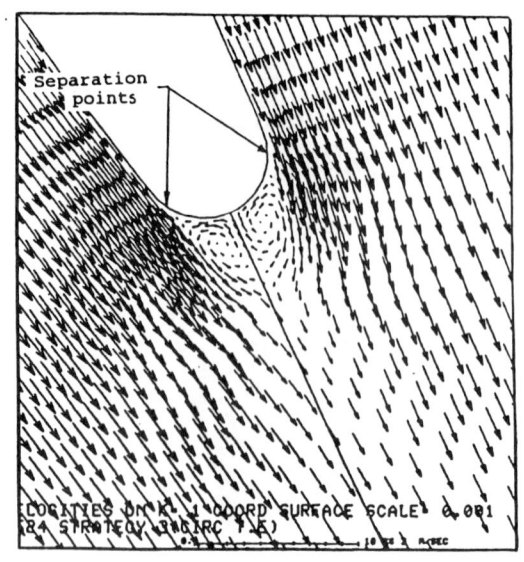 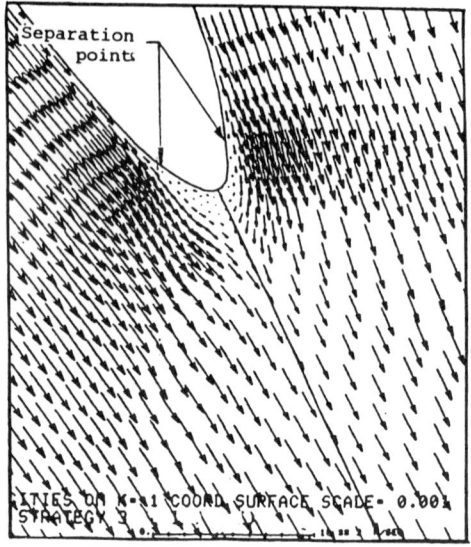

Circular t/e   Profiled t/e

5   HP turbine rotor - trailing edge base flow region.

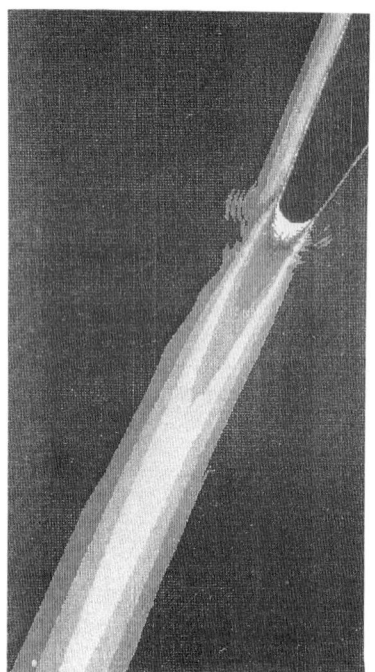 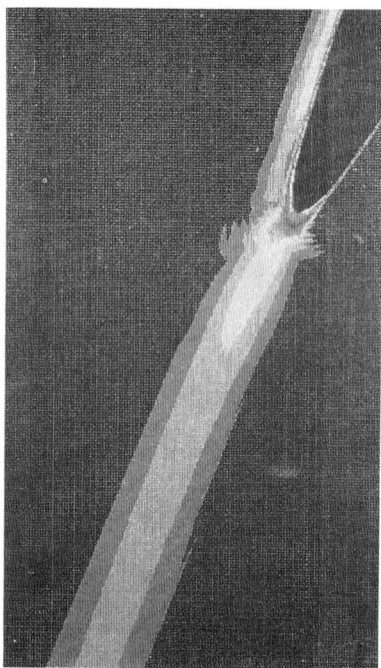

Strategy 3   Strategy 3
Circular T/E   Mnis=0.96   Profiled T/E

6   Total pressure contours in trailing edge/wake region.

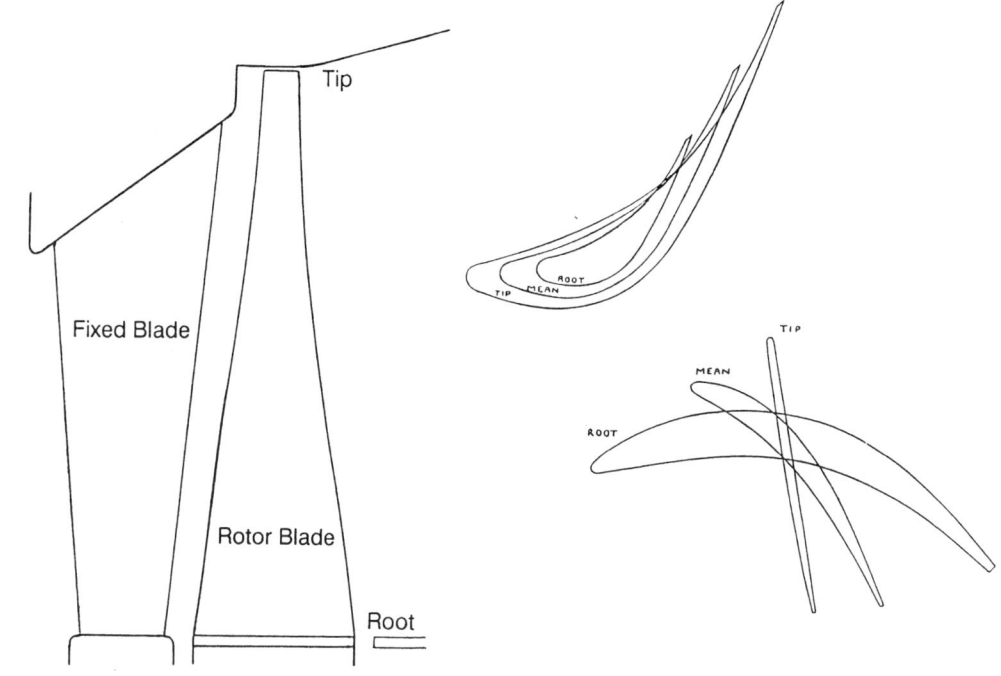

7 Steam turbine - last stage nozzle and rotor.

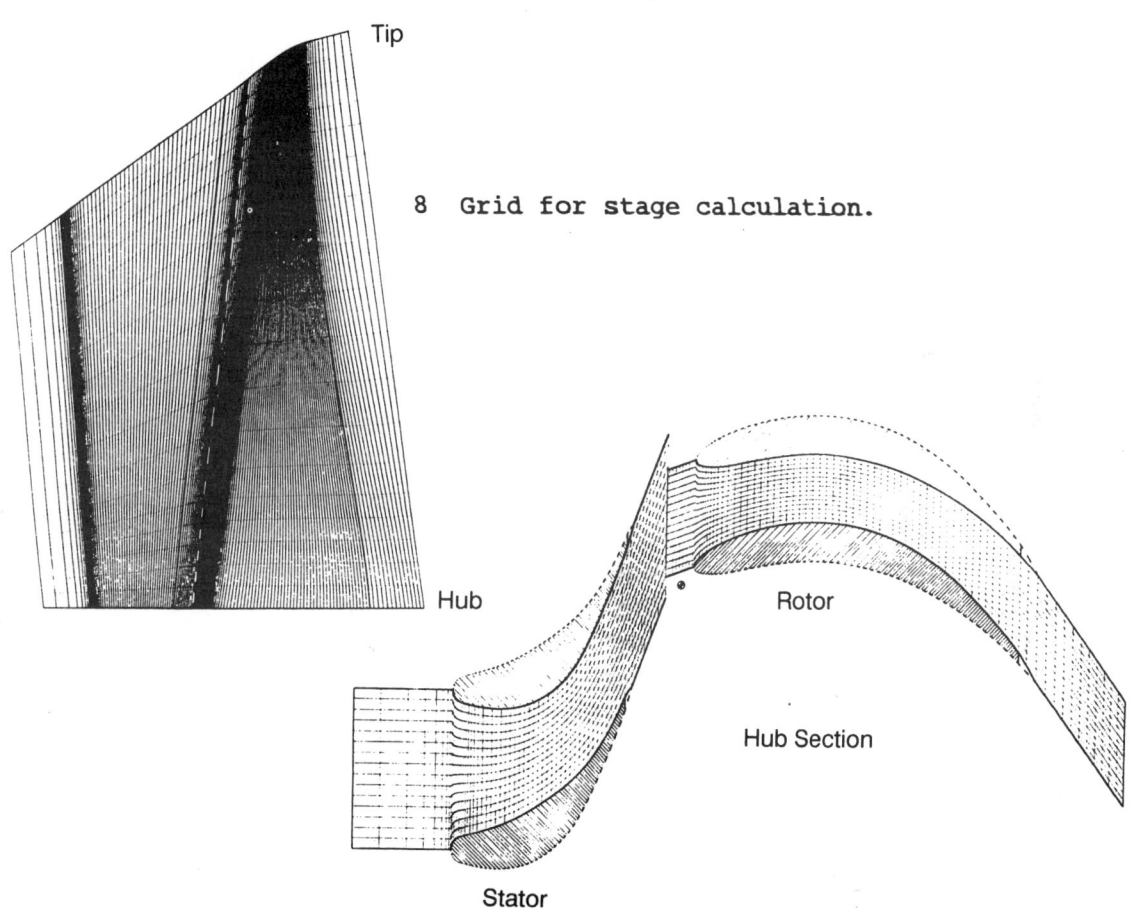

8 Grid for stage calculation.

9 Root aerofoil profiles and grid for stage calculation.

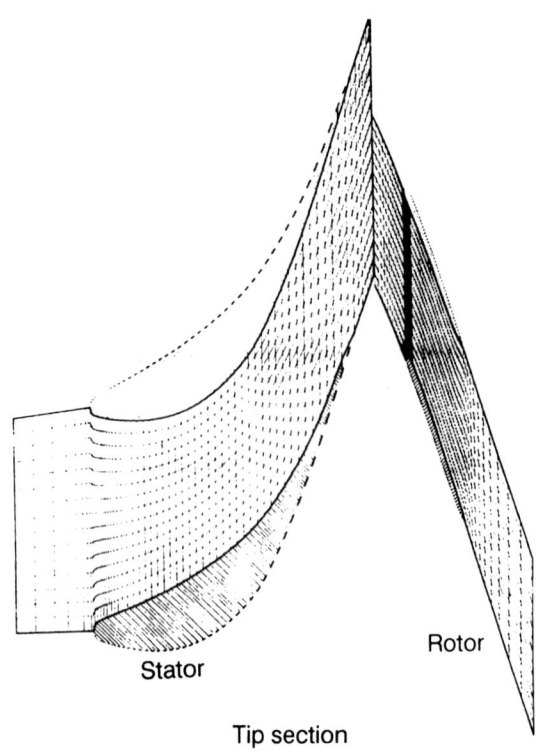

10   Tip aerofoil profiles and grid for stage calculation.

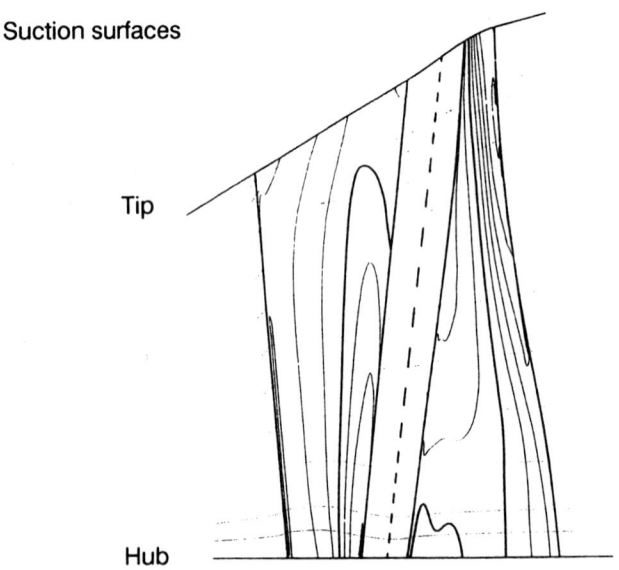

11   Mach number contours and aerofoil surface streamlines - suction surfaces.

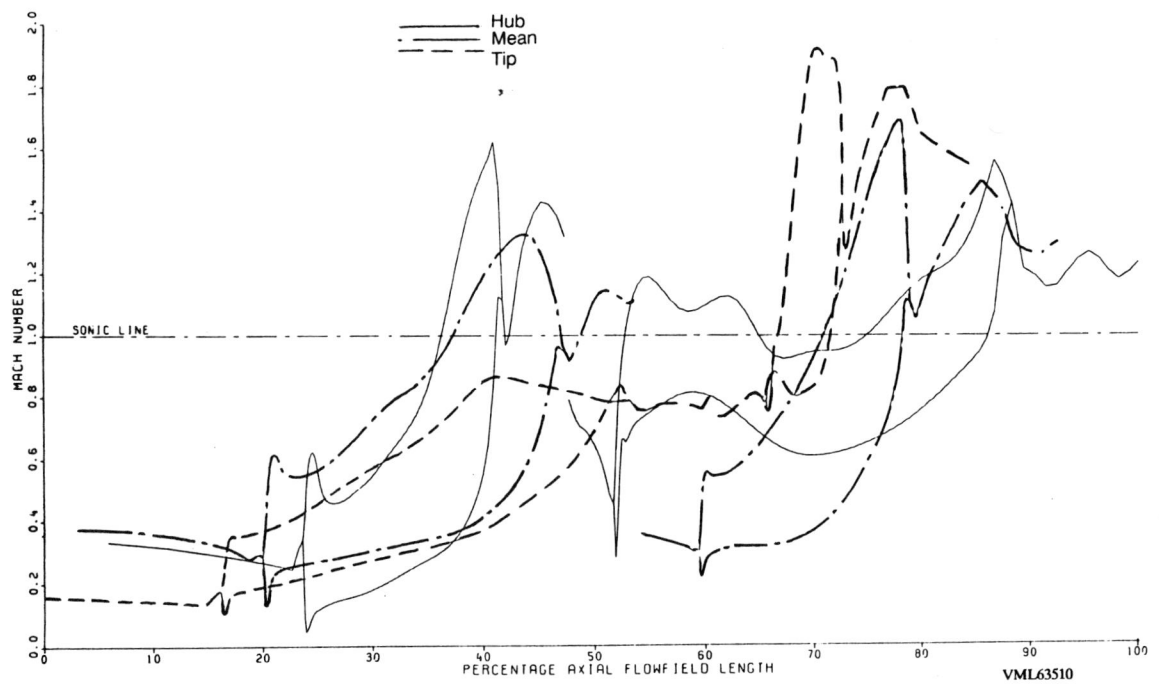

12 Aerofoil surface velocity distributions from stage calculation.

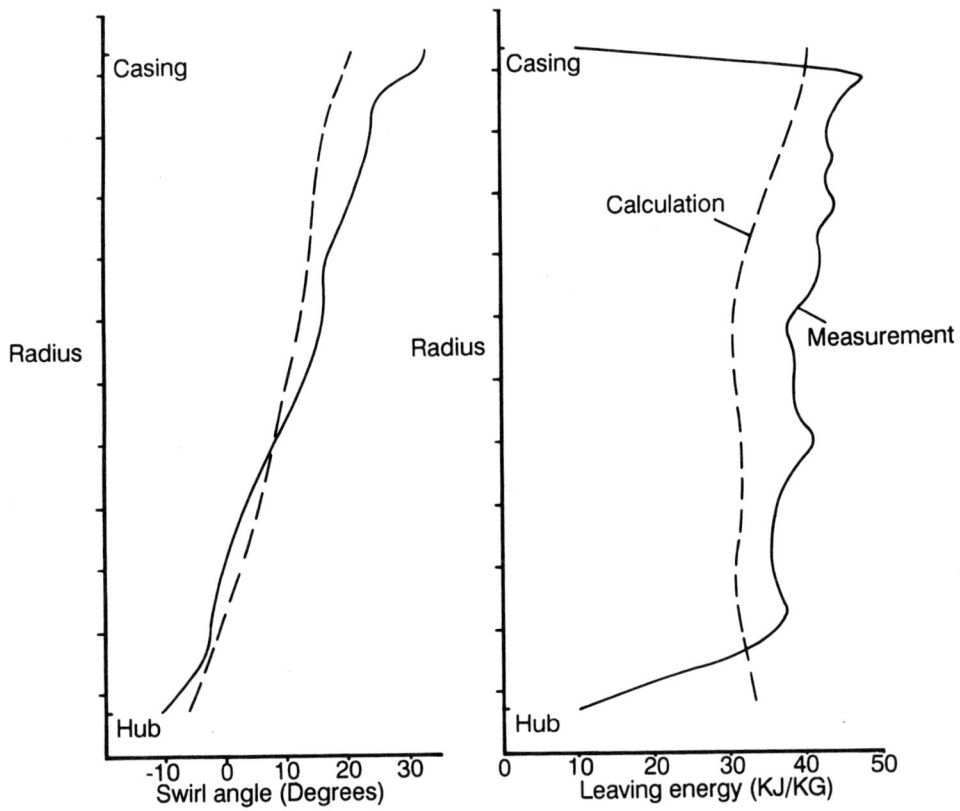

13 Comparison between measured and predicted radial variation of mean swirl angle and leaving energy.

C413/067

Prediction of smoke movement using CFD

N RHODES, BSc, MSc, CEng, MIMechE, I W CLARK, BSc, MSc, PhD, FRMS
and K ELSE, BEng
Mott MacDonald Limited, Croydon

SYNOPSIS   This paper describes the mathematical and physical basis of models used for fire and smoke movement prediction and their application for both design and validation. The models require the numerical solution of the basic equations governing three-dimensional, steady and transient flows with prescribed boundary conditionsand include the effects of turbulence, combustion, and radiation.

Predicted temperature and velocity distributions show very good agreement with data, and although gas concentrations appear to be less well predicted, the fundamental nature of field models permits their use in large-scale situations where simple models might be inapplicable.

NOTATION

g	acceleration due to gravity
h	enthalpy
k	turbulence kinetic energy
p	pressure
q	heat flux
u,v,w	velocity
x,y,z	cartesian co-ordinates
$C_{1,2,\mu}$	constants in turbulence model
G_b	generation term for kinetic energy
G_k	generation term for buoyancy
S_ϕ	source term
Γ_ϕ	exchange co-efficient
ϵ	turbulence dissipation
μ_{eff}	effective viscosity
μ_l	laminar viscosity
μ_t	turbulent viscosity
ρ	density
ρ_{ref}	reference density
σ_k	Schmidt number
Φ	dependent variable

1. INTRODUCTION

Computational Fluid Dynamics (CFD) is becoming increasingly used in the solution of design and analysis problems involving fluid flow, heat-transfer and combustion. The methods are conceptually quite simple. The flow domain is divided into numerous small volumes and the basic equations representing the conservation of momentum, energy, species concentration, etc., are solved numerically at a point within each one. The information obtained at every computed point in the flow domain builds up a complete picture of the flow situation.

This fine detail permits the use of more mechanistic representations of the processes that are occurring. For example, local heat release due to chemical reaction can be related to species concentrations, reaction kinetics, and temperature within a cell. Mathematical models of this kind therefore remove some of the need for empirical approximations used in simpler models and hence preserve a greater generality. The ability to alter the assumptions on which the physical processes are based and to test them with a mathematical model is also useful, and encourages a closer scrutiny of the relevant physics.

Thus, if engineers can gain confidence in the accuracy of the numerical solution of the equations, then the way is clear for exploration of the physics and chemistry of the processes of interest through systematic simulation and physical experiment. For example, such models might well provide a framework for investigating processes such as flashover and flame spreading.

This paper reviews the status of CFD or "field" models in terms of their mathematical basis, applications, and present level of validation. Field models are based on the solution of the fundamental equations governing all fluid-flow situations. They employ mathematical techniques that have been and continue to be developed in areas such as aerospace and nuclear energy. They are therefore well founded and benefit from advances being made in these areas, e.g. the use of distorted grids to enable easier geometry definition and improvements in the ease of use of computational fluid dynamics methods in general.

Inaccuracies in a mathematical model can arise from two sources, either numerical of physical. Numerical errors can occur, for example, when the grid size is too coarse, and gradients within the flow are not properly resolved. The cure is simply to refine the grid, but by how much? Since the answer is not known in advance it has become common practice to set up a problem using a relatively coarse grid, and then to refine the grid progressively until no significant change in the predicted values is observed.

Applications of field models range from the notable analysis of the King's Cross fire (1) which predicted unexpected flow behaviour and prompted a programme of experimental confirmation, to simpler validation exercises and pragmatic design studies. Validation studies have generally focussed on room or compartment fires with either natural or forced ventilation, but studies have also been undertaken for tunnel and aeroplane fires. Engineering design applications have been aimed at predicting smoke behaviour for buildings, particularly larger structures where the empirical content of zonal models may not be applicable. The example described below concerns the evaluation of smoke ventilation requirements in a railway repair depot. Studies of this kind provide an input into the assessment of the fire risks associated with the design, as well as quantitative information on the preformance of the system.

The next section of the paper provides an overview of the mathematical basis of field models, and briefly describes some of the physical models used for practical calculations, and those which are the subject of current research. Application studies are described in the following section, and the predictions are compared with data.

2. MATHEMATICAL ASPECTS OF FIRE AND SMOKE MODELLING

The equations that need to be solved for any fluid-dynamic problem can be written in the following general form:

$$\frac{\partial}{\partial t}(\rho \phi) + \text{div}\,(\rho \vec{v}_\phi - \Gamma_\phi \text{grad}\,\phi) = S_\phi$$

where

S_ϕ = source term
$\vec{v}_\phi$ = velocity vector
t = time
ρ = density
ϕ = the dependent variable
Γ_ϕ = diffusion coefficient

The computer solves a form of the above equation, which is obtained by integrating over a control volume. The solution methods are well established, reliable and widely used. Further details can be found in refernce (2). The dependent variable can represent the fluid velocities in each coordinate direction for single or multi-phase flows, enthalpies, species concentrations, turbulence quantities, radiation fluxes, and so on. The number of differential equations that can be used to represent a problem need not be limited when such a computational approach is adopted. The limitation is more likely to be in the degree of understanding of the relevant physics, whether suitable exchange coefficients and source terms can be formulated. Table 1 gives some examples of the values of Γ_ϕ and S_ϕ for equations commonly solved in field models.

The flow behaviour during a fire is usually three dimensional and is strongly influenced by turbulence and buoyancy effects. The simplest field models, therefore, need to solve equations for velocity in the three coordinate directions (u, v, and w) enthalpy (h), and pressure (p). They employ a fixed value of turbulent viscosity to represent the effects of turbulent mixing and a prescribed heat source to represent the fire. The limitations that are applied by assuming fixed values of these quantities can be relaxed by solving additional equations and representing the various phenomena on a cell-wise basis rather than

TABLE 1
Exchange Coefficient (Γ_ϕ) and Source Terms (S_ϕ) for Different Variables

ϕ	Γ_ϕ	S_ϕ
1	0	0 (Continuity)
u	μ_{eff}	$-\frac{\partial p}{\partial x} + \frac{\partial}{\partial x}(\mu_{eff}\frac{\partial u}{\partial x}) + \frac{\partial}{\partial y}(\mu_{eff}\frac{\partial v}{\partial x}) + \frac{\partial}{\partial z}(\mu_{eff}\frac{\partial w}{\partial x})$
v	μ_{eff}	$-\frac{\partial p}{\partial y} + \frac{\partial}{\partial x}(\mu_{eff}\frac{\partial u}{\partial y}) + \frac{\partial}{\partial y}(\mu_{eff}\frac{\partial v}{\partial y}) + \frac{\partial}{\partial z}(\mu_{eff}\frac{\partial w}{\partial y})$
w	μ_{eff}	$-\frac{\partial p}{\partial z} - g(\rho - \rho_{ref}) + \frac{\partial}{\partial x}(\mu_{eff}\frac{\partial u}{\partial z}) + \frac{\partial}{\partial y}(\mu_{eff}\frac{\partial v}{\partial z}) + \frac{\partial}{\partial z}(\mu_{eff}\frac{\partial w}{\partial z})$
h	$\frac{\mu_{eff}}{\sigma_h}$	$\dot{q}$
k	$\frac{\mu_{eff}}{\sigma_k}$	$G_k - \rho\epsilon + G_b$
ϵ	$\frac{\mu_{eff}}{\sigma_\epsilon}$	$\frac{\epsilon}{k}[(G_k + G_b)C_1 - C_2\rho\epsilon]$

$\mu_{eff} = \mu_t + \mu_l \qquad \mu_t = c_\mu \rho \frac{k^2}{\epsilon}$

$G_k = \mu_t \{ 2[(\frac{\partial u}{\partial x})^2 + (\frac{\partial v}{\partial y})^2 + (\frac{\partial w}{\partial z})^2] + [(\frac{\partial u}{\partial z})^2 + (\frac{\partial w}{\partial x})^2]$
$+ [(\frac{\partial w}{\partial y})^2 + (\frac{\partial v}{\partial z})^2] + [(\frac{\partial u}{\partial y})^2 + (\frac{\partial v}{\partial x})^2] \}$, and

$G_b = \frac{\mu_t}{\rho} g \frac{\partial \rho}{\partial z}$

a global one. Such refinements to this basic model might include:

- The use of a turbulence model. The two-equation model in which the kinetic energy of turbulence (k) and its rate of dissipation (ϵ) are solved has been used in many of the studies reported here (see Table 2). The turbulent viscosity in each cell is then calculated from:

$$\mu_t = C_\mu \rho \frac{k^2}{\epsilon}$$

where C_μ is one of the turbulence model constants.

- A combustion model, which requires the introduction of additional equations to predict the concentrations of reacting and inert species and source terms involving details of any kinetically controlled reaction rates.

The simplest combustion model assumes a diffusion-controlled, single-step reaction, which may be represented simply as fuel + oxidant → product. This implies that any fuel within a cell will react instantaneously with any available oxidant. A mixture fraction equation (f) is required in this case. If chemical kinetic influences are to be included, then a further transport equation for the mass fraction of fuel (m_{fu}) is required, and a prescription of the reaction rate. The most common formulation for the reaction rate is to take the minimum of the laminar Arrhenius expression or that deduced from eddy break-up concepts, (3) and (4).

A radiation model, which can enhance the treatment of heat transfer, but introduces further physical questions regarding the absorption and scattering coefficients of the medium and emissivities of the surfaces. Monte Carlo and Flux methods are available, (5). They have been developed nd validated for combustion and furnace modelling principally. They have not been used extensively for fire modelling, it being more common to use an enhanced wall heat-transfer coefficient to estimate the heat losses by convection and radiation. Similar assumptions to those used in zone models are frequently applied. See, for example, reference (6).

Finally, it is necessary to describe the details of the boundaries in the mathematical model. A no-slip boundary condition is applied on solid walls for velocity components. Fluxes of momentum and heat can be predicted from wall function relationships. Heat losses through the walls can be calculated from the wall conductivity and the local temperature gradients that are predicted.

On free boundaries, it is conventional to impose a fixed reference pressure. The mass inflows and outflows to the domain are then an outcome of the calculation. To ensure the validity of this, such boundaries should be sufficiently remote that they exert no unphysical effect on the solution. In fire-modelling applications, the boundary is usually fixed a little distance away from doorways in order to avoid such problems.

3. APPLICATION OF CFD MODELS

3.1 Typical design study

A recent study carried out for British Rail concerned the prediction of smoke movement in a railway repair depot. The objective was to assess the effect of different roof-mounted venting arrangements on the smoke removal rate. The Harwell FLOW3D code was used to predict the three-dimensional transient flow arising from a 10 MW fire in the building, the fire location and roof vent area being the primary parameters. Figure 1 shows some typical results from the analysis. In this case the fire is located in a corner of the building, and the figure shows a cross-section taken through the fire source. The contours of 1% and 10% smoke concentration are shown at several times during the transient. Plots of this kind provide an overall understanding of the smoke behaviour, whilst separate calculations from the model predictions give vent outflow rates and hence a quantitative basis for comparing different designs.

3.2 Validation studies

This section presents the results of studies in which mathematical models have been applied to predict particular experiments. The experimental cases include room fires investigated by the Swedish National Testing Institute (7), the National Institute of Standards and Technology (formerly National Bureau of Standards),(8), and a national laboratory, (9); a tunnel fire,(10); a nuclear reactor scenario, (11); a simulated hospital ward experiment carried out by the Fire Research Station(12), and a one-sixth scale sports hall (13). Table 2 summarizes the details of these experimental cases in terms of overall dimensions, the nature of the fire source and the heat release, whether steady or transient conditions are established, and some details of the modelling strategies.

Most of the experimental cases used a liquid or gaseous source of fuel for the fire. The main chemical reaction and the heat fuel release therefore can be calculated with reasonable accuracy, if complete combustion is assumed. The measurements reported tend to be broadly similar for most of the experiments. Thus, the information usually available for validation comprises:

- The mass balance for the system in terms of inflows and outflows.
- Gas velocity and temperature distribution at door-ways as functions of height.
- Point temperature measurements within a room.
- Thermocouple rakes that provide horizontal and vertical temperature distributions.
- Point gas concentration measurements e.g. CO_2 and O_2.

The field model can readily provide information to compare with these data, as well as giving both a broader understanding of the overall flow behaviour and detailed structure where required.

3.3 Temperature and velocity prediction

The comparison of model results with experiment for the Swedish test room, (13), is shown in Figure 2. The doorway velocity profiles at the symmetry plane are shown in Figure 2a, and the temperature profile in Figure 2b. The velocity is negative in the lower half, indicating entrainment into the room, and positive outflow occurs at heights above 1.0 m. The agreement with experiment is reasonable except at the ceiling, where the velocity is underpredicted by about 20%. The shape of the temperature variation at the door is well predicted, although some displacement from the experimental value can be observed.

Table 2 Summary of Experimental Cases and Mathematical Model Details

Experiment	Dimensions (LxWxH m)	Nature of Fire Source	Heat (kW) Release	Steady/ Transient	Turbulence Model	Radiation Model	Combustion Model
SNTI Compartment	3.6 x 2.4 x 0.8 0.8 x 2.0 door	Propane gas burner, rear wall	250	S	k-e	flux model	Kinetically controlled eddy break-up
NBS room fire	2.8 x 2.8 x 2.18 0.74 x 1.83 and 0.99 x 1.83 doors	Gas burner center of room 0.9 sq.m	31.6 to 158	S	k-e	none	Fixed heat sources
LLNL test cell	6 x 4 x 4.5 Outlet: 0.65m square duct 3.6m above floor Inlet: 2 x 0.12 slit; 0.1m above floor	Isopropyl alcohol: 0.91m diam. steel pan; natural pool fire	400	S	k-e	none	Kinetically controlled eddy break-up
Zwenberg tunnel	390 x 5 x 4 One end closed	Petrol fire: 200 l in 2.6 m sq. tray	14450 to 24950	S	fixed viscosity	none	Kinetically controlled eddy break-up
NRC/SNL/UL	6.5 x 4.25 x 3.0 1.2 x 2.4 door & 2.4 x 2.4 door	Heptane pool 0.3 x 1.5 m	600 to 900	T	k-e	none	Time-varying heat release
FRS hospital fire	7.85 x 7.33 x 2.7	0.45 x 0.5 polyurethane foam mattress	5 rising to 80	T	k-e	none	Kinetically controlled eddy break-up
Shimuzu construction Company	34 x 28 x 11.6	Pool fire	?	T	k-e	none	Kinetically controlled eddy break-up

Figure 3 shows doorway velocity and centre temperature profiles for the NBS room fire and the JASMINE model (15,16). Here again, the agreement is good, although with some discrepancy occurring between the hot and cold layers. The relatively coarse computational grid (13 by 12 by 12) may have had some influence (16). This may indicate weaknesses in the mathematical model in the area of turbulent mixing prediction.

The Swedish and NBS test cases are for naturally ventilated rooms. In contrast, the LLNL experiment employed forced ventilation, extracting 400 to 500 L/s of air from the room, and allowing inflow through a slit at floor level. Vertical thermocouple rakes have been used to measure the variation of temperature within the room, 1.5 m on either side of the fire tray, and on one of the walls and the ceiling. Although not shown here, the JASMINE predictions reported in (18) compare quite well with the model predictions. Larger errors occur at the floor and ceiling, and it is reported that this may be due to simplified wall heat-transfer assumptions, the coefficient having been fixed to 20 W/m^{2}K.

Figure 4 shows temperature predictions and experimental data for the tunnel fire prediction reported in (6). The comparison with data is not as good as the room fire tests described above, particularly in the region above the fire source. Errors are particularly significant in the natural convection case. The use of fixed turbulence viscosity and a simplified combustion model may have been the cause of the poor agreement for this case. Difficulty in converging the solution for the natural convection situation necessitated these simplifications. The forced convection cases show much better agreement probably because they employ more detailed models.

Prediction of the transient fire situations, caused by variable heat releases, which were investigated by the NRC, FRS and SCC, are reported in (12), (13) and (18). Point temperature measurements as functions at time was given for the NRC experiment in Figure 5. These show data at 1 ft, 2ft and 3ft below the ceiling and 3 ft above the floor. The rise and fall in temperature with time follows the variation in heat release during the experiment. It is interesting to note that the heat release was a combination of a heptane pool fire and burning cables. The predictions agree quite well with the experimental values except at the peak temperature; the maximum temperatures being under-predicted by about 15%. The predicted rise in temperature is not as sharp as the observed in the experiment, and this applies to the other simulated cases described in (18). The longer computation times associated with three-dimensional transient calculations may have necessitated a coarser grid (950 cells in this case) than is desirable for numerical accuracy.

The hospital ward, (12), and the sports hall, (13), both had an initial airflow distribution, the former caused by convection heaters on one wall and the latter by forced air fans. The initial conditions for these predictions were therefore obtained by performing a steady-state calculation prior to running the transient. Figures 6 and 7 show temperature data from these tests.

The hospital ward results, Figure 6, show the variation above the fire and at several locations in the room. All the results show reasonable agreement. Figure 7 shows similar results for the sports hall.

Taking the above results as a whole, the bearing in mind that each case is unique in its geometrical features, location, and strength of fire source, it can be concluded that the field models have been quite successful at predicting temperature within the test rooms and the conditions at exit. The major features of the flow structure must therefore be quire well predicted for this to be the case. There have been rather few grid studies performed, and most of those reported indicate only a small effect of grid size on temperature predictions.

3.4 Species concentration prediction

There are fewer measurements of gas concentrations in the literature. Time-varying CO_2 measurements are available for the hospital ward fire, and CO_2 and O_2 measurements for the tunnel fire and national laboratory room fire. Figure 8 shows the hospital fire case CO_2 variation at two locations with time, and the vertical variation after 12 minutes. It seems that the concentration is reasonably well predicted at "nose" height, but is poorly predicted at "bed" height. The vertical variation illustration the overprediction further, particularly at the lower levels.

The tunnel fire predictions (6) agree fairly well in most cases. However, the size of the experiment and the relative sparseness of measurements do not make it a particularly good case for validation purposes, and it is almost certainly the case that a finer computational grid would be desirable before quantitative comparisons could be made.

In the national laboratory test case, the measured exit concentrations of O_2 and CO_2 were 10.4% and 7.5%, respectively. The corresponding predicted values were 14% and 5.5%. This appears to be a similar level of accuracy to the other cases.

3.5 Mass inflows and outflows

Data are available for integrated mass inflows and outflows for three of the cases - the Swedish test room and the NBS and national laboratory experiments. Comparisons are made in Table 3 with predicted quantities. It can be seen that they compare well. The worst error is about 17% for the Swedish test cases; the other predictions are within 10% or less.

Table 3 Experimental and predicted total mass flow (kg/s)

	Measured		Predicted		Remarks
	in	out	in	out	(kW)
SNTI room	1.01	0.89	0.84	0.84	
NBS	0.446		0.474	0.476	32
	0.58		0.555	0.557	63
	0.624		0.617	0.622	105
	0.688		0.657	0.655	158
	0.677		0.683	0.684	629
LLNL	0.3	0.24	0.269	0.257	

4. CONCLUDING REMARKS

This paper has provided a general outline of the mathematical and physical basis of field models that have been used for the prediction of fire and smoke movements. These models are based on the numerical solution of the basic equations of fluid motion and therefore represent a fundamental approach to the problem of prediction. They require assumptions to be made by the user for the characterization of effects such as turbulence and combustion, and various models exist for incorporating into the framework of a field model.

The validation studies that have been performed indicate that the major flow characteristics resulting from the fire source and ventilation arrangements can be predicted by the models. Quantitative comparison is good or temperature and, where available, velocity data, but not as good for gas concentrations. Virtually all of the studies reported have used quite coarse computational grids. Grid studies have shown little dependence of temperature variation measurements have not been subject to the same examination. Therefore, no clear guidelines exist for the user of field models in terms of nodalization, with one exception. Gas concentration measurements have not been subject to the same examination. Therefore, no clear guidelines exist for the user of field models in terms of nodalization, and it may be the case that the choice of grid is dependant upon the geometry to such an extent that each case should be examined individually for numerical accuracy.

Long computer times, on the order of several hours, are an obvious disadvantage of field models. While this should not be a consideration where fire safety is concerned, it is nevertheless a factor in the choice of design approach. Before embarking on a field-modelling exercise, the user must be certain that the results will justify the additional complexities of the approach. The architectural application mentioned earlier was one in which zone models were not deemed to be suitable for the larger scale of building and, in order to obtain building permits, alternative justification of safety aspects were required. Given the level of validation described in this paper, the use of a field model does seem justifiable.

With the ever-increasing power of computers, it may soon become normal practice to use a field model for smoke-movement prediction and ventilation design. A more user-friendly model would be required and many safeguards would have to be incorporated before use of such a program by non-experts in computational techniques could be envisaged. An alternative role for the field model might be the development of design guidelines by numerical experiments rather than physical ones - the cost would certainly be less. The development of zone models might proceed in a similar way.

5. REFERENCES

(1) The Kings Cross Underground Fire: Fire Dynamics and the Organisation of Safety. I.Mech.E Seminar (June 1989).

(2) Burns, A.D. and Wilkes, N.S. A Finite Difference Method for the Computation of Fluid Flows inComplex Three-Dimensional Geometries. AERE R 12342.

(3) Spalding, D.B. 1971. "Mixing and chemical reaction in steady confined turbulent flames." Thirteenth Symposium (International) on Combustion, The Combustion Institute Pittsburgh, Pa.

(4) Spalding, D.B. 1976. "The influences of laminar transport and chemical kinetics on the time-mean reaction rate in a turbulent flame. "Proceedings of the 17th Combustion Symposium, Combustion Institute, pp. 431-439.

(5) Guilbert, P.W., Comparison of Monte Carlo and discrete transfer methods for modelling thermal radiation. AERE-R 13423.

(6) Kumar S., and Cox G. 1985 "Mathematical modelling of fires in road tunnels." 5th International Symposium on the Aerodynamics

and Ventilation of Vehicle Tunnels, Lille, France.

(7) Sundstrom, B., and Wickstrom, U. 1981. "Fire: full scale tests. Calibration of test room-part 1. "NORD-TEST Project 143:78, 2. Technical Report SP-RAPP 1981, 48 National Testing Institute, Bords, Sweden.

(8) Steckler, K.D.; Quintiere, J.G.; and Rinkinen, W.J. 1982. "Flow induced by fire in a compartment. "National Bureau of Standards, NBSIR 82-2520.

(9) Alvarez, N.J.; Foote, K.L.; and Pagni, P.J. 1984
"Characteristics of fire in a forced ventilated enclosure." Combustion Science and Technology.

(10) Fiezlmayer A.H. 1976. "Brandversuche in einem tunnel." Bundesminsterium fur Bauten und Technik, Heft 50, Vienna.

(11) Cline D.D.; Von Risesmann, W.A.; and Chavez J.M. 1983.
"Investigation of twenty-foot separation distance as a fire protection method as specified in 10CFR50, Appendix R." NUREG/CR-3192, SAND83-0306.

(12) Kumar S.; Hoffman N.; and Cox. G 1985. " Some validation of JASMINE for fires in hospital wards". Presented a the First International PHOENICS Users Conference, Dartford, Kent, England.

(13) Pericleous, K.A.; Worthington, D.R.E.; and Cox G. 1988. "The field modelling of fire in an air-supported structure." Second International Symposium of Fire Safety Science, Tokyo, Hemisphere, Washington.

(14) Markatos, N.C., and Pericleous, K.A. 1983. "An investigation of three-dimensional fires in enclosures." Proceedings of the 21st National Heat Transfer Conference, ASME/AIChE, HTD, Vol 25, pp. 115-124.

(15) Cox G.; 1983. "A field model of fire and its application to nuclear containment problems". Proceedings of the CSNI Specialist Meeting on Interaction of Fire and Explosion with Ventilation Systems in Nuclear Facilities, Los Alamos National Laboratory, New Mexico".

(16) Cox G., and Markatos N.C. 1984
"Hydrodynamics and heat transfer in enclosures containing a fire source" Physico-chemical Hydrodynamics Journal. Vol. 5, No. 1 pp 53-66.

(17) Cox G., and Kumar S. 1984. " The mathematical modelling of fire in forced ventilated enclosures." 18th DOE Nuclear Airborne Waste Management and Cleaning Conference Baltimore, MD.

(18) Boccio J.L.; Usher, J.L. Singhal, A.K.; and Tam, L.T. 1985.
"The use of a field model to analyze probable fire environments encountered within the complex geometries of nuclear power plants." Presented at the 23rd National Heat Transfer Conference Heat Transfer in Fire and Combustion Systems HTD-Vol 45, pp 159-166.

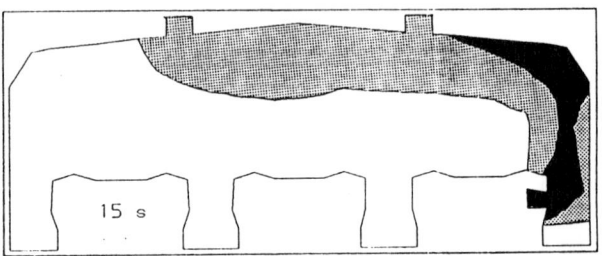

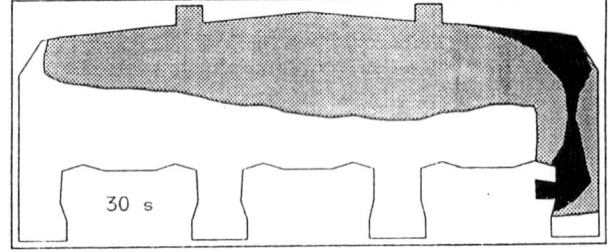

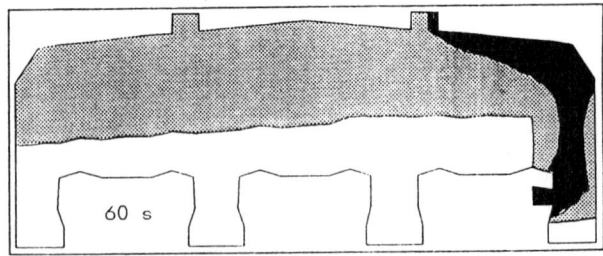

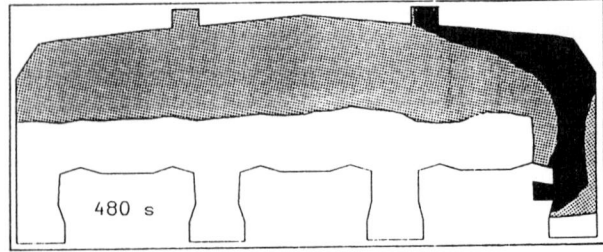

Fig 1 Variation of smoke concentration in a railway repair depot

Fig 2 Swedish test room:

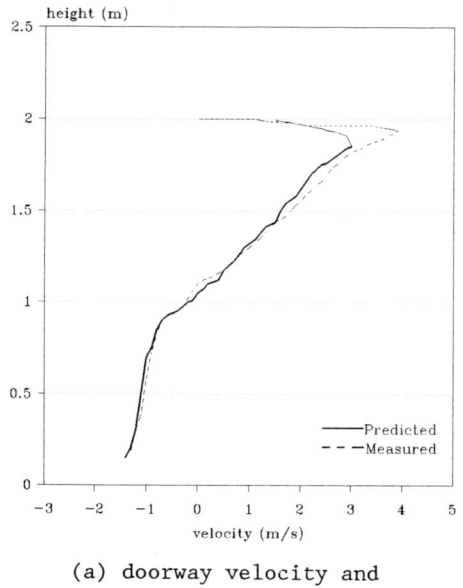

(a) doorway velocity and

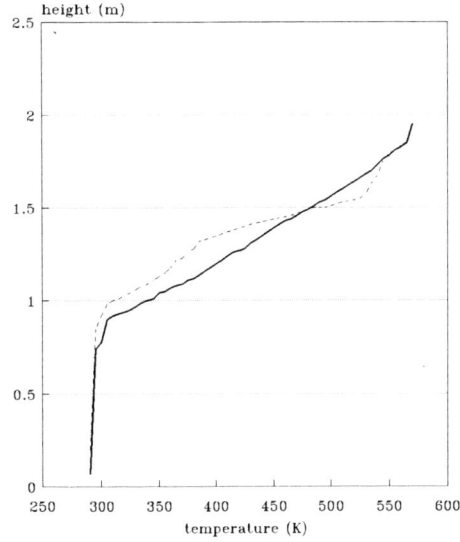

(b) temperature profiles

Fig 3 NBS room: doorway velocity and temperature profiles

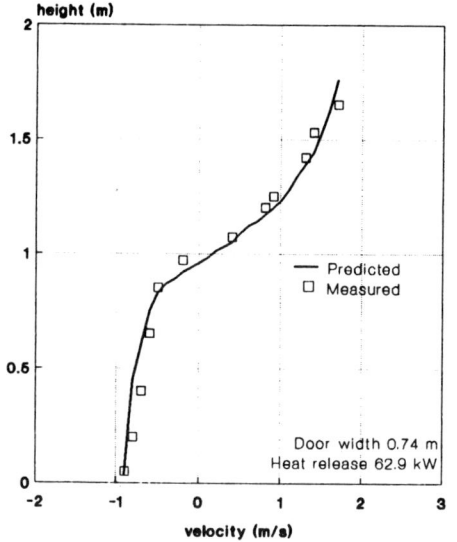

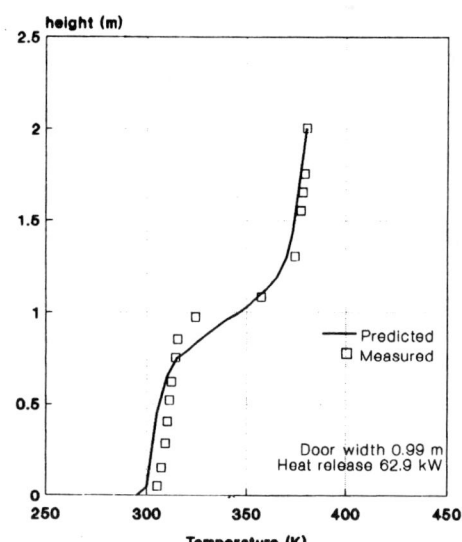

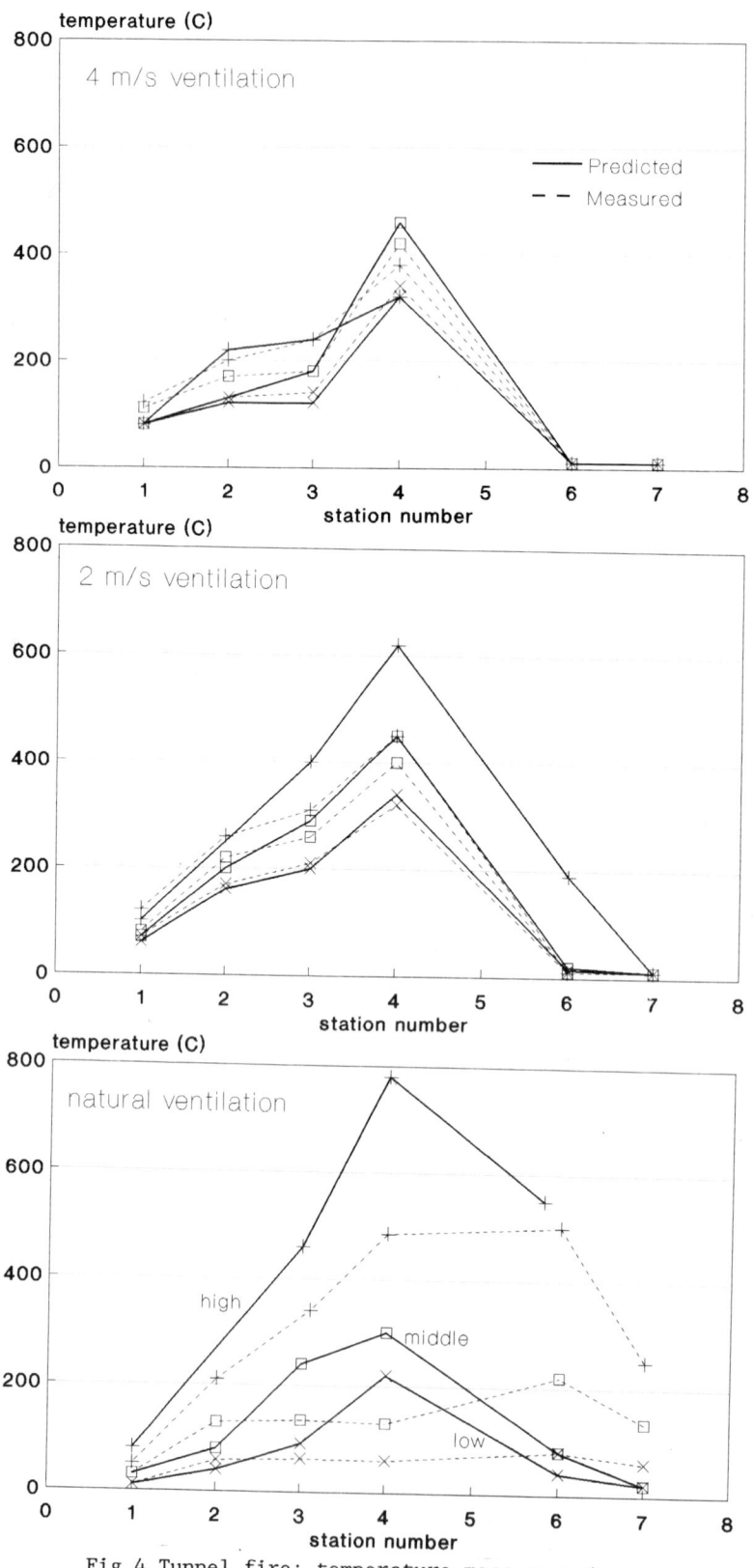

Fig 4 Tunnel fire: temperature measurements compared with prediction for natural and forced ventilation cases

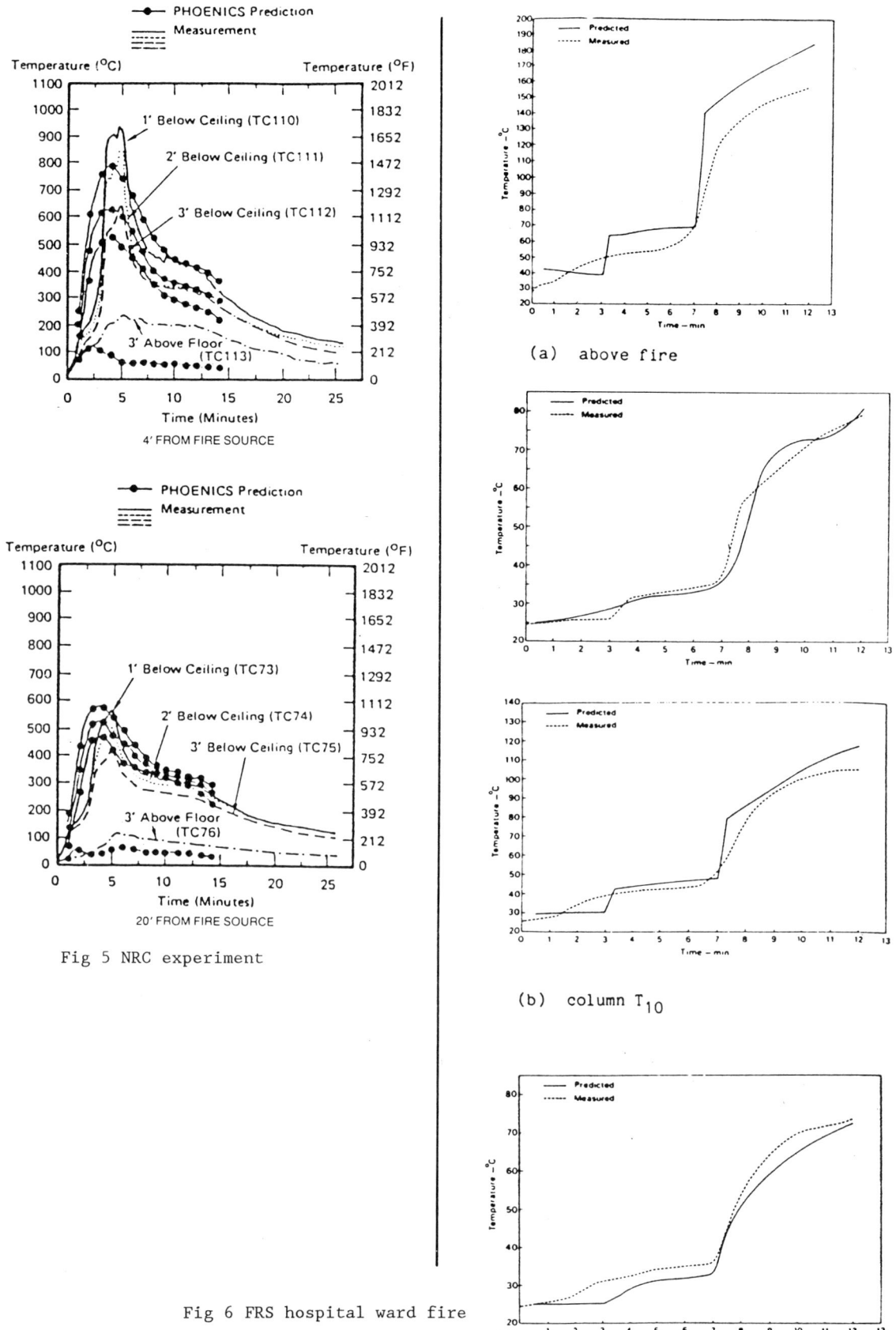

Fig 5 NRC experiment

Fig 6 FRS hospital ward fire

(a) above fire

(b) column T_{10}

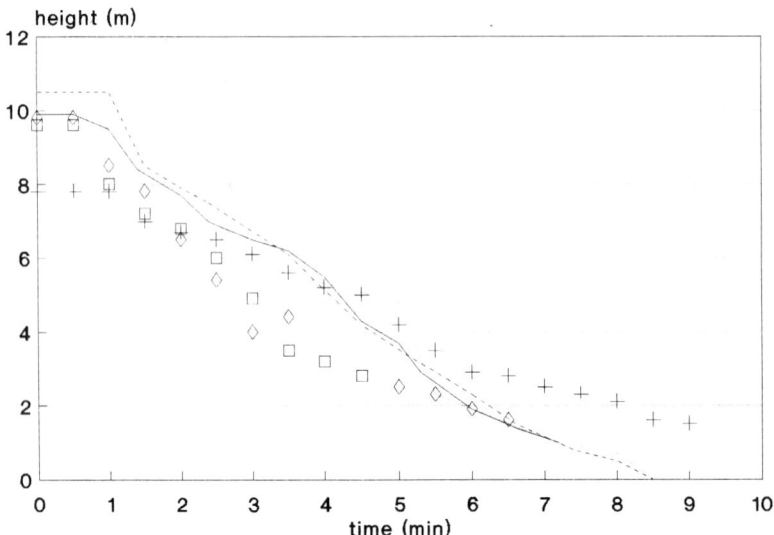

Fig 7 Sports hall fire: 10°C rise contour at three rates versus time

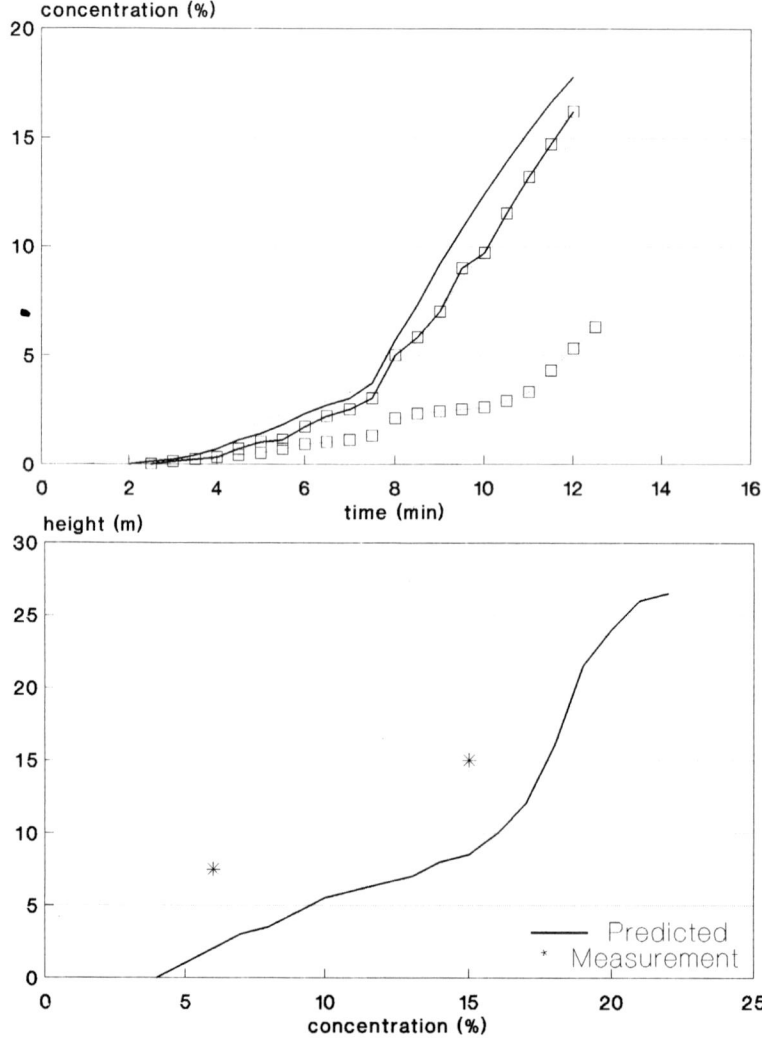

Fig 8 FRS hospital ward fire: concentration versus time at "nose" and "bed" height and concentration versus height

C413/059

Meeting vehicle pollution regulations by combustion technology

J CHAPMAN, Rover Group Limited, Lighthorne, Warwickshire
A C COLE, BSc, PhD, ERA Limited, Dunstable, Bedfordshire
S WALLACE, Rover Group Limited, Lighthorne, Warwickshire
J H WEAVING, BSC, PhD, CEng, FIMechE, FSAE, ERA Limited, Dunstable, Bedfordshire

SYNOPSIS

The preferred and perhaps only method to meet U.S. and imminent European Regulations is by 3-way catalyst with an engine running at stoichiometric air-fuel ratio. However, these regulations have neglected to cater for carbon dioxide which is a major contributor to the "greenhouse" effect. A suitable way of reducing all regulated pollutants and at the same time reducing carbon dioxide is by running engines under very lean burn conditions.

This paper presents an in-depth research on ten different combustion chambers for conventional vehicle S.I. engines performed in order to identify the preferred chambers for stable running. Data from stratified charge experimental engines are compared.

The four valve pent roof combustion chamber was selected by Rover for their new 'K' Series engine. Significant data are presented.

1. INTRODUCTION

No-one doubts the importance of reducing atmospheric pollution or that the motor vehicle makes a contribution to the total pollution of carbon monoxide (CO), nitric oxide (NO) and unburnt hydro-carbons (HC). Additionally, as with any combustion system for fossil fuel, it produces large quantities of carbon dioxide (CO_2).

The present legislation in the U.S. and Japan and proposed legislation for the EEC for petrol engines (1992) with existing technology, necessitates the use of a 3-way catalyst with feedback control of the fuel system to maintain the mixture at a stoichiometric air/fuel ratio. This latter is essential for a 3-way catalyst to both oxidize CO and HC and reduce NO.

This legislation is not necessarily the optimum as the CO_2 problem was not appreciated when the regulations were formulated; it is essential to have a clear view of the targets. These are; damage to health, damage to the environment, and the "greenhouse" effect. In view of this, further research on Air Quality Standards is highly desirable. For the purpose of this paper, however, it will be assumed that the above regulations have to be met but in addition, CO_2 will be reduced as much as possible.

As is well known, minimum fuel consumption is obtained weak of stoichiometric so there is a fuel penalty of about 10% with a 3-way catalyst, but what is equally important is that carbon dioxide is, in consequence, some 10% higher than it need be so contributing to the greenhouse effect. It is the objective of this paper to contribute to the basic knowledge of combustion in the internal combustion petrol engine and so help to minimise the pollution and point the way for optimum control.

All pollutants may be reduced with improved fuel consumption by a high compression lean-burn approach using conventional or stratified charge combustion engines (1). By burning weak (beyond 20:1 AFR) as is well known NO reduction is considerable but experiment shows there will be a residual amount of CO and HC which can be minimised with a catalyst. This approach requires, first the selection of the optimum combustion chambers and secondly, expedients to reduce further the remaining pollutants to meet the regulations. A large part of the paper deals with the first factor while for the second a management system will be required to programme the lean-burn approach.

Running at AFR's beyond 20:1 results in a loss of some 20% of the maximum engine power. It is proposed therefore that this is restored by means of a management system with a lean-burn sensor tuned to switch the mixture from lean-burn to stoichiometric for running near and at full throttle. A three-way catalyst will be needed to keep the NO in check under these conditions. This catalyst would also oxidise the residual HC and CO.

2. THE HIGH COMPRESSION LEAN BURN ENGINE (HCLB)

Lean burn reduces the flame speed with consequent loss of efficiency; this can be restored and indeed efficiently improved by increasing the compression ratio. In order to evaluate the capabilities of the various common combustion chambers when running under HCLB conditions, a series of comprehensive tests were performed by Rover Group using 500 cc Hydra single cylinder engines. The combustion chambers chosen are shown in Fig. 1, the objective being to select the best 4-valve and 2-valve engine for further evaluation in multi-cylinder form. For the high compression ratios 12:1 was chosen as above this figure improvements in efficiency are minimal. It is realised that with present premium fuel, some retardation will be necessary to prevent 'knock' but the gain in economy at part load is believed to justify this.

Combustion rate has a major effect on engine performance. The principle factors being burn velocity and flame area. Flame area is a function of the combustion chamber shape and spark plug position. Burn velocity is increased by higher temperatures, (due to increased compression) optimum fluid motion and turbulence intensity, but reduced by weaker mixture ratios.

2.1 Types of Combustion Chamber

The combustion chambers chosen for evaluation (Fig. 1) were:

1. 'O' Series, 9:1 compression ratio
2. 'O' Series, 12:1 compression ratio
3. 'O' Series High Swirl, 12:1 compression ratio (not shown)
4. Weslake, 9:1 compression ratio (WES 9:1)
5. Weslake, 12:1 compression ratio (WES 12:1)
6. 4-Valve Pent Roof, 12:1 compression ratio (4V)
7. 2-valve Hemispherical, 12:1 compression ratio (HEMI)
8. Compact chamber under exhaust valve, 12:1 compression ratio (MUE)
9. Compact Chamber under Inlet Valve, 12:1 compression ratio (CCI)
10. Bowl-in-Piston, 12:1 compression ratio (BIP)

The 9:1 heads were included to provide a baseline for comparison. The 'O' series has been a production cylinder head designed by the Rover Group and used in many production engines. To extend the baseline the compression ratio of the 'O' Series was increased to 12:1. To represent a high swirl chamber, a design due to Weslake, also used by Rover, was employed. The 9:1 Weslake was known to have a low octane requirement but little was known about it at 12:1 ratio (2).

Traditionally the 4-valve Pent Roof chamber has been associated with high performance design, but more knowledge was required on its lean-burn capability and burning characteristics. To differentiate chamber shape from dual inlets, a 2-valve version namely the 'Hemi' was included. The May combustion chamber has been promoted as a high compression design and is used on some Jaguar engines (3). To evaluate how much was due to the compactness of this chamber as distinct from its location, a similar head with the compact chamber under the inlet valve (as opposed to exhaust valve in the May) was designed. To compare these latter two chambers with the familiar bowl in piston, this too was included in the test programme.

Finally, to evaluate the effect of port induced swirl on burn rate to improve the 'O' Series lean-burn performance a special inlet port was designed, other features being kept the same.

2.2 Test Programme

The firing testing facility is centred around the Hydra engine configured to 89 mm stroke by 84.5 mm bore. Overhead camshafts are driven by toothed belt and have separate individual cam lobes to permit variations in valve timing.

The engine inlet air, oil and water temperatures are controlled by thermostatic control of heaters and coolers. The remotely operated engine controls for throttle, fuel injection, ignition timing, EGR and dynamometer speed are interfaced with the test bed computer. All of the measured parameters from the engine, temperatures, pressures, air and fuel flows, torque output, exhaust gas emissions analysis and EGR %, are displayed on digital readouts which are also interfaced with the test bed computer for data acquisition, calculation, plotting and analysis of results.

The test programme consisted of a comprehensive matrix of test points of varying loads and speeds. Air-fuel ratio (AFR) and ignition were varied over a large range at both full and part throttle conditions.

The programme was divided into three phases, which will be as far as is possible correlated;

Phase 1 -
Firing performance, including performance and emission data, pressure diagram measurement at selected points, together with ionisation gap data for monitoring flame travel.

Phase 2 -
Motored performance data. A second single cylinder engine of identical bore/stroke configuration was constructed to allow measurement, by hot wire anemometry, of air flow characteristics at specified locations within the combustion chamber, using one, two and three hot wires simultaneously as required.

Phase 3 -
Steady state rig tests of each head to determine flow and swirl characteristics.

Full throttle performance curves were carried out on three fuels, AF 2114 (RON 96), AVGAS (RON 105) and Trace Lead (RON 91) in the 1000 to 5500 rpm range with mixture and ignition loops at 1000 rev/min increments. Part throttle testing was carried out on AF 2114, at four conditions:

1. 2000 rev/min   2.0 bar
2. 3000 rev/min   5.5 bar
3. 1500 rev/min   1.0 bar
4. 2000 rev/min   4.0 bar

The chamber's response to three rates of Exhaust Gas Recirculation (5, 10, 15%) at 4000 rev/min, 5.5 bar and 2000 rev/min, 2 bar was assessed. Mixture strength was weakened from 14:1 until misfire occurred or HC became excessive. Idle tests comprised fuel and ignition loops at 900 rev/min.

In-cylinder pressure measurements and flame arrival times were measured by Kistler 6121 pressure transducer and ionisation probes respectively. Measurements were recorded at 1000, 2000 and 4000 rev/min full throttle together with 3000 rev/min and 5.5 bar and 2000 rev/min 2 bar part throttle conditions. Further measurements were taken at the 2000 rev/min 2 bar condition to evaluate the effects of 10% EGR and an 8° ignition retard on cylinder pressure and flame speed.

This programme allowed a wide selection of operating conditions and variables to be studied which are representatives of selected points on US Federal and European exhaust emission cycles.

3. ENGINE TEST RESULTS

For space limitations only some of the results of this comprehensive programme are presented. The authors have selected the full load and 2.0 bar 2000 rev/min conditions to be covered reasonably fully.

3.1 Full Load Tests

Fig. 2 shows the full throttle results; the compact chambers, 'May' (under the exhaust valve) and the similar chamber under the inlet gave almost identical BMEP so only the former has been plotted. At 12:1 C.R. all the heads were 'knock' limited before maximum power on 96 RON fuel. Most of the two valve chambers produced similar power and indeed had been designed to do this. As would be expected the 2 valve hemispherical head had better breathing and thus high speed performance. The Weslake (9:1 CR) had better power at lower speeds as it was knock free. The superior power of the 4-valve head at the higher speeds again was to be expected. The faster burning of the high turbulence heads is clearly indicated by the small ignition advance required for M.B.T.

It will be noted that although the 4-valve was knock limited at low speeds, it lost little against the 'O' series at 9:1 while being very superior at higher speeds. When run on 91 RON trace lead fuel similar trends were observed with the expected reduction in B.M.E.P. levels.

Looking at fuel consumption, the improvement due to CR increase from 9:1 to 12:1 will be noted at the higher speeds, but generally no improvement at low speeds due to the necessary retardation.

3.2 Part Load Tests.

It is at the part loads that the superiority of the high compression ratio is impressive (Fig. 3) and the capability of the 4-valve engine to run as lean as 23:1 AFR with considerable gain of efficiency from stoichiometric is also impressive.

Generally when comparing part throttle specific fuel consumption of the various heads, the degree of improvement decreases with an increase in speed/load condition. This was confirmed by a 23% improvement (not shown for space limitation) of the 4V when compared with 9:1 'O' Series at 1500 rev/min, 1 bar; gradually decreasing to a figure of 11.4% at 3000 rev/min 5.5 bar. However, when comparing the two 12:1 builds (4V and 'O') a 14% gain at 1500 rev/min, 1 bar of the 4V decreases to a minimal improvement of 3% at 3000 rev/min 5.5 bar. Apart from illustrating the expected gains in fuel consumption of 8% produced by the increase in compression ratio the graphs also highlight the higher burning velocities of the 4V and its ability to extend the lean burn limit up to 23:1 AFR.

3.3 Emissions

With the proposed system full load running would be at stoichiometric; and the 3-way catalyst would take care of NOx, HC and CO, therefore emission performance is focussed at part load lean mixtures. 2000 rev/min and 2.0 bar has been selected for illustration (Figs. 3 and 4) similar trends at 1.0, 4.0 and 5.5 bar were found.

At weak mixtures CO values are low and again would be further reduced with a 3-way catalyst, which acts well as an oxidising catalyst. This makes the NO pollution characteristics as the most important as no catalytic reduction is possible in the presence of excess oxygen. HC is also important as it is a sensitive indicator of misfire or partial burning due to flame quench.

Fig. 4 shows the NO results. As would be expected the high compression ratio compact chambers are the worse, 'May' being particularly high, with the 2 and 4 valve hemi chambers next, both being similar. However, at 22:1 with stable running these three chambers are as good as the 'O' Series at its lean limit of 19:1.

For hydrocarbons at 12:1 CR, (Fig. 3) the superiority of the 4-valve is very impressive. Next comes the 'May' and compact head under the inlet-valve which gave similar figures and secondly showed the capability of running very weak. The higher HC values compared with the 4-valve are no doubt due to the squish areas which are difficult to burn due to the close proximity of cooled surfaces. In general, the high compression heads have higher HC levels than the baseline 9:1 engines due largely to the increased surface to volume ratio.

3.4 Octane Requirements

Figure 5 shows the Octane requirements of the various heads under full throttle conditions. As only three fuels of rather differing specifications were used, these are clearly not exhaustive but are good enough to rate each head in order.

The Weslake 9:1 has by far the lowest requirement being significantly lower than the less active 'O' Series at the same ratio. At 12:1 CR all would require some retardation for 96 RON. The Weslake still has a slight advantage over the other heads tested, being knock free above 2000 rev/min. The 4-valve is clear of knock on AVGAS above 2000 rev/min, the compact chambers were more sensitive.

3.5 Exhaust Gas Recirculation (EGR)

The acceptance of EGR for further reduction of NO in the lean burn situation is an important requirement for good combustion chambers. The large number of tests at the several speeds and loads of this programme all gave an approximate linear reduction of NO with percentage EGR up to 15%. Some results for 15% EGR at 2 bar, 2000 rev/min and 18:1 air/fuel ratio are discussed below.

The 4V showed a 67% reduction of NO but with no detrimental effect on fuel consumption. The quiescent 'O' in contrast, for a 60% reduction had an HC increase of approximately 500% and a fuel penalty of 25%. There was minimal fuel consumption penalty for both the CCI and May and both gave 80% reduction in NO, with 60% increase in HC. The Weslake gave similar results to the 'May'.

It follows that a compromise has to be made to suit the total emission system proposed to meet specific regulations.

4. COMBUSTION MEASUREMENT

From the in-cylinder pressure measurements recorded against crank position the maximum pressure, maximum rate of pressure rise and angles of occurence can be taken directly from this data. By analysing the pressure - crank angle data, the nett heat released by the gaseous mixture can be calculated showing the shape and period of the heat release in the chamber. The pressure diagrams and the derived heat release data for the seven chambers tested are summarised in Tables 1 and 2.

Ionisation probes were inserted into the chambers to register the arrival of the ionised flame by the current pulse across the electrodes. This pulse, recorded on the high speed data acquisition system, was used to give mean 'equivalent' flame speeds.

4.1 Cylinder Pressure

At wide open throttle (WOT) the severe ignition retardation required to avoid 'knock' has the overriding effect. As the degree of 'knock' and retardation is reduced with increasing engine speed, the burn rate, maximum pressure and rate of pressure rise increase sharply. However, at 4000 rev/min the knock limited fast burning 12:1 builds still have lower maximum pressures and rates of pressure rise than the 9:1 builds. At 2000 rpm (Table 1) the 9:1 chambers exhibit much higher maximum pressures and rate of pressure rise than the heavily retarded 12:1 builds, though the 12:1 chambers, if optimised, have a potential that is 30% greater in terms of pressure. The optimum crank angle for peak pressure, irrespective of operating conditions or chamber designs, is 12-15° ATDC as shown by the optimised 9:1 Weslake (Fig. 6). The amount the 12:1 builds are retarded results in the peak pressure occurring as much as 20° later than the optimum. Similarly in cases of severe retardation, the rate of pressure rise is also retarded, giving a double peak in the pressure diagrams and reduced so much that the rate of rise due to compression can be higher than combustion. As ignition is near TDC at this condition, the delay period (defined as crank angle from ignition to 1% mass burnt) is very short due to high pressures, temperatures, gas activity and compact volume. The maximum rate of burning shows the same phenomena as the maximum pressure; the Weslake 9:1 being the only one at optimum giving the highest burn rate, nearest to TDC. At part throttle, the relative burning characteristics of the different chambers remained similar for all conditions of load, speed, SFC, EGR and ignition retard.   Results for 2000 rev/min at 2 bar are taken as typical (see Table 2). Ignition was optimised to give peak pressure at 12° ATDC for all chambers.

4.2 Flame Speed

At full throttle from Fig. 7a the effect of chamber shape on flame speed as measured by ionisation probes is apparent, though it is perhaps surprising that some of the chambers with considerably different internal gas flows (see Section 5) have similar flame speed characteristics. Certain generalities can be made, the first is that 'flame speed' is increased approximately linearly with engine speed and secondly that chambers with high internal gas motion (e.g. swirl) have up to double the flame speed of quiescent chambers ('O' series).

At part throttle (from fig. 7b) increasing AFR generally lowers the flame speed roughly linearly, this would be expected as laminar flame speed, to which turbulent flame speed is related, also reduces with increased AFR, (4). This drop in laminar flame speed at lean mixture is compensated by the increased turbulence in the active types of combustion chambers (e.g., 4 valve pent roof).

In most of the combustion chambers, several ionization probes were fitted and two examples are shown in Fig. 8a and b. The effect of the high axial swirl of the Weslake, (Fig. 8a) positions P and M are equally spread from the plug but 'flame speed' to P in the direction of swirl is roughly double that to M against the swirl. The speed across the chamber is also high showing general activity probably generated by the 'tip' between inlet and exhaust valves which is known to have a considerable effect on combustion. With the 4V chamber (Fig. 8b) at part load, the random nature of the 'flame speed' to the various gap positions is difficult to explain, but it clearly has fast burn in all directions. The situation for the May head (not shown) gave very rapid acceleration across the hot exhaust valve, while the combustion chamber over the inlet valve had a more uniform burn averaging the maximum and minimum of the 'May'. In the 'O' series chamber, flame speeds were roughly half those of the chambers at 12:1 and some 25% lower than the others at 9:1.

5. INCYLINDER GAS MOTION

Gas motion within the combustion chamber is of crucial importance to the combustion efficiency of the engine design.   Within this study the flow field of the principal combustion chambers has been evaluated through both hot wire anemometry techniques (on a motored engine) and a steady state gas flow rig.

5.1 Hot Wire Anemometry

Hot wire anemometry provides a useful and relatively simple means of understanding the nature of gas motion at the point of measurement within the chamber of a motored engine. Measurements at the normal spark gap position are particularly useful for correlation with burn rate data, but of course, it cannot give precise information about the complete gas flow field within the cylinder.

Within this study five of the ten original chamber designs were evaluated in terms of gas motion.   Referring to Figure 1 they were;

1) O series, 9:1
5) Weslake, 12:1
6) Four valve, Pent Roof 12:1
9) Compact, Chamber under Inlet, 12:1
10) Bowl-in-Piston, 12:1

The data was measured using a Dantec 55M CTA system with a 10 μm PtRh triple wire probe.   The wires have an aspect ratio (l/d) of between 225 and 250 and operated at a mean temperature of 800 K.   The calibration function of the wire followed the work of Davies and Fisher (5) and incorporated the Collis and Williams temperature correction (6), i.e.;

$$Nu \, (T_f/T_m)^{-0.17} = A + BRe^n$$

(Nu is Nusselt No. and Re Reynolds No.)
where T_f is the mean film temperature, $(T_m + T_w)/2$
T_w is the wire temperature
T_m is the instantaneous measured gas temperature.

The calibration constants; A B and n, are determined from a steady flow high temperature (650K) and pressure (14 bar) calibration rig.

Data was acquired from 50° BTDC to 50° ATDC about TDC compression for 100 cycles at half degree intervals and stored for subsequent processing.   From the raw data, both ensemble average and standard deviation curves are produced for each wire orientation.   The standard deviation curve is related to turbulence intensity but of course includes cyclic variability.   Table 3 attempts to condense the data by presenting for each test condition and chamber shape the maximum, minimum and average velocity and standard deviation (irrespective of wire).

Examination of this table shows many interesting facets.   First it is clear from the mean velocities measured that the 4-valve head and the Weslake are the most "active" heads followed by the compact chamber under inlet valve.   The 'O' series and 'Bowl-in-Piston' are more quiescent.   Secondly the actual levels of velocity and standard deviation are apparently speed and throttle dependent rising with increasing engine speed and reducing with decreasing throttle opening.

A third observation, that despite the great differences between the heads, there is no significant differences on the velocity standard deviation.   From this we may assume that, with this method, true turbulence is being masked by the "turbulence" caused by cycle-to-cycle variation and as such does not correlate with the activity of the head as was hoped.   The likelihood of ordered small scale turbulence in such a highly transient situation is small, so that the chances of characterising it are remote.   Even the three wires will not give the absolute direction of flow, as a wire even when lying along the direction of flow does not give a zero signal. For the purpose of comparison the wire giving the maximum value at the plug is plotted in Fig. 9 as being the most relevant.

It will be seen that the two 'O' series combustion chambers (curves No. 1 and 3) show low 'activity' in that during compression and expansion the gas motion remains reasonably constant. This ties up with slow burn and the need for large ignition advance, Fig. 2. Comparing the 4-valve (No. 6) and the Weslake, (No. 5) both have high compression velocity but the 4-valve drops to a lower value at TDC and continues to fall slowly after TDC. The Weslake, though dropping considerably at TDC tends to rise afterwards, indicating that the high axial swirl, though attenuated, still remains. The chamber under inlet valve, where the induction swirl is considerably less relies on compression to create the axial swirl and after TDC this swirl remains, though as with the Weslake, it is greatly attenuated.

5.2 Steady Flow

In order to obtain both quantitative and qualitative air flow data, a steady air flow rig with axial swirl measurement capability was used. Fig. 10 highlights the excellent air flow of the four valve configuration with little or no axial swirl. This is in direct contrast to the Weslake which produced a high degree of swirl to the detriment of air flow. The shrouding effect between inlet valve and combustion chamber wall on the Weslake is not so prominent at the higher valve lifts on the 12:1 due to the shallower combustion chamber; this would also account for the differences in swirl levels. The combustion chamber under the inlet valve design with its inherent masking effect also shows poor air flow characteristics at the higher valve lifts and is further compounded by the relatively small inlet valve diameter. Air flow restriction was not, however, reflected by the overall chamber performance to the degree that would have been expected. Swirl levels proved to be of a consistent value over the valve lift range with a respectable swirl ratio of 0.96. (Swirl ratio = swirl rate x bore/inlet velocity).

It is well known that swirl can be induced by inlet port design and this was demonstrated by modification of an 'O' Series inlet port to create a High Swirl 'O' Series (HSO), without compromising the comparatively good air mass flow characteristic of the standard 'O' Series port. It can be seen from the results that swirl increased with valve lift, this being attributed to port contours having more effect at the high gas velocities. Also seen is the increase in port restriction as indicated by the air flow at valve lifts greater than 6 mm.

The 2-valve hemi configuration has relatively small inlet valve diameters in order to accommodate three spark plug positions within the chamber resulting in relatively poor air flow performance. The extra two plugs, fired simultaneously, were used to compare against the single central plug position. As there was no detrimental effect, in future designs, a central plug could be eliminated this allowing larger valves.

6. CORRELATION OF TEST DATA

Gas motion is obviously a key element of the HCLB concept. The steady flow results confirm the well known trade off between induction generated axial swirl and flow capability, the Weslake design being the best example of this. Other designs increase the gas motion during compression, i.e. 4-valve and May, and generally do not show high rates of induction axial swirl.

The 4-valve design, in particular, shows very good flow capability over a wide valve lift range indicating port restriction limits the effective cross-sectional area on other designs (e.g. high swirl types). Although the 4-valve design shows very little axial swirl, it is known that barrel vortices are created during induction (7). The incylinder gas flow measurements on a motored engine confirm this gas motion by showing an increase in gas velocity during the early stages of compression (as the barrel vortices accelerate) This is followed by a significant decline as the vortices collapse. Apparently similar gas activity is recorded in the Weslake design, but here it is an axial swirl motion that is accelerated and then partly broken down to turbulence by the design's characteristic 'tip' in the combustion chamber. In contrast to the 4 valve it is worth noting in Fig. 8 that the Weslake and May chambers show increases in gas motion after TDC indicating the regeneration of ordered, probably axial, gas motion at the start of the expansion stroke. Whether this is true for the fired, as opposed to motored engine, is unknown, though it is tentatively concluded that the 4 valve swirl breaks down into turbulence effectively.

The part load combustion measurements tend to echo the gas motion results. The difference between 'active' and non-active designs in terms of burn rate is considerable at both 9:1 and 12:1 compression ratios. The best example is the 1% to 99% burn duration of the 'O' series design, in that it is almost twice that of the Weslake at both 9:1 and 12:1 compression ratio, Table 2. The flame speed measurements give further information about the nature of the gas motion and the flame growth sensitivity to air-fuel ratio. The Weslake design with its axial gas motion and off-centre plug position shows considerable flame speed variations within the chamber. The 4-valve shows better flame speed uniformity and typically a 30% reduction in the magnitude for a weakening of five air-fuel ratios (Fig. 8).

Part load emissions and fuel economy are heavily influenced by amongst other things, combustion rate, heat transfer and combustion temperatures. Compression ratio improves cycle efficiency but also increases NOx emissions (linked to combustion temperature), similarly rapid combustion from active chamber designs also produce higher gas temperatures and hence NOx. However, as can be clearly seen by operating at leaner air-fuel ratios, the fast combustion systems result in lower NOx levels without any significant hydrocarbon or fuel economy penalties. The exception to this is the under-valve chamber designs, which although they are known to be rapid burning, the large squish areas cause considerable hydrocarbon formation at the leaner air fuel ratios.

Full load performance is seen to be flow and ignition limited. The considerable ignition retarding of the high compression designs at low speed suggests that 12:1 is too high for even the fast burn chambers. The 4-valve is distinctive by a considerable torque increase (10%) between 3000 and 4000 rpm. The Weslake design shows its reduced high speed breathing capability, confirmation of the steady flow tests.

In conclusion therefore, it is clear that of the chamber designs considered the 4-valve compact chamber offers the most potential in terms of emissions and performance within a lean burn multi-cylinder application. Section 8 of this paper discusses this application further. Of the two valve designs it is clear that the Weslake is superior apart from its poorer high speed breathing capability.

7. COMPARISON OF STRATIFIED CHARGE ENGINE (SCE) WITH HCLB

Some 15 years ago (c. 1975) considerable research work was performed with the SCE in the hope of meeting the stringent U.S. regulations. The outcome of this work showed that NO could be very greatly reduced but that HC emissions were too high. The Honda Civic was indeed marketed in the U.S. (8) with HC reduction achieved using a thermal reactor. Increase in the severity of the regulations resulted in most companies dropping this approach. However, in view of the CO_2 "greenhouse effect" it seems timely that this work should be reviewed and a comparison with the HCLB is being attempted below.

The basic advantage of the SCE, first researched by the late Sir Harry Ricardo, (9) is that power is reduced by weakening the mixture rather than throttling. Theoretically, this gives a considerable improvement in part-load fuel consumption due to lower pumping losses and, due to the higher value of the ratio of the specific heats of the charge, a higher thermal efficiency. From the pollution angle the lower temperatures during combustion give much lower NO and the improved efficiency gives lower CO_2.

There have been a variety of different combustion systems which are reviewed in (1). These divide into two classes the Open Chamber and the Pre-Chamber types. The principal exponents of the Open Chamber have been the U.S. Ford Motor's Proco (10) and Texaco (11) while Honda, Rover (Fig. 11a), Porsche and others have researched the Pre-Chamber type. Opinions have been expressed that the Open Chamber types are bound to be more efficient than Pre-chamber configurations because of higher heat losses as the gases expand from the Pre-Chamber to the main chamber but, as the Pre-Chamber may be as small as 5% of the volume of the main chamber, this effect can be minimised. With a stratified charge engine it has been found that some throttling is necessary and Fig. 11b for the Rover work shows the situation. Throttle setting is indicated by manifold depression and the curves show the effect on NO as the mixture is weakened from rich to weak. It will be noted that it was possible to run to nearly half power before some throttling became necessary to prevent misfire. (Curves end at point of misfire.) With 127 mm Hg depression, ¼ load can be obtained with correspondingly low NO. An engine running near stoichiometric would have NO values close to the peaks shown, while the stratified charge engine can be tuned to follow the wide-open throttle curve until it is necessary to move to the 127 mm Hg depression curve at 4 bar.

It is not possible, with the data at present available to make a precise comparison because different single cylinder engines were used and different performance programmes were adopted, however, quite a useful insight can be obtained. It will be seen from Fig. 3 that the fuel consumption of the Porsche (curve No. 11) and VW (No. 12) are not as good as the 4-valve but it has to be borne in mind that the CR was only 8.5:1. The VW was further penalised by the high friction due to driving a Lanchester damper, a high penalty at this low load. Figure 4 shows the NOx values from which it is apparent that the Porsche is superior. The Rover results are similar, at 12:1 the compression ratio would be a little higher. The NO response of several SCE over the load range are shown in Fig. 11c.

A 4-cylinder Rover experimental SCE (1.85 litre) was evaluated in a vehicle and produced the results shown in Table 4 when tested to the US-CVS Chassis Dynamometer test schedule. It will be noted that the SCE without a catalyst was poor on HC but this, as with CO, can easily be handled by an Oxidising catalyst. NOx without any assistance from the catalyst was below the US Federal standard but did not reach the severe Californian requirement for NO. It is probable that with EGR and a 3-way catalyst, the latter would be achievable.

TABLE 4

CONDITION	POLLUTANT g/mile			
	HC	CO	NOx	MPG (US)
Without Catalyst	8.0	6.0	0.7	24.0
With Catalyst	0.35	1.0	0.8	24.2
U.S. Federal	0.4	3.4	1.0	
U.S. Californian	0.4	9.0	0.4	

8. FOUR VALVE ENGINE IN MULTI CYLINDER FORM

Rover Group, as many automobile manufacturers, recognised 4-valve technology as a potential solution to the demands of new combustion systems. The good overall performance of the four valve configuration identified by the high compression lean burn research programme described above confirmed that certain inherent properties of the design were beneficial. Ignition advance was lower than most, suggesting a faster burn chamber, similarly the central location of the sparking plug also suggested that flame propagation paths were at a minimum, so reducing the tendency of end gas 'knock'. An additional consideration was the capability of downsizing the power unit whilst maintaining customer requirement for performance outputs. Other advantages were seen to include reduced fuel requirement (reduction of CO_2 emission), reduced mass emissions, improved packaging capabilities, reduced weight and the possibility of basing an engine family on a common short engine leading to possible cost reductions.

For a specific engine design, combustion requirements must be considered within the context of many principal design parameters; engine weight and size, power requirement, rated speed, bore to stroke ratio, etc. With respect to cylinder head design the basic layout must consider three primary features, namely aspect ratio (height of chamber to bore), barrel swirl ratio (in-cylinder activity) and spark plug position.

A new combustion system was developed utilising 4-valve technology based on the following design criteria:

- effective degeneration of barrel swirl vortices
- minimum activity for the required burn rate incurring less port restriction.
- achieving high compression ratios without piston protruberances.
- favourable surface/volume ratio for good thermal efficiency.
- narrow head design from small included valve angle giving good underbonnet packaging.

Data from this new combustion system indicated that increasing the burn rate by utilising induced in-cylinder activity that degenerated before TDC, improved combustion stability, lean burn capability, combined HC and NOx emission levels and fuel economy.

The improved lean burn capability also provided a wider tune window for engine management system calibration, Fig. 12. This figure also indicates the economy potential (reduced CO_2) of the higher activity chamber even when running at stoichiometric AFR's.

Specific consideration was given to the manufacturing aspects of such a combustion system. New technology benefits can only be used for volume manufacture if a feasible, reliable and insensitive design package can be produced. This rationale has been used throughout the design and development of Rover Group's new family of engines designated "K" (12).

To illustrate the superior characteristics of a 4-valve combustion system a comparison has been drawn with a 2-valve port injected production engine. Performance is measured against a single point injection version of the new 4-valve engine.

Specific power output, in terms of BMEP, of the 4-valve system shows a 15% improvement over the 2-valve port injected design. Figure 13 illustrates part throttle operating characteristics. Benefits in emission levels and fuel economy are demonstrated showing an improvement of 13% and 12% respectively. For calibration of the engine management system it is also apparent that the tune "window" for the proposed EEC (1992) requirements is greater and performance is better at any individual AFR.

9. CONCLUSIONS

1. Ten combustion chambers, many of radically different design, have been evaluated in considerable depth, the particular objective being their capability to burn lean mixture, the final aim being to meet the present and proposed legislated pollution regulations and at the same time to reduce the carbon dioxide emissions which contribute to the "greenhouse" effect.

2. The outstanding chamber for 1 above has been identified as the 4-valve pent-roof design, well known for its good power characteristics. This programme has shown that it is exceptionally good for lean-burn operation.

3. The 4V lean-burn high compression chamber at part load and 23:1 AFR has lower specific NOx production than any of the other chambers, thus enabling a good contribution to CO_2 reduction.

4. Due to the high compression ratio, NOx at full throttle is high necessitating the use of a 3-way catalyst for maximum power and acceleration. The high CR gives improved efficiency.

5. The best 2-valve combustion chamber is more difficult to judge. For relatively low octane fuel requirement (91 RON), the Weslake stands out at 9:1 CR. However, at 12:1 there is little to choose between the Weslake and the compact chambers, May and CCI. The 'Hemi' however, has the best lean-burn capability and with larger valves would give good power.

6. EGR reduces the NOx for all chambers but only the 4V has no fuel consumption penalty.

7. Strategies to meet the objectives of this paper must still be open ended. No doubt 3-way stoichiometric systems will still be favoured by many as it is a well developed technology. However, to enable further CO_2 and fuel consumption reductions to be achieved engine friction and weight together with transmission and vehicle aerodynamics have to be addressed. Rover vehicles with the 'K' engine are good examples of this system.

8. Potentially the lean-burn engine with a 3-way catalyst and an engine management system having a lean-burn feed-back sensor will give lower carbon dioxide and better fuel consumption than a stoichiometric tuned engine. With the capability of full throttle performance, the 3-way catalyst for controlling NOx, no power need be sacrificed. Engine management will, however, be difficult.

9. The stratified charge engines also have great potential for reducing carbon dioxide as they virtually have no problem in burning lean and are thus less sensitive to precise control by a sensor of a management system. Examples have demonstrated possible improvement, even in comparison with the 4-

valve HCLB system but final proving of this still remains to be demonstrated.

10. Further research is necessary to discriminate between the alternative concepts when applied in actual vehicles equipped with the most up-to-date management system.

ACKNOWLEDGEMENTS

The Authors would like to thank the Directors of Rover Group for permission to publish this paper, the D.T.I. for their support in helping to fund the project and all those directly involved in the work.

REFERENCES

1. WEAVING, J.H. "Internal Combustion Engines".
   - Elsevier, 1990.
2. ANON. "The New Austin 7"
   - Automobile Engineer, October 1952.
3. MAY, M.G. "The High Compression Lean Burn Spark Ignition Engine", - I.MECH.E Conf.(A.D.) "Fuel Economy and Emissions of Lean Burn, Engines", June 1979.
4. BRADLEY, D., HYNES J., LAWES M & SHEPPARD C.J.W., "Limitations to Turbulence - enhanced burning rates in lean burn engines". International Conference on Combustion in Engines. I.MECH.E., London, 1988, P.17.
5. DAVIES, P.O.A.L. and FISHER, M.L. "Heat Transfer from Electrically Heated Cylinders" - Proc. Royal Soc. 1964 (SER A) Vol.280, pages 486-526.
6. COLLIS D.C. and WILLIAMS M.J. "Two-Dimensional Convection from Heated Wires at Low Reynolds' Numbers" - J. Fluid Mech. 1959, Vol.6, No.3.
7. BENJAMIN, S.F. "Development of the GTL Barrel Swirl Combustion System" - (p.203) I.MECH.E. Conference "Combustion in Engines", 1988.
8. YAGI S, FUJII M., NISHIKAWA M., AND SHIRAI H., "A Newly Developed 1.5 litre CVCC Engine", SAE 800321, 1980.
9. RICARDO, H.R. "Recent Work on the Internal Combustion Engine" - SAE Journal, Vol.10, May 1922.
10. SIMKO, A.O., SCUSSEL, A. and WADE, W., "The Ford Proco Engine Update, - SAE Paper 780699, 1978.
11. TIERNEY, W.T., "United Parcel Service applies Texaco Stratified Technology to Power Parcel Delivery Vans". - SAE Paper 801429, 1980.
12. CHAPMAN, J., DRAPER, A., FAIRHEAD, G.S., AND WALLACE, S., "Optimisation of Combustion Chamber Design" - I.MECH.E., Strasbourg 1989,

Fig. 1a & 1b Combustion Chambers of the cylinder heads tested.

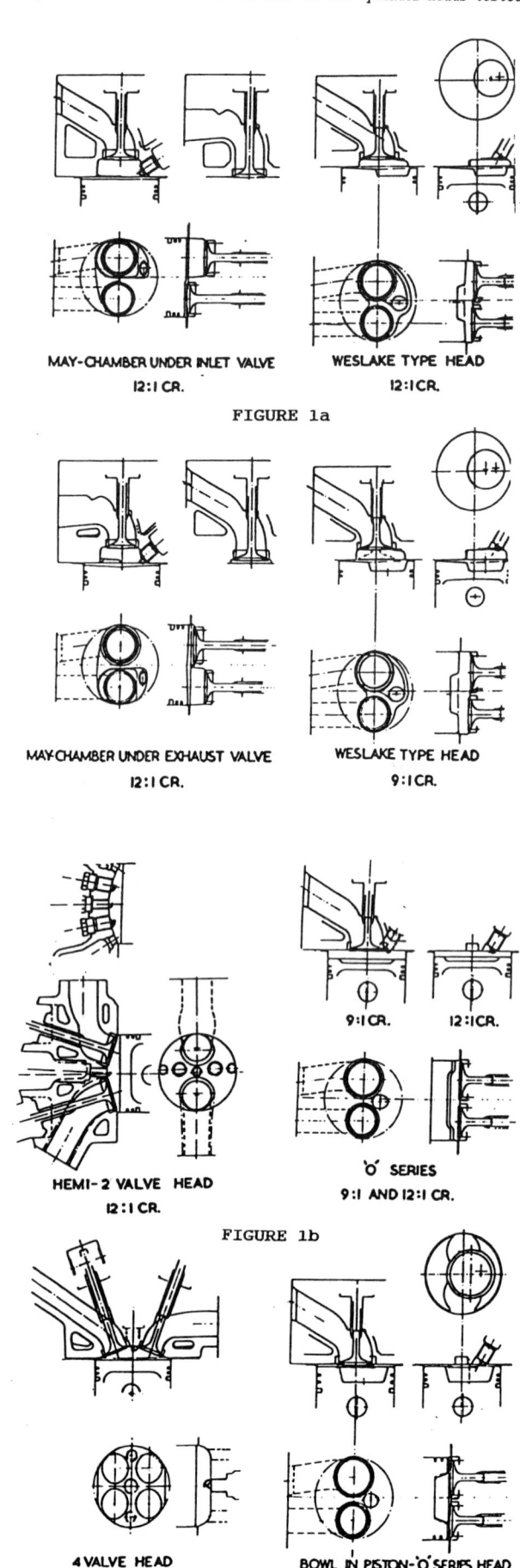

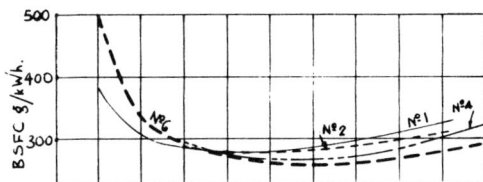
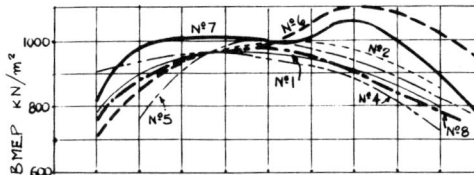
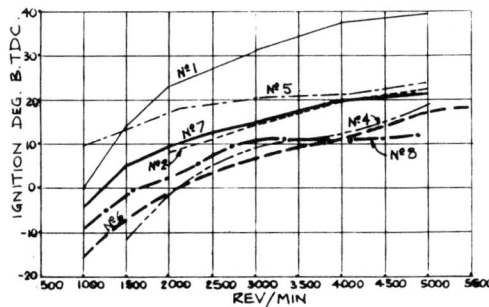

Fig. 2

Full load performance, BMEP, Fuel Consumption and Ignition setting (borderline 'knock') of several heads evaluated.

Curve 1   "O" Series 9:1 Compression Ratio (C.R.)
    2   "O" Series 12:1 C.R.
    3   "O" Series 12:1 C.R. High Swirl
    4   Weslake, 9:1 C.R. (WES. 9:1)
    5   Weslake, 12:1 C.R. (WES.12:1)
    6   Pent Roof 12:1 C.R. 4-Valve (4V)
    7   Hemispherical 12:1 C.R. (HEMI)
    8   Compact Chamber under Exhaust Valve 12:1 C.R. (MAY)
    9   Compact Chamber under Inlet Valve 12:1 C.R. (CCI)
   10   Bowl in Piston 12:1 C.R.
   11   Porsche Stratified Charge Engine SCE
   12   V.W. Stratified Charge Engine

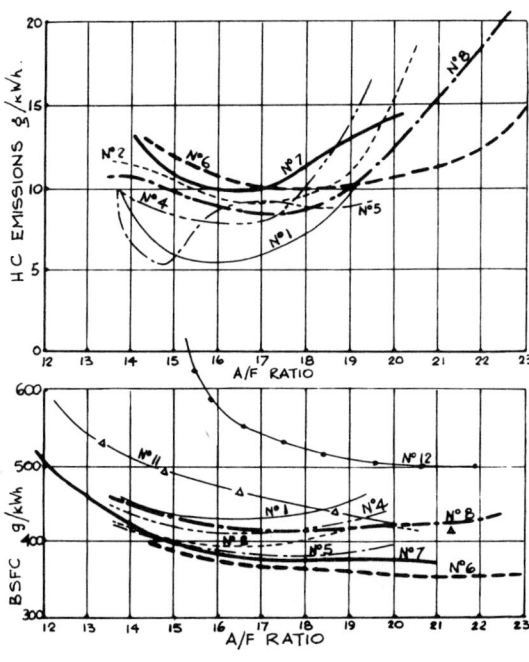

Fig. 3

Part throttle performance 2000 rev/min, 2 bar BMEP showing specific fuel consumption and HC emissions. Key as Fig. 2.

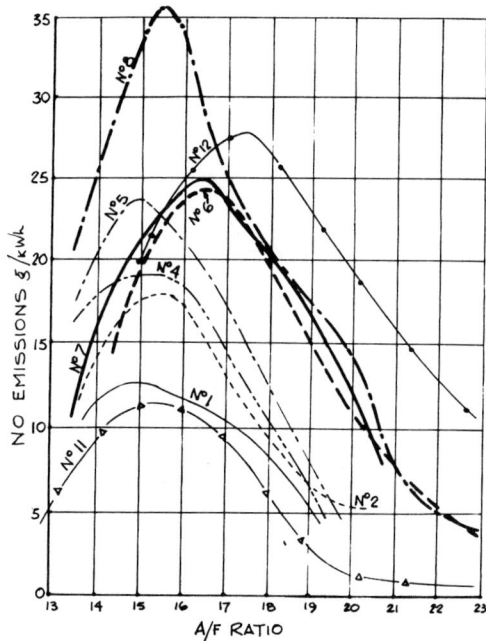

Fig. 4

Part throttle NO emissions 2000 rev/min, 2 bar BMEP. Key as Fig. 2.

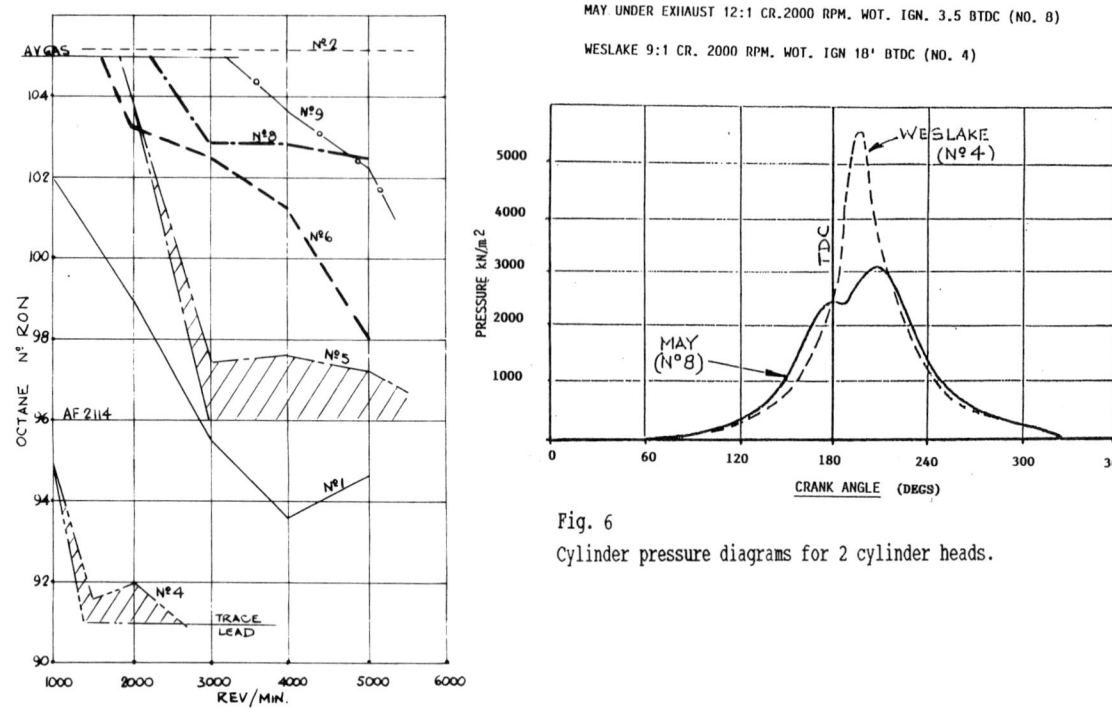

Fig. 6
Cylinder pressure diagrams for 2 cylinder heads.

Fig. 5
Fuel octane (RON) requirements for several heads. Key as Fig. 2.

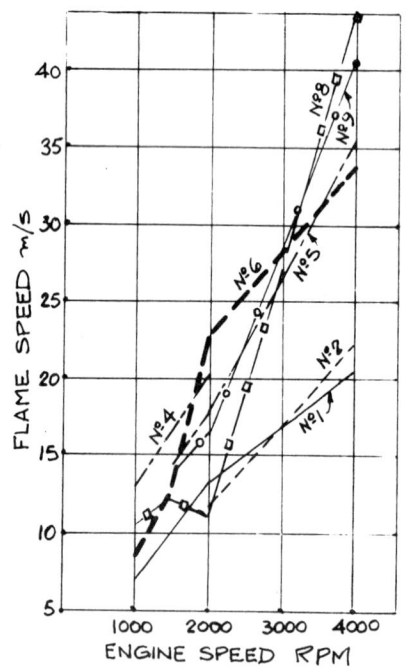

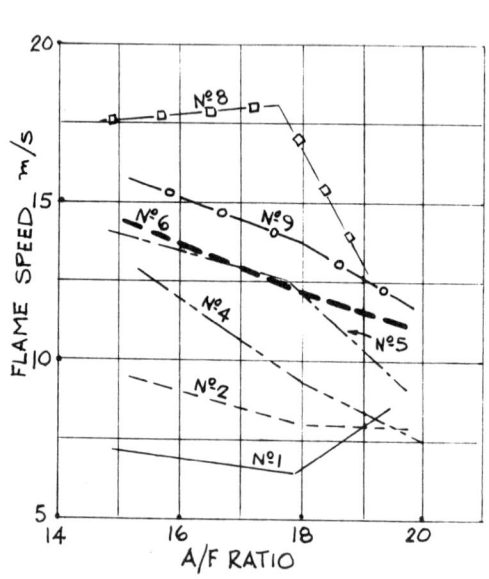

Fig. 7a
Flame speed - Full throttle. Key as Fig. 2.

Fig. 7b
Flame speed - Part throttle. 2000 rev/min 2 bar BMEP. Key as Fig. 2.

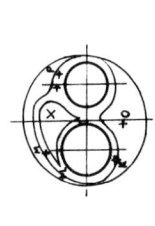
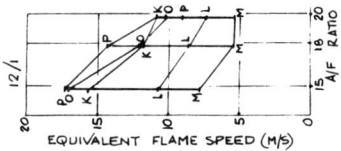

Fig. 8a
Flame speed "Weslake" 9:1 and 12:1, 2000 rev/min 2 bar BMEP.

Flame speed "4 Valve" 12:1 2000 rev/min 2 bar BMEP.
Fig. 8b

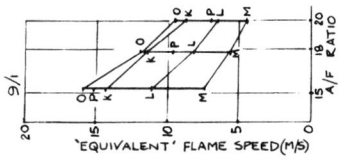
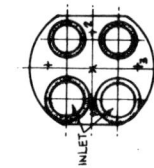
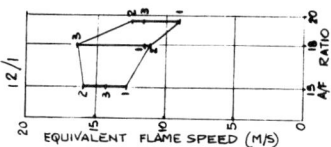

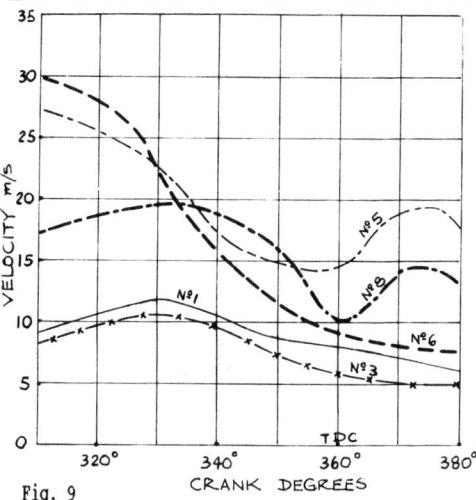

Fig. 9
Velocity measurements by Hot Wire Anemometry.
Full throttle, 2000 rev/min.
Highest reading from triple wire probe at sparking plug.
Key as Fig. 2.

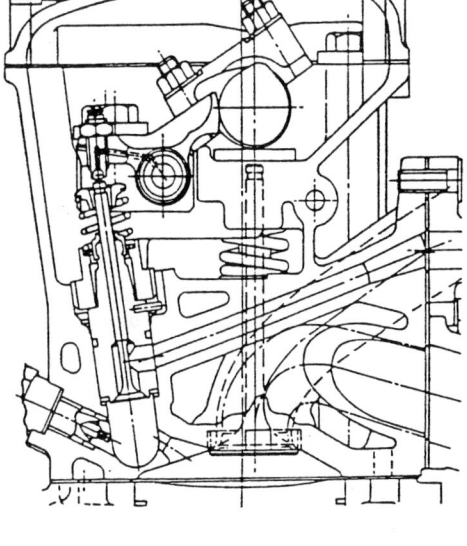

Fig. 11a Rover SCE cylinder head.

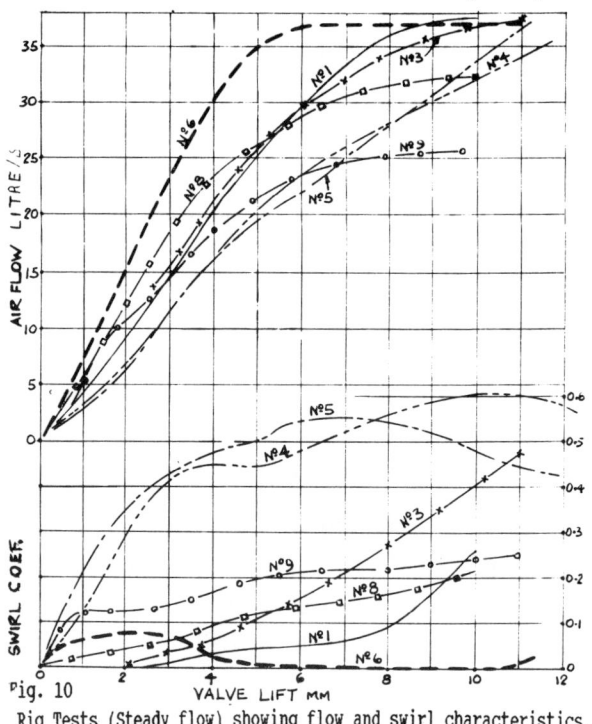

Fig. 10
Rig Tests (Steady flow) showing flow and swirl characteristics.

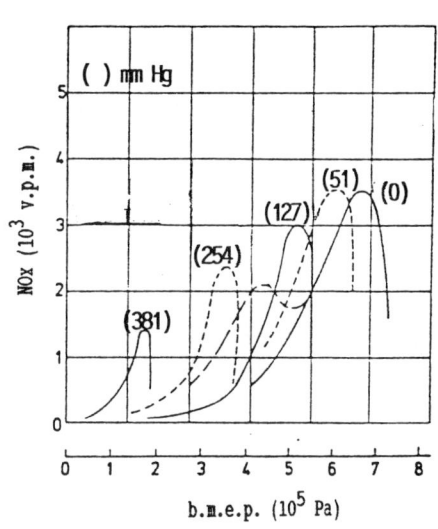

Fig. 11b NOx characteristic of SCE at varying throttle settings of main carburetter.

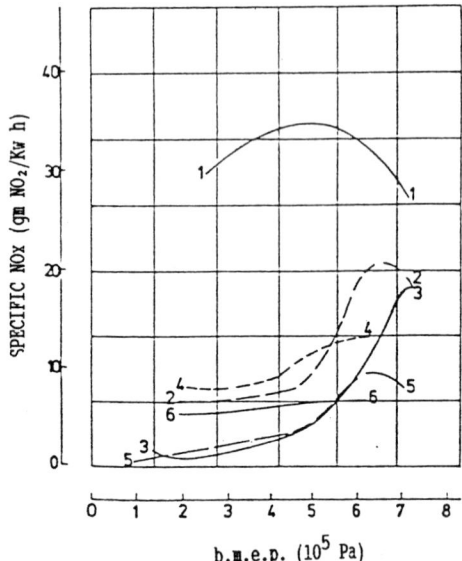

Fig. 11c A comparison of specific NO emissions of the Rover SCE and "O" series engines with other SCE.

1. Rover 1.8L "O" series.
2. Rover SCE (economy tuned).
3. Rover SCE (low NO tuned).
4. VW - SCE.
5. Porsche SCE.
6. Texaco TCCS engine.

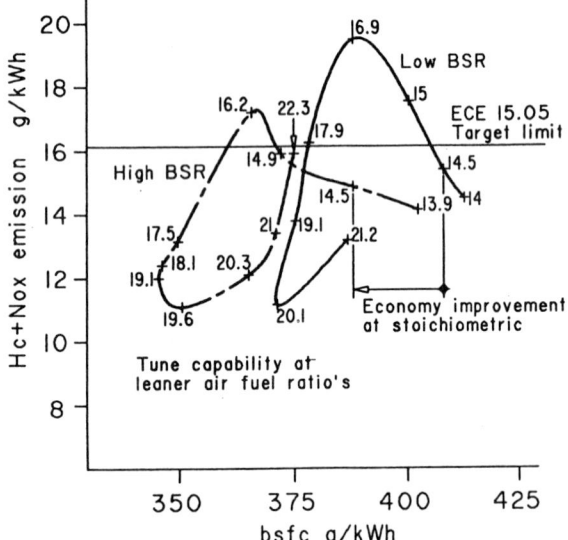

Fig. 12 Rover 4 cylinder experimental 4 valve engines. Comparison of High Barrel Swirl (BSR) with low BSR.

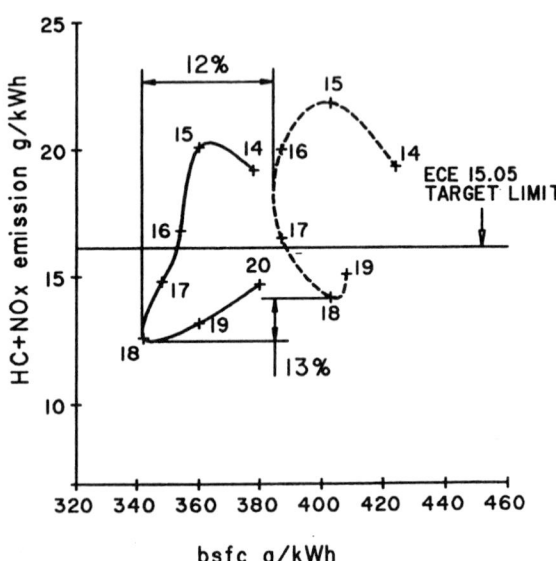

Fig. 13

Rover 4 cylinder 4 valve experimental engine. Comparison with 2 valve engine of equal cylinder capacity.

C413/039

Experimental and theoretical study of the radiation of gases containing dust particles

J STASIEK, MSc, PhD and M W COLLINS, MA, PhD, DSc, CEng, MIMechE, MIEE, MINucE
Thermo-Fluids Engineering Research Centre, The City University, London

SYNOPSIS The attenuation of thermal radiation within a dilute cloud of pulverised coal and ash are investigated experimentally and theoretically, for different ranges of particle size. An empirical expression is developed for obtaining the absorptivity and emissivity of a coal/ash cloud. A new nomogram is also presented on the basis of this expression.

1. INTRODUCTION

Many modern and practical engineering systems require the knowledge of the radiative properties of a particulate medium. Some examples that involve highly concentrated small particles are fluidized and pocked-bed combustors, pocked-bed catalytic reactors, microsphere insulations, combustors with deposited soot on the walls, M.H.D. generators, diesel engines and furnaces laden with aglomerated particles. The radiation transfer through atmospheric aerosols is of considerable interest in atmospheric transmission calculations. In these systems the absorption and scattering of thermal radiation by particles plays an important role in the overall energy transfer, and its understanding is central to the prediction and evaluation of system performance.

Within a suspension of particulate matter, thermal radiation can be either enhanced or attenuated, depending upon the size and concentration of particulates, the temperature distribution, and the radiative properties of the matter. Even for a pure material such as carbon, the specification of the pertinent radiative properties is not straightforward. The crystalline structure of the carbon and the surface condition can greatly affect the spectral emissivity [8] [23]. For a heterogeneous substance such as coal, the difficulties are multiplied since the particulate cloud can be made up of varying fractions of unreacted coal, char, ash and soot, each of which has its own optical properties dependent upon wavelength, temperature and particle size. Stull and Plass [17], Buckius and Hwang [5], Viskanta et al. [22], Grosshandler and Monteiro [8], Tien [20], Mackowski et al [14], Kumar and Tien [13] have examined the importance of some of these parameters (mean particle radius, size distribution, temperature, back-scatter) for polydispersions of various coals and fly-ash using experimental and theoretical models (Lorenz-Mie or Rayleigh theory) with average or precise spectral properties. Whitson [23] has compiled comprehensive data from more recent measurements of the refractive index and emissivity of graphite and carbonaceous material at high temperatures. The results of over twenty-five references were summarised in graphical and tabular form for the spectral region between 0.4 and 12.5 μm. Blokh [1] [2] [3] has reviewed the results of many experiments on bulk coal, soot, flame, and pulverised coal. Using Mie's theory, the absorption and scattering coefficients of different coals were calculated as a function of particle size. The emissivities of twenty different dusty materials have been measured and interpreted by Bierman and Vortmeyer [4]. The radiative characteristics of polished specimens of several coals have been determined by Foster and Howarth [12]. They found significant differences between the spectral refractive index for coal and polycrystalline graphite.

Recently, Kumar and Tien [13] have presented a complete model for predicting the absorption and scattering characteristics of homogeneous dense particulate sytstems by including the near-field interparticle effect (where the internal field of each particle is affected by the presence of the others) and that of coherent addition (i.e. taking into account the constructive/destructive interference of the scattered radiation by each particle in the far field, which is manifested by a change in the scattered field. Engineering radiative heat transfer in absorbing-scattering-emitting media have been presented by many authors and have been extensively discussed a.m. by Van de Hulst [21[; Hottel and Sarofim [9], and Siegel and Howell [16].

This study is undertaken to determine the extinction coefficient, absorption and emissivities coefficients of different coals and ash, under conditions similar to the early pyrolysis zone of a pulverised coal furnace. Several different coal and ash types have been experimentally investigated.

---
* On leave of absence from Technical University of Gdansk, Poland

The dependence of the total emissivity and absorptivity on mean particle diameter, size distribution, fractions of unreacted coal, temperature, chamber furnace and place of extraction of fuel are all demonstrated graphically. The theoretical background on thermal radiation attenuation is also briefly reviewed in this paper. From this background and using the considerable experimental data of Blokh [2] [3], Bierman and Vortmeyer [4] and Stasiek et al [18] [19] a new nomogram is designed and equations are formulated. These may be used for the determination and calculation of emissivity and absortivity of dust clouds. Its optical properties are dependent on wavelength, temperature, particle size and concentration. Particle diameters are typical of furnace operation, covering two ranges of 10 to 200 μm and 0.2 0.6 mm, and temperatures ranging from 300 to 2200K.

2 THEORETICAL BACKGROUND

Computation of the transport of thermal radiation in the particulate system requires an accurate knowledge of the extinction and the scattering coefficients. This is evident by considering the propagation of radiation within an absorbing, emitting and scattering medium which is governed by the equation of transfer [15] [16]:

$$\frac{dI_\lambda(S,\omega)}{dS} = -K_\lambda I_\lambda(S,\omega) + a_\lambda I_{\lambda,b}(S,\omega)$$

$$+ \frac{\sigma_{s,\lambda}}{4\pi} \int_0^\infty I_\lambda(S,\omega') \Phi_\lambda(\omega' \cdot \omega) \, d\omega' \quad (1)$$

Where $K_\lambda = a_\lambda + \sigma_{s,\lambda}$ is the extinction coefficient and in general is a function of position S.

The three terms on the right-hand side of the transfer equation represent respectively (a) the attenuation of intensity due to absorption and scattering, (b) the gain due to emission, and (c) the gain due to the scattering in the direction ω from all other directions. The intensity I_λ is defined as the energy per unit area per unit solid angle per unit wavelength and the scattering phase function $\Phi_\lambda(\omega' \cdot \omega)$ is a specification of the radiation intensity scattered from the direction ω' to the direction under consideration, normalised by the isotropic scattered radiation intensity. i.e., $\Phi(\omega' \cdot \omega) = 1$ for isotropic scattering [13] [15] [16]. Scattering and absorption characteristics of a particle are governed by three factors: the particle shape, the particle size relative to the wavelength of the incident radiation ($p = \pi d/\lambda$), and the optical properties of the particle and the background medium. The last factor is represented by the complex refractive index m defined as $(n-i\kappa)$ where n is the index of refraction and κ is the index of absorption. For particles not too closely spaced (distance between centres greater than 3r [9]), the effect of a cloud of particles can be found by summing over all individual particles [9] [13] [20] [21].

The radiative properties for a polydisperse cloud of spherical particles are expressed in terms of the extinction efficiency, scattering efficiency and phase function for a single particle. The extinction and scattering coefficients are given as

$$K_\lambda(m,N) = \int_0^\infty \pi r^2 N(r) Q_e(p,m) \, dr \quad (2)$$

$$\sigma_\lambda(m,N) = \int_0^\infty \pi r^2 N(r) Q_{sc}(p,m) \, dr \quad (3)$$

and

$$\Phi_\lambda(\omega' \cdot \omega) = \frac{1}{\sigma_\lambda(m,N)} \int_0^\infty \pi r^2 N(r) Q_{sc}(p,m) \Phi_\lambda(\theta,r) \, dr$$

$$(4)$$

where

$\Phi_\lambda(\theta,r)$ is the phase function for individual particles of size r and $(\omega' \cdot \omega) = \cos\theta_0$ [15]

The absorption coefficient, $a_\lambda(m,N)$ is the difference between the extinction and scattering coefficients. The functional dependence of the radiative properties on wavelength, index of refraction and particle size distribution is explicitly denoted. Dimensionless radiation properties are obtained from considerations of the limiting expressions in the size parameter p. As for any participating medium the spectral attenuation factor (Q_e, Q_{sc}, Q_a) of particles can be determined from the general equation [1] [7] [21] as:

$$Q_e = -\frac{2}{\bar{m}^2 p^2} \text{Re} \sum_{\nu=1}^\infty i\nu (1+\nu)(-1)^\nu (c_\nu - b_\nu)$$

$$(5)$$

$$Q_{sc} = \frac{2}{\bar{m}^2 p^2} \sum_{\nu=1}^\infty \frac{\nu^2(\nu+1)^2}{2\nu+1} (|c_\nu|^2 + |b_\nu|^2)$$

$$(6)$$

in which the peak factors of the electrical and magnetic fluctuations are complex functions of size parameter p and the complex refractive index $\bar{m}$ (the background medium) and m (the participating medium).

Also in equations (5) and (b), c_ν and b_ν are:

$$c_\nu = c_\nu(p,\bar{m},m)$$
$$b_\nu = b_\nu(p,\bar{m},m) \quad (7)$$

where

ν - frequency of incident radiation

p - size parameter = $\pi d/\lambda$ (d = 2r)

d - diameter of the particles

λ - wavelength of incident radiation

The sizes of m and m̄ in their turn can vary with the radiation wavelength λ. Dimensionless radiation properties (5) and (6) are usually obtained from considerations of the limiting expressions in the size parameter, p. [1] [7] [9] [13] [17] [21]

The size parameter may be viewed as having three ranges, small (p < 0.6) intermediate (0.6 < p < 5) and large p > 5. Because the results have been discussed by many authors ([1] 17] 19] [13] [17] [21]) the calculations for the relevant expressions will not be presented here. However, simple expressions for carbon spherical particles result in attenuation efficiencies for a range of size parameter 0 < p < 20 as shown in Fig. 1. It is clear from these curves that the strongest dependence of the attenuation factors Q_e, Q_{sc} and Q_a on p occurs in the range. As wavelength λ increases so the maximum attenuation is displaced to the region of low values of p and where the true absorption has a considerable influence on the attenuation.

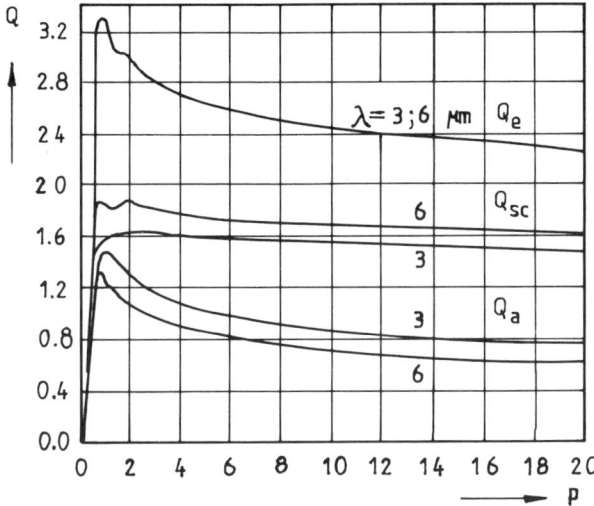

Fig 1. Spectral attenuation efficiencies for radiation from carbon particles depending on size parameter p at different wavelength λ.
$Q_e = Q_{sc} + Q_a$

The scattering power of carbon particles increases sharply with increasing p to a value p ≈ 1. At higher values of this parameter the scattering power of the carbon particles ceases to depend on p and is wholly governed by the optical constants n and κ. In this case, as λ increases so does the attenuation factor due to scattering Q_{sc}. In contrast to scattering the true absorption of carbon particles is stabilised and ceases to depend on p at appreciably higher values of the latter, while, with increase of λ the attenuation factor due to absorption Q_a decreases. At large values of p when diameter of the particle d becomes appreciably greater than the wavelength λ the spectral attenuation factors Q_a and Q_{sc} cease to depend on p and tend towards specific asymptotic values which depend on the wavelength λ, or, more accurately, on the complex refractive index m(λ). Following from theory and computational results [3] [7] [20] if p becomes sufficiently large (i.e. p ≫ 1) the overall attenuation efficiency (extinction efficiency) for all substances becomes independent of the absolute value and m and equals:

$$Q_e = Q_a + Q_{sc} = 2 \qquad (8)$$

Attenuation of a Particulate System

Attenuation (absorption plus scattering) characteristics of many particles in a close system can be obtained from the single-particle characteristics. The procedure depends on the scattering regime to which the system of particles belongs. An independent and dependent scattering regime map was presented recently by Tien [20].

The independent theory is based on the assumption that each particle in the assembly scatters and absorbs radiation in a manner unaffected by the presence of other particles. Thus the extinction and scattering of energy by the system is expressed by a simple algebraic addition of the energy extinguished and scattered by each primary particle. The cross section for the system of N particles is the sum of the cross section for each particle, the individual particles being assumed to scatter and absorb radiation independently of the others [1] [9] [16].

The corresponding attenuation coefficient for the medium of monodisperse spherical particles of diameter d is obtained by:

$$K_\lambda = Q_e \frac{\pi d^2}{4} N \qquad (9)$$

or

$$K_\lambda = Q_e A B = \frac{3}{2} \frac{f_v Q_e}{d} \qquad (10)$$

where

$A = \dfrac{3}{2d\rho}$ [m²/kg] - free surface of particle

ρ [kg/m³] - density of solid material

B [kg/m³] - density of cloud particles

$f_v = \dfrac{B}{\rho}$ - solid volume fraction

$N = \dfrac{6}{\pi d^3} \dfrac{B}{\rho}$ [1/m³] - number of particles in unit volume

According to the data of [1] [2] and [3] the attenuation efficiencies are given by the formula

$$Q_e(p) = \frac{D_p}{\sqrt[3]{\pi}} p^{1/3} \qquad (11)$$

where D_p is a factor depending on the type of fuel. Values of the factor D_p are given in Table 1 for a number of fuels. [1] [2] [3] [4]. If solid particles (for p ≫ 1) are suspended in a transparent gas, the emissive and absorptive powers of the medium, depend for a given temperature and layer thickness, only on the concentration, dimensions and physical properties of the particles. In this case the total attenuation efficiencies of the dust particles can be calculated from the formula [1] [11]

$$k = M(m)\phi(a) \left(\frac{d}{\lambda_{max}}\right)^{1/3} = D_p \left(\frac{d}{\lambda_{max}}\right)^{1/3} \quad (12)$$

where

$M(m)\phi(a)$ - is a function expressing the absorbtivity of the dispersed substance as a function of its physical properties and the wavelength of the incident radiation.

k- is the total attenuation efficiencies of the dust treated as a grey gas.

λ_{max} - is the wavelength corresponding to the maximum value of the spectral black body radiation at an assigned temperature. ($\lambda_{max} T = C_3$; $C_3 = 2897.8$ mK - Wien's constant).

In general the attenuation coefficients of cloud particles can be calculated using the equation [1] [4]:

$$K = k A B = D_p \left(\frac{d}{\lambda_{max}}\right)^{1/3} A B =$$

$$= D_p (\lambda_{max})^{-1/3} (3/2\rho)^{1/3} A^{2/3} B =$$

$$= D_p (T/C_3)^{1/3} (3/2\rho)^{1/3} A^{2/3} B \quad (13)$$

Taking Eq. (13) into account, we can write the following expression for absorptivity or emissivity of a dusty gas [1] [4] [19]:

$$a_s|\varepsilon_s| = 1 - \exp[-k A B L] = 1 - \exp[-K L] \quad (14)$$

Equation (14) gives either the absorptivity, if the T of equation (12) is a source of incident radiation, or the emissivity, if T is the cloud temperature.

3. Experimental Investigation and Results.

The purpose of this experimental study is to obtain more comprehensive attenuation data further to that already published and to design a nonogram for funding the absorption or emission coefficient of dust-laden streams.

The radiative characteristics of coal and ash clouds are determined experimentally by the following sequence of experiments.

Experimental Series: No 1

Experimental investigation of the large particle coal cloud's extinction coefficient K_λ was based on attenuation of the visible radiation emitted by the standard-strip lamp. Calculation of the extinction coefficient was done based on the Lambert-Bourger law. The experiment was carried out in a stainless steel chamber 1000mm high, 60mm deep and 40mm wide. The measurement of radiation attenuation was made 600 mm down from the place where pulverized coal and air was fed and mixed in the rectangular channel. (Fig. 2). The main parts of the experimental stand are: 1 - test channel 2 - standard-strip lamp 3 - monochromator, 4 - photomultiplier, 5 - digital voltmeter, 6 - high-voltage source supply, 7 - feeder, 8 - compressor, 9 - flowmeter 10 - low-voltage source supply. The experiment investigated two ranges of coal particle d, namely from 0.2 to 0.3 mm and 0.5 to 0.6 mm. Coal dust was fed into the test channel via the disc feeder by the following amounts:

(i) $0.2 < d < 0.3$ mm $m_c = 3.65; 7.80; 11.7$ and 15.56 kg/h

(ii) $0.5 < d < 0.6$ mm $m_c = 3.4; 7.35; 11.0$ and 14.6 kg/h

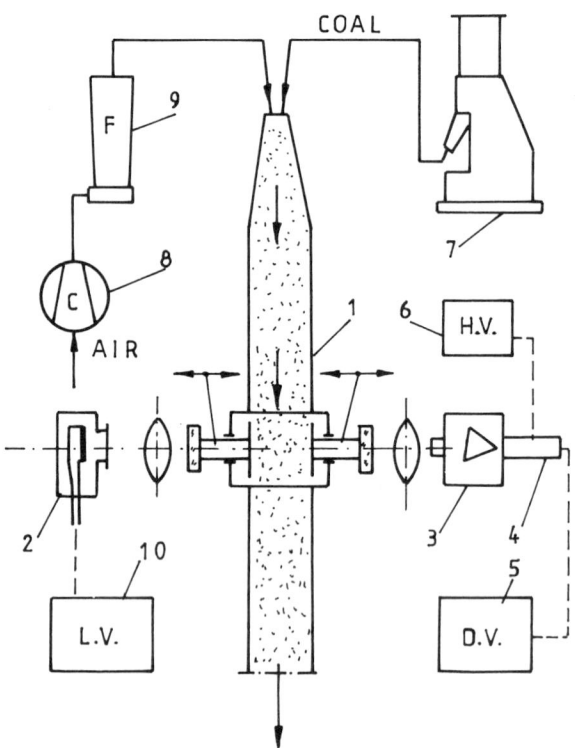

Fig. 2 The experimental stand configuration. 1. Experimental channel 2. Walfram strip lamp 3. Monochromator 4. Photomultiplier 5. Digital voltmeter 6. High-voltage source supply 7. Feeder 8. Air compressor 9. Flowmeter 10. Low-voltage source supply.

The transporting air volume was $V_1 = 21$ m^3/h and $V_2 = 28$ m^3/h respectively. In order to determine volume concentration of coal dust a quantity of 10000 particles of each size range was weighed. The results were:

for $0.2 < d < 0.3$ mm, the weight is 0.455g, and for $0.5 < 0.6$ mm it is 1.30717g. The measurement was made for wavelength $\lambda = 0.6$ μm. Fig. 3 demonstrates the experimentally determined decrease of radiation intensity (spectral transmittance) as a function of path length L, particle diameter and concentration N, and volume air velocity. The experiment proved that for large particles (like coal and ash), that is for $\lambda \ll d$, the scattering cross section approximately equals $2\pi d^2/4$ and the extinction efficiency $Q_e = 2$ (i.e. Fraunhofer and Fresnel diffraction plus reflection). Also Fig. 3 compares the experimental data (point values) with Fraunhofer and Fresnel theory (lines). The agreement is generally very good for the complete range of parameters.

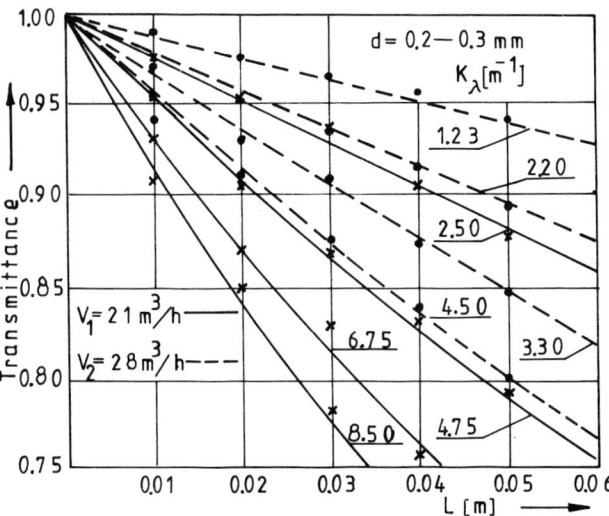

Fig. 3 The transmittance for coal particles as a function of layer thickness, particle diameter and concentration, and volume air velocity (——, ——— independent scattering).

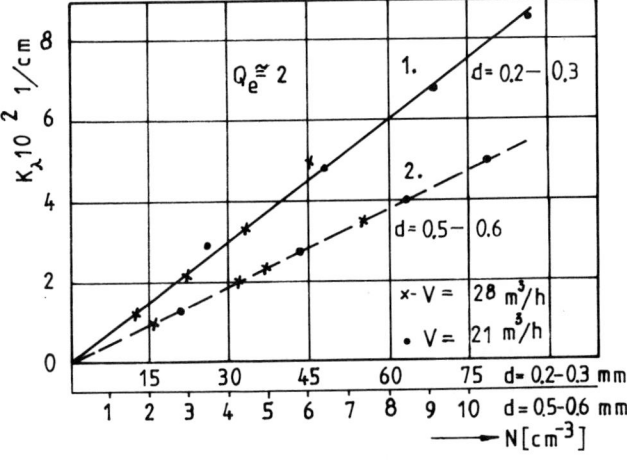

Fig. 4 Dependence of extinction factor is a function of concentration, and coal diameter.

Experimental Series: No 2

Experimental investigations were also performed for the determination of total emissivity and absorptivity of coal ash and coal dust for particle size parameter $5 < p < 90$. A schematic diagram of the experimental facility is shown in Fig. 5. The main element of the stand was a ceramic tube containing 12 individually cooled elements of internal diameter $D = 100$ mm and length $L = 1200$ mm. The tube was instrumented with Pt - PtRh thermocouples so that the surface and flow medium temperatures could be measured. The pulversied coal ash and air was heated by a 3-phase 30 kW generator plasma. Pulversied coal ash was fed into the spouted-bed mixer and introduced into an optically active medium. The equipment for measurement of the absorptivity and emissivities of the air-nitrogen-ash mixture is shown in Fig. 5b. The system is assembled from a calibrated light source, a radiation pyrometer with chopper, lens and cool background. The particle cloud was prepared from coal-ash

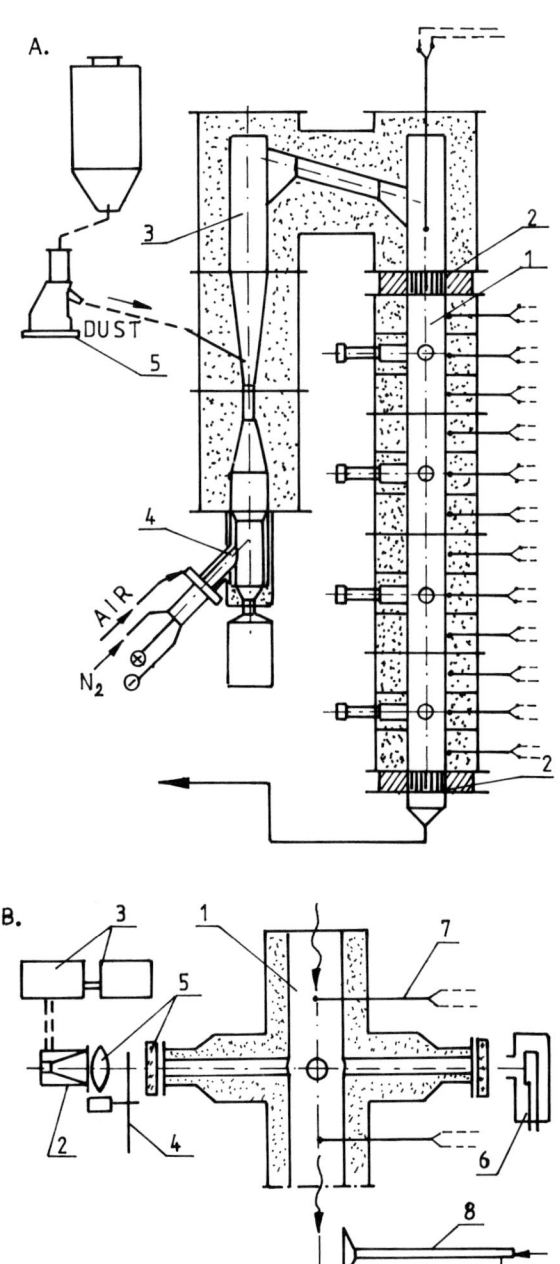

Fig.5 Schematic diagram of the experimental stand for determining attenuation factor.
A.) 1. ceramic channel 2. inlet and exit head 3. spouted bed mixer 4. generator plasma 5. feeder.
B.) 1. ceramic channel 2. detector 3. alternating current amplifier 4. chopper 5. optical system 6. wolfram strip lamp 7. thermocouple 8. cool black background

giving a range of size as small as 20 to 200 μm; and temperatures ranging from 300 to 2200 K. Temperature variations of less than ± 10K were encountered, thus allowing the assumption of isothermal conditions to be made. The total radiation pyrometer was calibrated by a black body light source. Fig. 6 shows the experimentally determined total dimensionless attenuation coefficients of the particles as a function of particle diameter and the wavelength λ_{max} from Wien's Law.

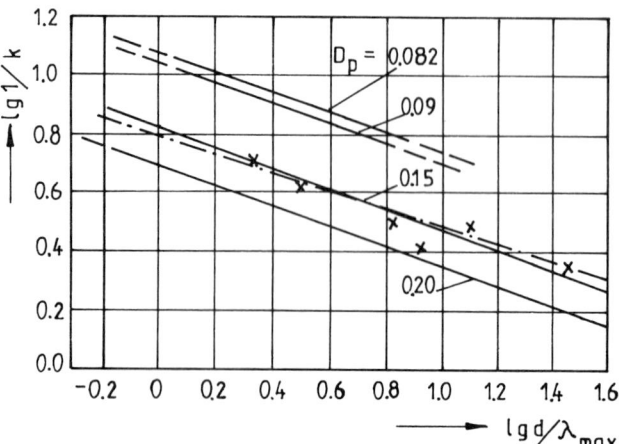

Fig. 6 Dependence of the attenuation coefficients for coal and ash and d/λ_{max} from Blokh (*-*-* = authors experiment).

According to the data of Ref. [1] [2] [3] and [4] the total attenuation efficiencies of the dust can be correlated by a function of the general form:

$$k = D_p \left(\frac{d}{\lambda_{max}}\right)^{1/3} \quad (15)$$

where

D_p = from 0.15 to 0.28 for coal

D_p = from 0.08 to 0.15 for ash

Table 1 presents examples of values of the factor D_p, for a number of fuels determined experimentally by Blokh [1], and Bierman and Vortmeyer [4].

Table 1
Values of the Factor D_p in Formula (13) and (15) [1] [4]

Type of fuel	ρ kg/m³	d μm	D_p -
Pechora coal 1 Ash dust coal dust	2000 1460	2.8÷87.0 5.8	0.20 0.28
Donets gas coal 2 Ash dust coal dust	2150 1800	8.0 3.5	0.08 0.15
3 Estonian shale	2600	15.6÷125	0.15
4 Ruhr-Westerholt	2110	6.3	0.28
5 Saar (Ash dust)	2050	15.6	0.17
6 Kalkstein	2700	14.4	0.09

Determination of the absorptivity (a_s) or emissivities (ε_s) for ash cloud and coal can be carried out using the monogram of Fig. 7. The new monogram is designed using formula (19) and (20) and the experimental investigation of Blokh, [1] Bierman and Vortmeyer [4] and authors. More information about the conditions and the results may be found in [1], [2], [3], [4], [18], [19] and [24].

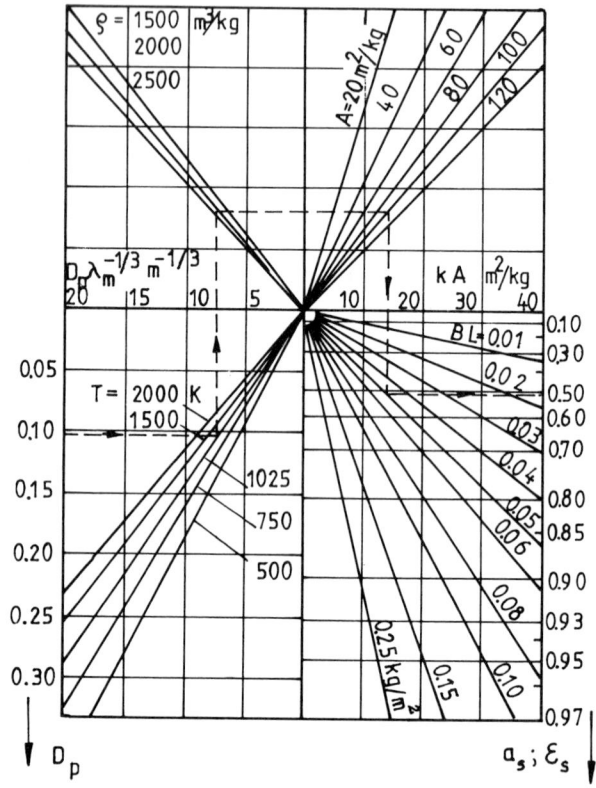

Fig. 7 Nomogram for funding the absorption or emission coefficient of a dust-laden stream.

4. CONCLUSIONS

The influence on radiative properties of temperature, type of fuel, density of solid material, particle diameter and concentration has been investigated for particles of small optical thickness: specifically coal and coal ash clouds. From the above data and that of other authors a nomogram has been constructed which is intended to be used both in engineering design and for parameters for theoretical calculations. For these purposes it is estimated that up to maximum values for B = 0.1 kg/m³, d = 110μm and 500K < T < 1500K, the overall accuracy in a_s and ε_s is ± 12%.

Beyond these values the nomogram is formed by extrapolation, and the error is likely to be higher.

REFERENCES

1. Blokh A.G. The Basis of Heat Exchange by Radiation. Energiz Moscow 1962

2. Blokh, A.G. Radiation of a Luminous Sooty Flame. Teploenergetika, 1964, 11, No. 4, 26-30

3. Blokh A.G. Radiation of Carbon Particles in a Flame, Teploenergetika, 11, No 7, 16-19.

4. Bierman P., Vortmeyer D. Radiation of Gases Containing Dust Particles Warme-und Stoffubertragung, 1969, 2, 193-202

5. Buckius R.O., Hwang D.C. Radiation Properties for Polydispersions: Application to Coal. Journal of Heat Transfer, 1980, 102, 99-103

6. Berne B.J; Pecora R. Dynamic Light Scattering, 1976, J. Wiley and Sons Inc., N. York.

7. Diermendjian, D, Electromagnetic Scattering of Spherical Polydispersions, 1969, American Elsevier Publishing, N. York.

8. Grosshandler W.L, Monteiro S.L.P., Attenuation of Thermal Radiation by Pulverized Coal and Char. Journal of Heat Transfer. 1982, 104, 587-593.

9. Hottel H.C., Sarofim A.F. Radiative Transfer. 1967. McGraw-Hill Book C. N. York.

10. Howell J.R. Thermal Radiation in Participating Media: The past, the Present and Some Possible Futures. Journal of Heat Transfer. 1988, 110, 1220-1229.

11. Kutateladze S.S. Fundamentals of Heat Transfer. 1963. Edward Arnold (Publishers) Ltd. London.

12. Foster P.J., Howarth C.R. Optical Constants of Carbons and Coals in the Infrared. Carbon 1968, 6, 719-729.

13. Kumar S. Tien C.L. Dependent Absorbtion and Extinction of Radiation by Small Particles. Journal of Heat Transfer, 1990, 112, 178 - 185.

14. Mackowski D.W., Altenkirch R. A., Menguc M.P. A Comparison of Electromagnetic Wave and Radiative Transfer Equation Analyses of a Coal Particle Surrounded by a Soot Cloud. Combustion and Flame, 1989, 76, 415-420.

15. Ozisik M.N. Radiative Transfer and Interactions with Conduction and convection 1973. John Wiley and Sons Inc., N. York.

16. Siegel R., Howell J.R. Thermal Radiation Heat Transfer. 1981 Hemisphere Publishing Corporation, N. York.

17. Stull V.R. Plass G.N. Emissivity of Dispersed Carbon Particles. Journal of the Optical Society of America, 1960, 50, 121-129.

18. Stasiek J. Determination of Extinction Factors and Scattering Cross Section in a Diffuse Medium: Air-Coal Dust under Flow Conditions. Flow Visualization III Ed. by W. J. Yang. 1985, 808-812, Hemisphere Publishing Corporation, New York.

19. Stasiek J., Collins M.W., Radiant and Convective Heat Transfer for Flow of an Optically Active Gas in a Cooled Tube with a Grey Wall. I.H.T.C. 9, 1990, Jerusalem, Israel (to be published).

20. Tien C.L., Thermal Radiation in Packed and Fluidized Beds. Journal of Heat Transfer, 1988, 110, 1230-1242.

21. Van de Hulst H.C. Light Scattering by Small Particles 1957 J. Wiley and Sons Inc. N. York.

22. Vistanta R, Ungan A, Menguc M.P. Predictions of Radiative Properties of Pulverised Coal and Fly-Ash Polydispersions. 20th National Heat Transfer Conference. ASME paper No 81-HT-24, 1981, Milwaukee, U.S.A.

23. Whitson M.E. Handbook of the Infrared Optical Properties of Al_2O_3, MgO and ZrO_2, Vol I. AD/A-103 722, 1975, The Aerospace Corporation, El Sugundo, California.

24. Stasiek J; Collins M.W., Vortmeyer D. Experimental and Theoretical Study on the Radiation of Gases Containing Dust Particles Warme und Stoffubertragung (in preparation)

C413/065

Swirl centre precession under steady flow conditions

S NADARAJAH, BSc (Eng), M J TINDAL, PhD, CEng, FIMechE and M YIANNESKIS, BSc, MSc, PhD, DIC
Department of Mechanical Engineering, King's College, London

SYNOPSIS Precession of the swirl produced downstream of the inlet valve of a production diesel engine was investigated experimentally under steady flow conditions at three different flow rates. Two techniques were employed: flow visualization using laser-sheet illumination and both still and video cameras, and velocity measurement by laser-Doppler velocimetry. Both ensemble-averaged and time-resolved velocities were determined; the latter were analysed to obtain additional information about the frequency content of the flow.

There was clear evidence of precession of the swirl centre with a dominant band of low frequencies which scaled with flow rate. The flow visualization provided useful information about the trajectory of the swirl centre and its range of movement in the cylinder.

1 INTRODUCTION

The performance of diesel engines is dependent on the rate of combustion which is largely controlled by the mixing process between air and fuel; this in turn is affected by the air swirl. Swirl will be influenced by precession of the swirl centre during the engine cycle, particularly in regard to cycle-to-cycle variations in the flow. Precession which develops during the induction stroke will persist into the compression stroke where it may be modified.

Swirl centre precession has been observed in quite simple symmetrical configurations, such as steady swirling flow in a circular pipe [Kito (1)] and flow through an axisymmetric port [Arcoumanis et al (2)]. The phenomenon seems to be inherent in any swirling flow, especially where the rate of swirl is high [Gupta (3)], so it is not surprising that it should be observed in engine cylinders. Chanaud (4) reported that a periodic motion occurs in some swirling flows, and that this motion can be described in terms of an oscillator that derives its energy from a hydrodynamic instability of the swirl. Evidence of precession of the swirl centre has also been observed in measurements of steady flows in the cylinders of engines fitted with both helical and directed inlet ports [Coghe et al, (5)].

The object of the work reported in this paper was to investigate precession of the swirl centre in the flow downstream of the inlet port from a production diesel engine. The study was carried out on a transparent model of the port, valve and cylinder under steady flow conditions. It began with a programme of flow visualization aimed at providing a general picture of the flow and continued with a series of time-resolved and ensemble-averaged measurements at a range of locations. By studying swirl precession under steady flow conditions, rather than in a running engine, it is possible to avoid all the complications introduced by cycle-to-cycle variations in the flow.

The presence of this precessional motion introduces uncertainties as to the true value of the turbulence velocity component. In steady flows, the instantaneous velocity U equals the sum of the mean velocity U_m and the velocity fluctuation, u.

$$U = U_m + u$$

However in the presence of a time-dependent variation such as swirl centre precession, having a mean velocity of its own, U_t, the instantaneous velocity is given by

$$U = U_m + U_t + u_{true}$$

In processing velocity measurements by ensemble averaging it is usual to characterise the turbulence levels by the r.m.s. value, taken as the square root of the difference between the squares of the instantaneous and mean velocities. Consequently if no alowance is taken of the swirl centre precession, the turbulence levels will be over-estimated.

2. APPARATUS AND EXPERIMENTAL TECHNIQUES

A refractive index matching technique was employed to gain access for the laser beams in the test section. The test section used was an acrylic replica of the helical port of a Perkins Prima direct-injection diesel engine (Figure 1) and the working fluid was a mixture of tetraline and oil of turpentine in the ratio 30.2 : 69.8 by volume. The matching refractive index was 1.489 at 21.72°C. The main features of the steady flow rig and method used to construct the replica have been described previously [Tindal et al, (6)]. The temperature of the liquid was controlled within ± 0.02°C throughout the experiments. The density and kinematic viscosity of the liquid were 894 kg/m^3 and 1.71 x 10^{-6} m^2/s respectively. Time-resolved measurements were made at several locations and flow visualization was carried out in several planes.

The laser-Doppler anemometer optical and signal processing systems are shown in figure 2; they comprised a 10 mW Helium-Neon laser, a rotating diffraction grating and associated optics, and a frequency counter (TSI model 1990B) for processing the Doppler signals. The ensemble-averaged data were obtained from the digital output of the counter which was interfaced to an Apple-IIe computer, and the mean and the r.m.s. velocities were calculated for a pre-determined number of samples (6144). The time-resolved data were obtained from the analogue output of the counter which was connected to an analogue to digital converter and signal acquisition unit (Cambridge Electronic Design 1401) interfaced to a Tandon microcomputer. The output voltage from the counter was offset by 5 volts in order to match the range of the signal acquisition unit and was sampled at a rate of 20 kHz for 1.6 seconds. To obtain a velocity trace with the least amount of drop-out, the data rate was maximised before the signal was recorded. The data rates in the experiments varied from 1.2 - 1.5 kHz. The principal characteristics of the velocimeter are listed in Table 1 below.

As the output voltage was inversely proportional to the frequency measured, the sampled signal was first converted into a signal proportional to velocity. The data was subsequently transfered to a Vax mainframe computer for statistical calculations and plotting. The mean and r.m.s. values of time-resolved velocity were checked against the ensemble-average values obtained with the Apple computer to confirm that the data transfer was correctly performed. The errors in the ensemble-averaged measurements are estimated at 5% and 10 % for the mean and the r.m.s. measurements respectively. As the mean and r.m.s. values obtained from the time-resolved measurements were nearly identical to those obtained from the ensemble-average measurements, the errors in the time-resolved measurement might be expected to be similar. A Fast Fourier Transform (FFT) package was used to transform the time-resolved results into frequency-amplitude spectra .

Table 1:
Principal characteristics of the laser-Doppler velocimeter.

Half angle of beam intersection in air (degrees)	3.9
Frequency to velocity conversion constant (m/s/MHz)	4.65
Intersection volume diameter at $1/e^2$ intensity (μm)	60
Intersection volume length at $1/e^2$ intensity (μm)	1320
Number of fringes in intersection volume without shifting	13
Fringe separation (line-pair spacing) (μm)	4.65

Flow visualization was carried out using both still and video cameras. The flow was visualized using a sheet of light from a 2W Argon-Ion laser. Minute air bubbles of around 60μm mean diameter were introduced into the flow as tracer particles. The size of the bubbles was such that their rise velocity was only 0.5% of the bulk velocity. This is sufficiently small to cause negligible deviation from the actual flow path.

3. EXPERIMENTAL RESULTS AND DISCUSSION

Figure 3 shows typical ensemble-averaged velocities in the form of vectors, for a valve lift of 9 mm. The vectors were drawn from measurements of the axial and tangential velocity components in the Y = - 5 mm plane. This set of results is shown to illustrate the pattern of the flow as it leaves the valve passage. The main part of the jet leaving the valve extends to the right hand corner of the region shown, with a recirculation region underneath the valve. Another part of the jet, coming out from the left hand side of the valve, feeds the part of the recirculation region underneath the valve on that side. Both of these recirculations form parts of an asymmetric ring vortex, whereas the jets are parts of the non-uniform annular jet issuing between the valve and the valve seat. Flow visualization showed two more recirculation regions between the jet and the cylinder head, a small one at the left hand side and a larger one at the right hand side. These recirculations make up another ring vortex that extends round the periphery of the cylinder head. The results shown in this figure were obtained with a flowrate of 1.85 litres/s, and those in Figures 4 - 6 with flowrates of 1.00 - 1.74 litres/s. It has been previously shown that the flowrate does not affect the flow patterns significantly [Cheung et al, (7)], and thus it may be expected that the flow pattern for all flow rates at which LDA measurements were made would be similar to that shown in Figure 3.

Figure 4 shows the locations of the precession centre which were obtained from about 250 frames of the video recordings at two z-planes: (a) z = 20 mm and (b) z = 48 mm. The instantaneous positions of the swirl centre are indicated in the Figure as dots, which often overlap each other. The valve lift and flow rate for these tests were 7.5 mm and 1.74 litres/s respectively. The sequence of pictures stored from the video recordings were studied, frame by frame. Thus it was possible to estimate the path of the swirl centre with time. It was not possible to record the swirl centre at every frame, because of the poor resolution caused by the relatively high video frame rate; in some locations the period of observation was too short for the tracer bubbles to form streaks of sufficient length. This limited the ability to follow the path of the swirl centre continuously. But the locations of the swirl centre that could be identified gave a good indication of the spatial extent of the precessional motion. For present purposes, it is useful to define the 'mean' centre of swirl as the average location obtained from a large number of instantaneous recordings. From the recordings obtained at the two planes, the precessional motion of the swirl centre appears to be more organised in the plane further downstream from the cylinder head (Figure 4 (b)) than in the plane closer to it (Figure 4 (a)). It was observed from the video recordings that: (1) the spread of the positions of the instantaneous swirl centre at z = 20 mm was greater than that at z = 48 mm; and (2) in certain instances when the instantaneous centre of swirl at z = 20 mm deviated significantly from the mean swirl centre position, the swirl motion momentarily broke up, but reformed almost immediately. In the plane further downstream (Figure 4 (b)), the range of movement of the swirl centre is almost the same in both x- and y-directions. In the plane nearer the valve, there is a greater excursion in the y-direction than in the x-direction. The number of occurrences of the instantaneous centre of swirl location near that of the mean centre is also greater for the station further away from the cylinder head.

Once the locations of the swirl centre were obtained from the video recordings, the control volume of the velocimeter was positioned close to the position of the estimated 'mean' swirl centre. Figures 5 (a) - (c) show the velocity traces for the tangential component at that location for three flow rates (1.00, 1.52 and 1.74 litres/sec), at a valve lift of 7.5 mm. It can be seen from all three velocity traces that the mean veclocity varies through a range which considerably exceeds the amplitude of the turbulent fluctuations. This is thought to be caused by the precession of the swirl centre. All three curves show regions of abrupt change of magnitude. This suggests that the centre of swirl could be changing position from one instant to another in an irregular and rapid fashion: this finding is supported by observations of the video recordings.

The ensemble-averaged results were studied to establish whether bi-modal velocity distributions were present in the flow, as indicated by the observations made in (9) for swirl centre precession in an axisymmetric swirling flow and in (5) for the flow downstream of a helical port. There was no evidence to support these findings.

The precessing vortex rotates about the mean swirl centre at a rate which is about twice that of the velocity of the fluid in the jet issuing from the valve curtain area. This is in accordance with the observations of (9).

Figure 6 shows the frequency-amplitude distributions for the three flow rates at the same location. The distributions were obtained from the FFT's of the velocity traces of Fig. 5.

The amplitude scale is in arbitrary units. The d.c. component has been subtracted from the results shown in the figure.

It has been suggested that the frequency of precession increases linearly with an increasing flowrate (2) or with increasing engine speed [Yianneskis et al, (8)]. Chanaud (4) also suggested that the frequency of oscillation in steady swirling flows will depend on the angular (swirl) velocity of the fluid near the boundary of the jet and the recirculation region. However there is no evidence of a single distinct frequency in the present results: instead, a band of dominant frequencies are apparent at the lower end of the spectrum. This can be expected for a complicated three-dimensional flow. Because of this, it was considered inappropriate to obtain an average frequency for the comparison of frequency distributions at different flow rates. In any case, it has been shown that in bounded axisymmetric swirling flows the period of rotation of the precessing vortex around the axis of the containing pipe can vary from cycle to cycle by up to 25% (9). Coghe et al (5) also found that the oscillations recorded in steady flows through a helical port did not exhibit any well-defined periodicity.

A comparison was made of the dominant frequencies recorded at different flow rates, a dominant frequency being defined as any frequency which had an amplitude greater than that of any other frequency above 50 Hz. By this means it was intended that frequencies of lower amplitude representing turbulent fluctuations were eliminated. The results of Fig. 6 indicate that the dominant frequencies increase with flowrate. About 85 percent of such frequencies encountered in the spectrum obtained with the 1.52 litres/s flowrate (Figure 5 (b)) scale with those recorded at the higher and/or lower flowrate. At the lower end of the spectrum the two higher flow rates were found to show additional dominant frequencies which were not apparent at the lower flow rate. Thus a complex form of scaling with flowrate is present, extending over a range of, rather than a unique, frequency. It has been previously found that in steady axisymmetric flows the rate of precession around the axis of symmetry is dependent on the flow rate (9), in agreement with the present findings.

The precession is found to involve a complicated three dimensional motion as suggested in Figure 7. The curved line shown in this Figure is a three-dimensional representation of the path of the mean precession centre down the cylinder, extracted from the video flow visualization. The mean axis of precession seems to originate at the edge of the recirculation region behind the valve, where rotational motions are present in all three planes. This is in agreement with the observations made for steady axisymmetric swirling flows in (9), where the precessing vortex core was found to precess round the axis of symmetry near the boundary between the forward and reverse flows downstream of a swirl generator. Chanaud (4) suggested that for precession instability to occur in swirling flows, a reverse flow is necessary in order to generate a region of zero axial velocity and a positive dU/dr gradient (in the centre of the recirculation region, near the edge of the jet issuing from the valve in the present flow); also that this motion will originate at the same region, where the axial velocity is zero. In the present highly three-dimensional flow, the precession shows up as a fairly complicated time-dependent motion. The centre of swirl originates near the valve lip in quadrant 3, moving in its path down the cylinder to quadrant 4 and finally to quadrant 1, as indicated in Figure 7.

When the conditions for instability suggested by Chanaud (4) occur in a swirling flow with a recirculation region, precessional oscillations may be triggered by other instabilities, such as intake jet flapping (8). Such disturbances travel downstream and are fed back to the region of zero axial velocity through the reverse flow in the recirculation zone, as proposed by Chanaud (4). This explanation accounts for the observed persistence of the precessional motion in steady flows and in reciprocating engines during the induction stroke, but not during the compression stroke (2, 8) where reverse flows are not normally present after about 90 degrees BTDC.

The presence of the precessing swirling vortex, as a three-dimensional structure rotating about the mean centre of swirl, has important implications for the turbulent flow structure in an engine cylinder, as in steady flows. In swirling flows, whether axisymmetric (9) or asymmetric, as in the present case, precession increases the levels of turbulence and may cause oscillations of the flow. This can be expected to have a significant effect on cycle-to-cycle variations in the flows in engines.

The investigation is currently being extended to study flows through axisymmetric ports with varying levels of swirl, in order to identify the influence of the intensity of swirl on the extent and frequency of the precessional motion. It should be noted that there is already evidence that the level of swirl affects the rate of precession in steady axisymmetric flows downstream of swirl generators (9).

4. CONCLUSIONS

The following conclusions can be drawn from this study:
1) In the steady flow leaving a production helical port, there is a precessional motion of the swirl centre which follows a complicated helical path as it moves down the cylinder. This precession will produce an apparent increase in the turbulent velocities measured.
2) The region over which this precession extends is more uniformly spread about the mean swirl centre at planes further away from the cylinder head. This suggests that the precessional motion is less organised nearer the cylinder head.
3) There are several dominant frequencies in the precession; these scale with increasing flowrate. The results suggest that in complex engine geometries there may be no unique frequency of precession and it may not be possible to characterise the motion by a single "average" frequency.
4) The 'mean' centre of precession of swirl appears to originate neat the centre of the recirculation region underneath the valve.

ACKNOWLEDGEMENTS

The authors acknowledge financial support from the Science and Engineering Research Council for the work presented in this paper. Thanks also go out to Dr. H. Fu and Mr. K. O. Suen for many useful discussions.

REFERENCES

1. Kito, O., "Axi-symmetric character of turbulent swirling flow in a straight circular pipe", Bulletin of JSME, Vol. 27, pp. 537 - 555, 1984.

2. Arcoumanis, C., Hadjiapostolou, A., and Whitelaw, J. H., "Swirl precession in engine flows", SAE Paper No. 870370, 1987.

3. Gupta, A.K., Lilley, D.G., and Syred, N., "Swirl flows", Abacus Press, 1984.

4. Chanaud, R.C., "Observations of oscillatory motion in certain swirling flows", J. Fluid Mech., vol. 21, part 1, pp. 111-127, 1965.

5. Coghe, A., Brunello, G. and Tassi, E., "Effects of intake ports on the in-cylinder motion under steady flow conditions", SAE Paper No. 880384, 1988.

6. Tindal, M.J., Cheung, R.S. and Yianneskis, M., "Velocity characteristics of steady flows through engine inlet ports and cylinders", SAE Paper No. 880383, 1988.

7. Cheung, R.S., Nadarajah, S., Tindal, M.J. and Yianneskis, M., " An experimental study of velocity and Reynolds stress distributions in a production engine inlet port under steady flow conditions", SAE Paper No. 900058, 1990.

8. Yianneskis, M., Tindal, M.J. and Suen, K.O., "A comparison of cycle-resolved and ensemble-averaged velocity variations in a diesel engine", SAE Paper No. 890617, 1989.

9. Claypole, T.C., Evans, P., Hodge, J. and Syred, N., "The influence of the precessing vortex core on velocity measurements in swirling flows", Proc. 3rd Int. Symposium on Applications of Laser Anemometry to Fluid Mechanics, Lisbon, 7-9 July 1986.

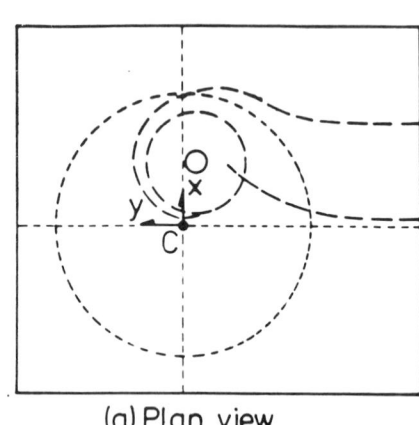
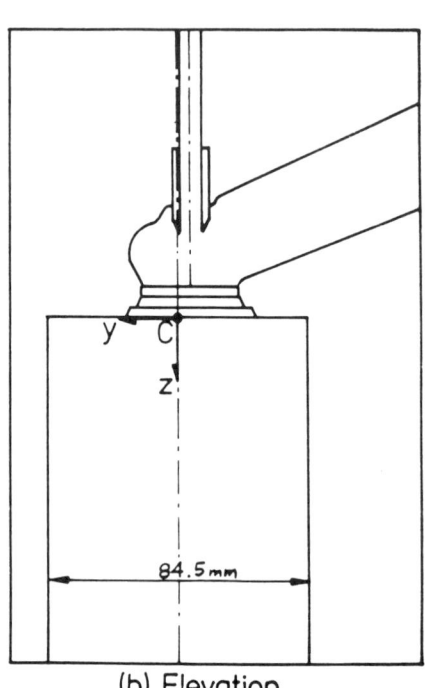

Figure 1 Plan view (a) and elevation (b) of the test section.

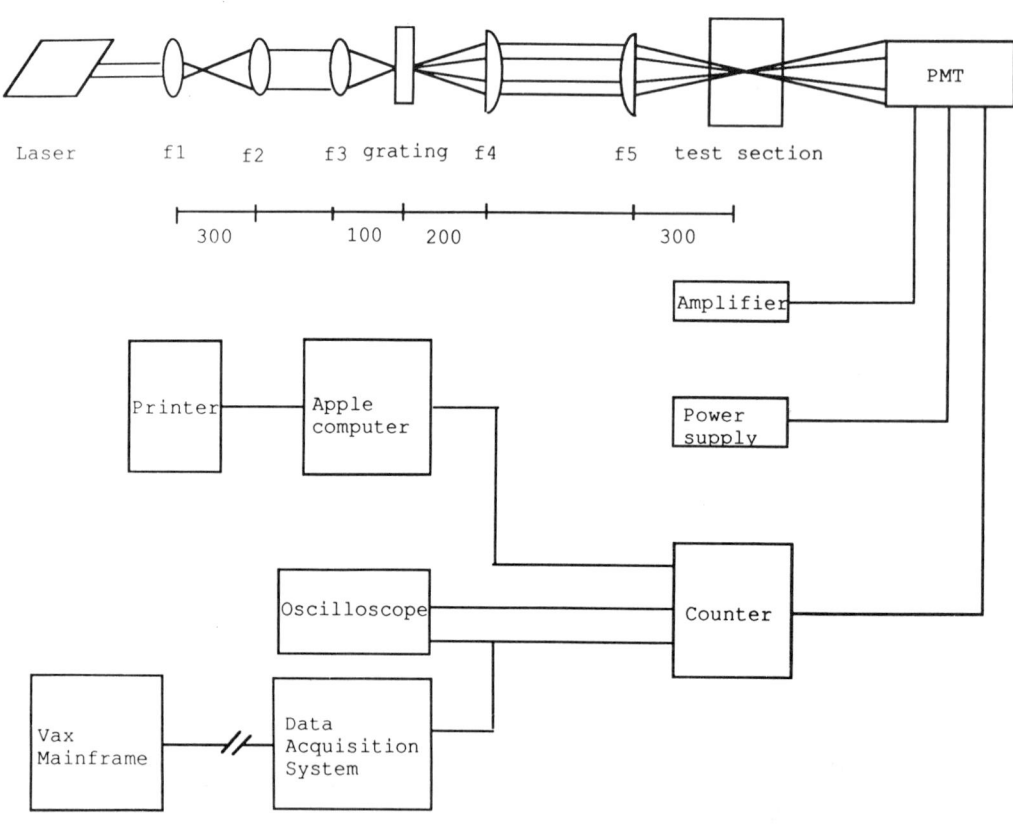

Figure 2. The laser-Doppler anemometer optical and signal processing systems. Distances in mm.

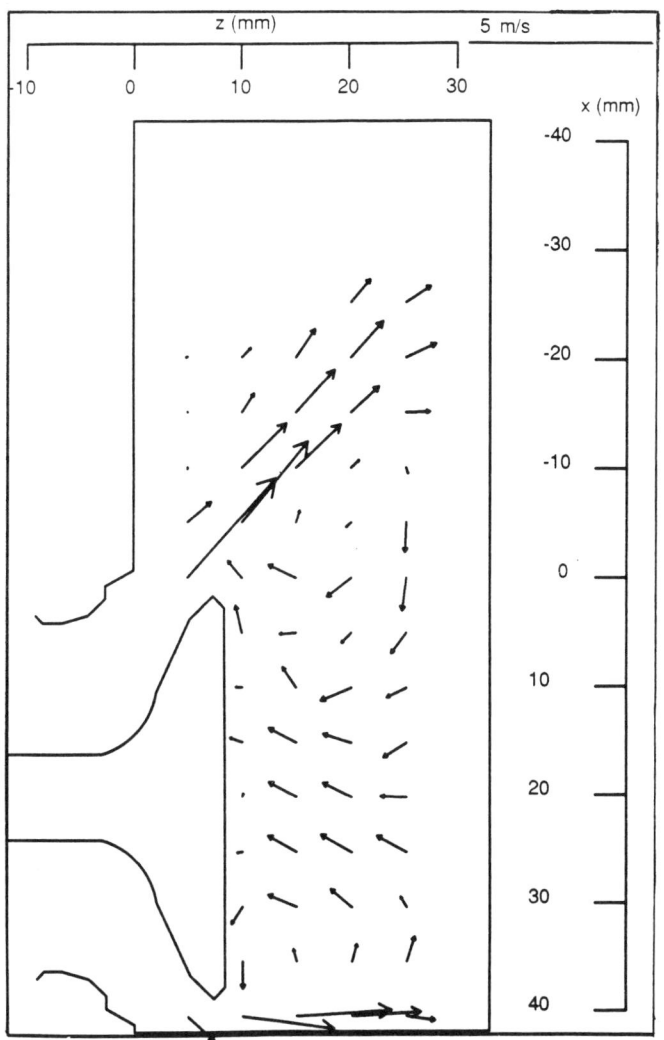

Figure 3. Velocity vectors in the y = - 5 mm plane. Valve lift 9 mm, Re = 34500.

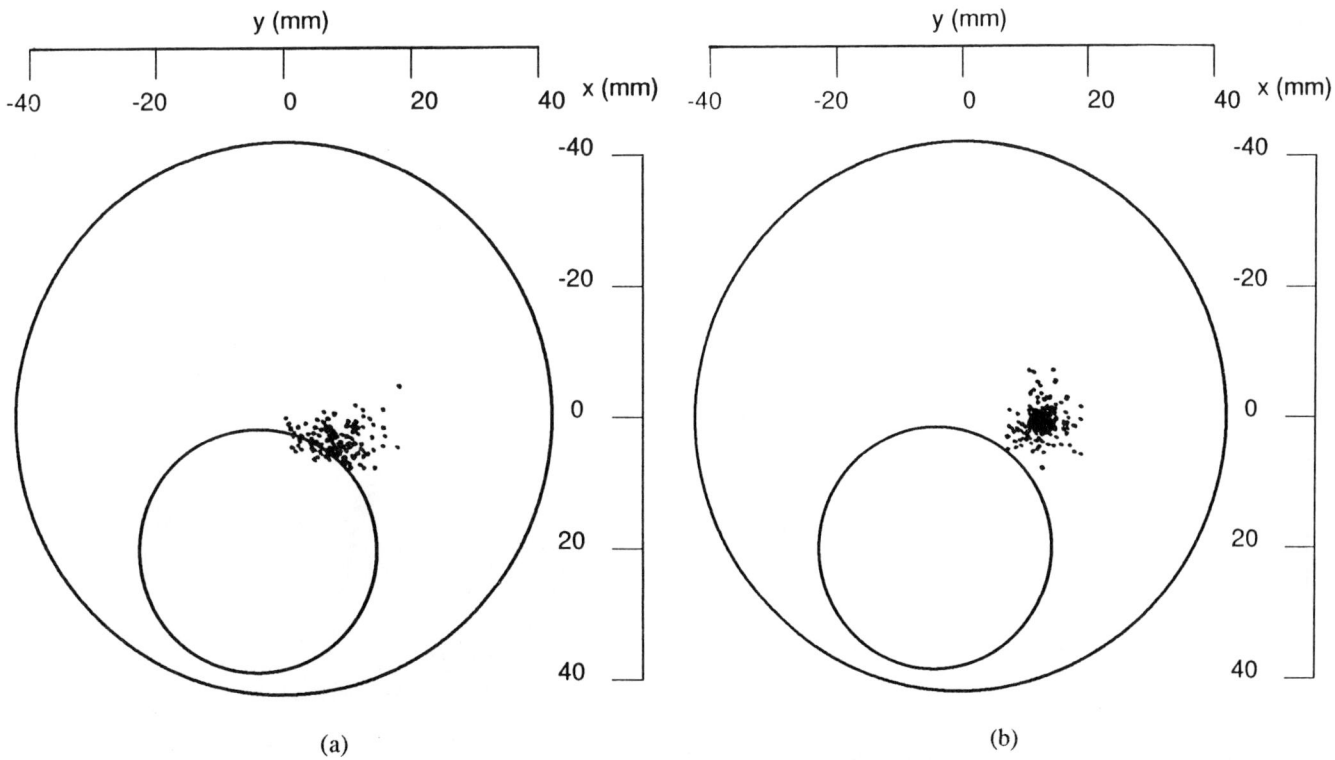

Figure 4. Instantaneous locations of the swirl centre obtained from 250 video frames: (a) z = 20 mm; (b) z = 48 mm.

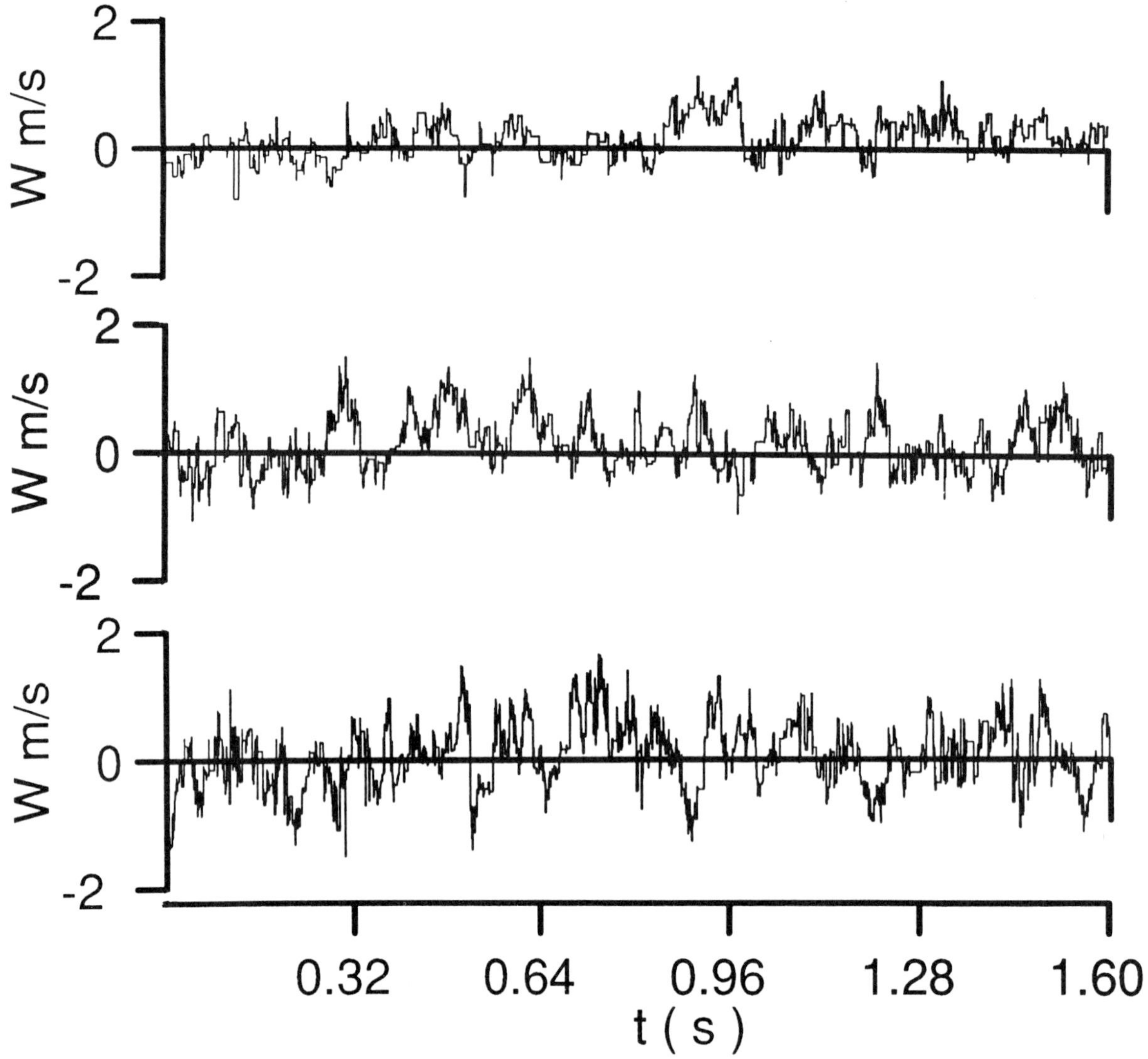

Figure 5. Recordings of time-resolved velocity variation at x = 6 mm, y = 11.5 mm, z = 48 mm. Valve lift 7.5 mm. (a) Re = 18900; (b) Re = 28500; (c) Re = 32900.

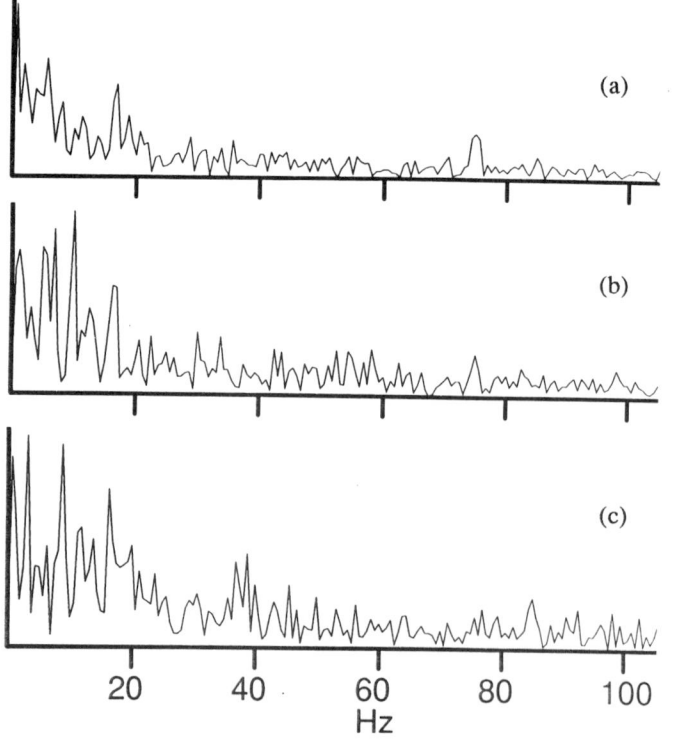

Figure 6. Amplitude-frequency spectra of the velocity recordings shown in Fig. 5. (a) Re = 18900; (b) Re = 28500; (c) Re = 32900.

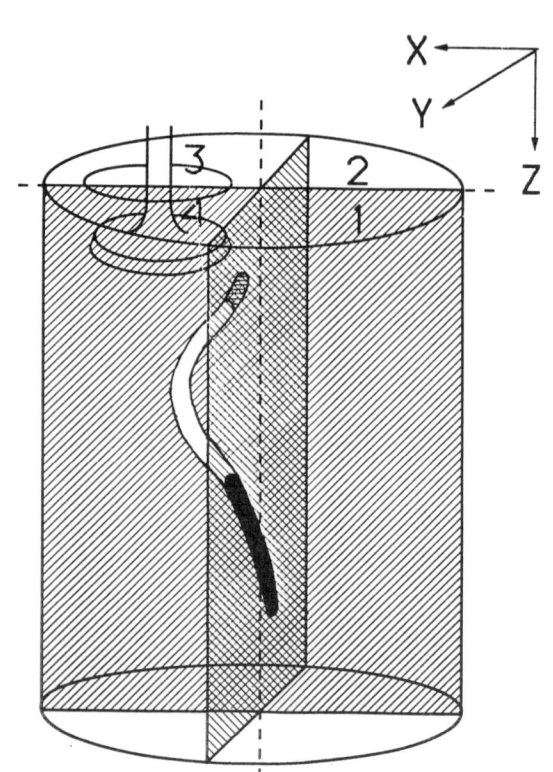

Figure 7. Three-dimensional representation of the path of the mean centre of swirl precession in the cylinder.

C413/009

Estimation of the differences of swirl and charging characteristics between cylinders in a multi-cylinder diesel engine

Y ISSHIKI, MEng, MJSME, MJSAE, Y SHIMAMOTO, PhD, MJSME,
MJSAE and T WAKISAKA, PhD, MJSME, MJSAE
Department of Mechanical Engineering, Kyoto University, Japan

SYNOPSIS Swirl ratio and volumetric efficiency of a 6-cylinder production engine having a complicated intake system have been estimated under firing conditions by means of a numerical method. It has been shown that the swirl ratio in each cylinder of the engine varies considerably with engine speed and that no small differences exist in swirl ratios between cylinders, though the configurations of all intake ports are the same. These differences in swirl ratios between cylinders are due to the unequal length of intake passage. Investigations of the factors causing such variations or differences in swirl ratios have revealed that the important factor is not only the volumetric efficiency, but also the temporal variation of mass flow rate through each intake port.

1 INTRODUCTION

In four-cycle high-speed direct injection diesel engines, the volumetric efficiency and the induction swirl generated by intake ports are important factors influencing engine performance and exhaust emissions. In designing intake ports, it is necessary that they generate suitable swirl with less pressure drop. As a measure of the quality of an intake port, swirl-generating and breathing characteristics measured with a steady flow rig equipped with a vane anemometer (1,2) or an impulse swirl meter (3,4) are usually used. However, it seems desirable to utilize the swirl and charging characteristics measured under actual operating conditions of engines.

Over several decades now, there have been many studies on swirl characteristics with test engines under motored or firing conditions (5-11), but it is only very recently that the influence of the intake system on the swirl characteristics under motored conditions has been investigated (12). It is still a difficult task to measure the swirl characteristics of intake ports under firing conditions of production engines that have complicated intake systems. Therefore, the authors have developed a simplified numerical method for easily estimating both the swirl ratio and the volumetric efficiency under actual operating conditions of production engines (13-15).

A large number of studies (3, 6, 16-18) have revealed that fuel consumption and smoke density are strongly influenced by induction swirl intensity. Therefore, efforts have been made in manufacturing intake ports to limit the deviation of their swirl ratios from the prescribed value, for example, within plus or minus 0.1. In multi-cylinder engines, differences may exist in swirl ratios and volumetric efficiencies between cylinders, due to the difference in the transitional behavior of gas flow in each intake port. Since such uneven swirl ratios and volumetric efficiencies exert undesirable influences upon engine performance and exhaust emissions (19), it is necessary to investigate these differences between cylinders under firing conditions.

In this study, the swirl ratio and volumetric efficiency of a 6-cylinder production engine under firing conditions are estimated by means of a numerical method which has been developed by the authors. Furthermore, the effect of transitional behavior of intake flow on the swirl ratio is investigated both experimentally and numerically using a single-cylinder test engine in order to analyze the swirl characteristics in a 6-cylinder engine.

2 METHOD OF ANALYSIS

It has already become possible to estimate the volumetric efficiency in multi-cylinder engines, which have complicated intake and exhaust systems including an air cleaner, manifold and muffler, under firing conditions by one-dimensional calculation of the gas exchanging process in intake and exhaust systems (20). For estimating the swirl ratio under actual operating conditions of engines, a three-dimensional analysis is necessary of the gas flows in the intake port and also in the cylinder along with the gas exchanging process in the intake and exhaust systems. Computation time to simulate the gas flow in the whole region by three-dimensional calculation is very long. In this study, therefore, the swirl ratio in a 6-cylinder engine under firing conditions is estimated in the following simplified way:

(1) The gas exchanging process in the intake and exhaust systems of the engine is calculated under firing conditions by the one-dimensional characteristic method in order to obtain the temporal variations of the total pressure and total enthalpy at the intake port inlet and also the temporal variation of the pressure in the cylinder along with the volumetric efficiency.
(2) The unsteady gas flow in the intake port alone is calculated three-dimensionally using the above-mentioned state quantities as boundary conditions at the port inlet and outlet.
(3) The swirl ratio at the time of intake valve closing is calculated by integrating the angular momentum flux (with respect to the cylinder axis) through the valve curtain area in the period of intake valve opening (see Appendix 1).

Thus this method can easily estimate the swirl generating ability of intake ports under actual operating conditions, though it is not possible to predict the influences of cylinder wall and combustion chamber on the swirl ratio.

To solve the unsteady gas flow in the intake port, the finite volume method (13) with Cartesian coordinates (x, y, z) is used. The gas is treated as a compressible viscous fluid. It is assumed, for convenience sake, that the velocity distribution at the cross-section of intake port inlet is uniform and its direction is perpendicular to the section. For treating wall boundaries that have complicated configurations, a porosity approach is applied to the control volumes which are adjacent to walls (14). It has already been confirmed that the calculated values of the discharge coefficients and swirl ratios of helical ports by means of this method almost agree with the experimental ones under steady flow conditions (15).

3 ESTIMATION OF THE SWIRL RATIO AND VOLUMETRIC EFFICIENCY IN A MULTI-CYLINDER ENGINE

A 6-cylinder four-cycle production diesel engine is investigated in this study. The main specifications of this engine are; bore = 108 mm, stroke = 113 mm, compression ratio = 18.9, intake valve opening = 20° BTDC, closing = 48° ABDC and intake port diameter = 46 mm. The schematic diagram of the intake and exhaust systems is shown in Fig. 1(a). The average of the volumetric efficiencies of six cylinders, η_{va}, has been obtained under firing conditions by one-dimensional calculation of the gas exchanging process. The variation of η_{va} with engine speed is shown in Fig. 1(b) together with the experimental data (21). Since the calculated results agree well with the experimental ones over a wide range of engine speed, it seems appropriate to use the results of one-dimensional calculation as the boundary conditions in three-dimensional calculation.

In order to investigate the effects of transitional phenomena on the swirl characteristics under the condition of the same port configuration, all intake ports in the above-mentioned engine are replaced by a helical port shown in Fig. 2(a). The size of the intake manifold is a little modified. Calculations have been carried out under firing conditions in this 6-cylinder engine to obtain the swirl ratio and volumetric efficiency of each cylinder.

The variations of swirl ratio S_C and volumetric efficiency η_v of each cylinder with engine speed N_e are shown in Fig. 2(b). The definition of S_C is described in Appendix 1. It is found that the swirl ratio of each cylinder is not constant, but varies considerably with engine speed. Of course η_v varies with engine speed. The variations of S_C and η_v of the No. 1-3 cylinders show a different tendency from those of the No. 4-6 cylinders. These phenomena are supposedly due to the difference in transitional behavior of gas flow, which is caused by the unequal lengths of the pipe elements P-3 and P-4 in the intake manifold as shown in Fig. 1(a). The degree of unevenness in S_C, that is, the ratio of the difference between the maximum and minimum values to the averaged value over the six cylinders is larger than the degree of unevenness in η_v; the degree of unevenness in S_C is 7.8% at its maximum, while that in η_v is 3.6% at its maximum. It is inferred from these results that in actual engines differences may exist in swirl ratios between cylinders, and such differences may be so large as to affect the fuel consumption and smoke density.

Thus the present numerical method, that is, the hybrid method which employs the one-dimensional and three-dimensional approaches makes it possible to estimate the swirl characteristics under actual operating conditions of a multi-cylinder engine.

Figure 2(b) shows that the swirl ratio S_C increases with an increase of engine speed N_e, in the same way as the volumetric efficiency η_v, within the range of engine speed from 1000 rpm to 2000 rpm. This result agrees with the experimental result in reference (12), which reported that the swirl ratio increases with an increase of volumetric efficiency under motored conditions of an engine. In the range of a higher engine speed than 2000 rpm, however, the variation of S_C with N_e is different from that of η_v. This fact suggests that there can be other factors influencing the swirl ratio in addition to the volumetric efficiency. These factors are investigated in the next section using a single-cylinder engine along with the 6-cylinder engine.

4 INVESTIGATION OF THE FACTORS INFLUENCING THE SWIRL RATIO

4.1 Investigation using a single-cylinder engine

A single-cylinder test engine with a simple intake system is used. The main specifications of this engine are; bore = 130 mm, stroke = 150 mm, compression ratio = 15.6, intake valve opening = 18° BTDC, closing = 52° ABDC and intake port diameter = 52 mm. This engine is equipped with a helical intake port with a configuration similar to the port shown in Fig. 2(a).

As it is not easy to measure the swirl ratio under actual operating conditions of the test engine, experiments have been carried out under the condition of intermittent flow by using an experimental apparatus which has the same intake port and cylinder bore as the test engine as shown in Fig. 3(a). This apparatus is equipped with a cam-operated intake valve which moves in the same way as the intake valve of the test engine, but has no exhaust valve. The cam shaft is rotated at a speed of 900 rpm (corresponding to an engine speed N_e = 1800 rpm) and the average pressure difference Δp between the atmosphere and the cylinder is kept constant during experiments investigating the effect of intake passage length l_p on the swirl characteristics. Here, l_p is the center line length of the intake passage measured from the intake pipe inlet to the intake port outlet. The average mass flow rate of intermittent flow is measured by an orifice. A conventional vane anemometer (2) is installed in the cylinder at a distance of 1.1 times the cylinder diameter away from the cylinder head. The average rotation speed of the vane is measured in order to know the characteristics of swirl generation qualitatively, though it is not possible to measure the instantaneous swirl speed under the condition of intermittent flow by the vane anemometer.

Figure 3(b) shows the average mass flow rate $\bar{m}$ and average vane speed $\bar{N}_v$ measured under the condition of Δp=11.2 kPa. It is found from the experimental results that $\bar{m}$ varies largely with the intake passage length l_p due to the dynamic effect of the intake passage, and also that $\bar{N}_v$ varies largely with l_p. It is inferred that the swirl intensity is influenced by the average mass flow rate from the fact that the variation of $\bar{N}_v$ roughly accords with the variation of $\bar{m}$. Also shown in Fig. 3(b) are the calculated values of $\bar{m}$ and $\bar{N}_v$ by means of the present hybrid method; the values of $\bar{m}$ are three-dimensionally calculated ones and the values of $\bar{N}_v$ are obtained by a similar manner to the method for obtaining the swirl ratio (see Appendix 2). Good agreement of the calculated values of $\bar{m}$ with the experimental values shows the validity of this calculation method. It is found that the tendency in the variation of calculated values of $\bar{N}_v$ with l_p is similar to that of the experimental values, though the level of the calculated values is higher than that of the experimental ones since the decay of swirl is not taken into account in the calculation.

From the above-mentioned experiment, it is inferred that the swirl intensity is influenced by the average mass flow rate. In order to examine this inference, experiments have been carried out under the condition of a given average mass flow rate in spite of using the intake pipes of various lengths in the experimental apparatus. The average pressure difference Δp has been set as shown in Fig. 3(c) for each l_p to maintain the average mass flow rate $\bar{m}$ at 0.0307 kg/s. The measured and calculated values of average vane speed $\bar{N}_v$ are shown in Fig. 3(c) together with the calculated values of Δp. The calculated values of Δp agree very well with the experimental values, and further, the calculated result of $\bar{N}_v$ predicts the trend in the variation of $\bar{N}_v$ measured at constant $\bar{m}$ relatively well. The experimental result shows that there still exists the variation of $\bar{N}_v$ with l_p at constant $\bar{m}$, though the magnitude of the variation is smaller compared with that in Fig 3(b). Therefore, it is obvious that in the case of unsteady flows there are other factors influencing the swirl intensity to be generated by an intake port in addition to the average mass flow rate.

Next, calculations have been carried out on the single-cylinder test engine under motored conditions using the intake passages having a length $l_p=0.2$ m or 1.4 m. The calculated values of swirl ratio S_C and volumetric efficiency η_v are plotted in Fig. 4(a) versus engine speed N_e. In the case of the short intake passage ($l_p=0.2$ m), that is, in the case where the intake system consists of an intake port alone, the variations of S_C and η_v with N_e are small, since the dynamic effect of the intake passage is weak. In the case of the long intake passage ($l_p=1.4$ m) however, the variations of S_C and η_v with N_e are very large. This is obviously due to the strong dynamic effect. In this case, the trend in the variation of S_C with N_e is very different from that of η_v. To examine the cause of such a difference, the temporal variations of mass flow rate are investigated.

4.2 Influence of the temporal variation of mass flow rate on the swirl ratio

The variations of m_{deg} (mass flow through the intake valve curtain area per unit crank angle) with crank angle Θ in the single-cylinder engine are shown in Fig. 4(b) at various engine speeds N_e for each case of $l_p=0.2$ m and 1.4 m. It is found that in the case of $l_p=0.2$ m, the form of the m_{deg} curve changes a little with N_e, while in the case of $l_p=1.4$ m it changes largely with N_e. From Figs. 4(a) and 4(b), it follows that the swirl ratio varies largely with engine speed in the case where the form of mass flow rate curve changes largely with engine speed. Comparing the swirl ratio at 1400 rpm with that at 2000 rpm in the case of $l_p=1.4$ m, the former swirl ratio is much larger than the latter, though the volumetric efficiencies at these two engine speeds are nearly the same. As shown in Fig. 4(b), the forms of m_{deg} curves at these two engine speeds for $l_p=1.4$ m are largely different from one another. It is found that high swirl ratio is brought about when m_{deg} curve has a sharp peak in the period of high valve lift, i.e. takes a parabola-like form ($N_e=1400$ rpm), while low swirl ratio is brought about when m_{deg} curve has no peak, and even has a drop in the period of high valve lift, i.e. takes a trapezoid-like form ($N_e=2000$ rpm). It is concluded from these facts that the swirl ratio is influenced by the form of mass flow rate curve. The cause for this phenomenon is explained as follows: The swirl ratio is dominated by the ratio of the integration of angular momentum flux to the integration of mass flow rate in the period of intake valve opening, where the angular momentum flux is dominated by the product of two velocity components v_r and v_{xy} according to its definition (see Appendix 1), while the mass flow rate is dominated by a velocity component v_r. Therefore, it follows that the swirl ratios at two engine speeds are not the same when the patterns of the temporal variations of intake flow velocity are different, even if the masses of intake gas per one cycle at the engine speeds, i.e. the values of the integration of mass flow rate in the period of intake valve opening are the same. Consequently, it is confirmed that the pattern of the temporal variation of mass flow rate is one of the important factors influencing the swirl intensity.

4.3 Investigation using the 6-cylinder engine

The temporal variation of mass flow rate in the 6-cylinder engine in Sec. 3 is investigated. Figure 5(a) shows the variations of m_{deg} with crank angle Θ at 2000 rpm and 3000 rpm with respect to the No. 2 cylinder. The volumetric efficiencies in these two cases are nearly the same. It is found that the swirl ratio at 2000 rpm, where m_{deg} curve is a parabola-like form, is much larger than the swirl ratio at 3000 rpm, where m_{deg} curve is a trapezoid-like form. Such a relationship between the swirl ratio and the form of the m_{deg} curve is the same as that in the single-cylinder engine mentioned previously. Consequently, in the case of the multi-cylinder engine also, the variation of swirl ratio with engine speed is explained by the difference in the pattern of the temporal variation of mass flow rate.

The swirl ratio S_C and volumetric efficiency η_v of each cylinder at 2500 rpm, for example, are shown in Fig. 5(b) by re-plotting them from Fig. 2(b) to examine the differences in swirl ratios between cylinders in the 6-cylinder engine. The swirl ratios S_C of the No. 1-3 cylinders are smaller than those of the No. 4-6 cylinders, though the volumetric efficiencies η_v of the No. 1-3 cylinders are a little larger than those of the No. 4-6 cylinders. Since the lengths of the pipe elements P-3 and P-4 of the intake manifold are different from one another as shown in Fig. 1(a), the transitional behaviors of flows in the intake ports are different between the group of No. 1-3 cylinders and that of No. 4-6 cylinders. Therefore, it is concluded that the differences in S_C and η_v between the two groups are due to the differences in the transitional behaviors of flows.

5 INVESTIGATION OF THE DEVELOPMENT OF FLOW IN THE INTAKE PORT

It has been found from the above-mentioned investigations that the temporal variation of mass flow rate is an important factor in swirl generation. As a next step, an attempt is made to investigate the development of flow in the intake port (22), i.e. helical port shown in Fig. 2(a) by comparing the swirl ratios under engine operating conditions with the swirl ratios under quasi-steady conditions, which are calculated as follows:

Firstly, steady flows in the present intake port have been calculated three-dimensionally at some fixed lifts of intake valve by prescribing the mass flow rates and state quantities (the total pressure and enthalpy at the intake port inlet) which are equivalent to the mass flow rates and state quantities calculated at each corresponding valve lift under firing conditions in the 6-cylinder engine. Next, the angular momentum flux (with respect to the cylinder axis) through the valve curtain area has been obtained from this quasi-steady state calculations at some valve lifts, and then the swirl ratio at the time of intake valve closing has been estimated by integrating the obtained angular momentum flux in the period of intake valve opening (see Appendix 3). This swirl ratio is designated the quasi-steady swirl ratio.

The quasi-steady swirl ratios calculated with respect to the No. 1 and No. 6 cylinders at various engine speeds are shown in Fig. 6(a). The swirl ratios already shown in Fig. 2(b), which are designated the unsteady swirl ratio, are also plotted in Fig. 6(a). It is found that in both No. 1 and No. 6 cylinders, the quasi-steady swirl ratios lie near the unsteady swirl ratios within the range of engine speed from 1000 rpm to 2000 rpm, however, they become smaller than the unsteady swirl ratios in the range of higher engine speed than 2000 rpm. This fact implies that in the range of low engine speed, the velocity field in this helical port under engine operating conditions is nearly the same as that under quasi-steady conditions, namely, the flow is nearly in the fully developed state. On the other hand, in the range of high engine speed, the velocity field under engine operating conditions seems to be different from that under quasi-steady conditions. Then, the velocity distributions around the intake valve at the highest speed within the engine speed range analyzed, i.e. at 3000 rpm, are examined.

In Fig. 6(b), the velocity distributions at the valve curtain area under the quasi-steady condition are compared with those under the unsteady condition at

$N_e=3000$ rpm with respect to the No. 1 cylinder. The velocity distributions are expressed by the normalized magnitude $|v_{xy}|/\bar{v}_r$ and direction γ_{xy} of the velocity vector projected on the plain perpendicular to the valve axis as shown in Fig. 6(b), where $|v_{xy}|$ is the magnitude of vector v_{xy} and $\bar{v}_r$ is the average of radial velocity component v_r at the valve curtain area. Shown in this figure are the velocity distributions at the middle of each valve lift of No. 1 cylinder at crank angles $\Theta=60°$, 104° and 148° ATDC.

According to Fig. 6(b), the values of $|v_{xy}|/\bar{v}_r$ under unsteady and quasi-steady conditions are almost the same at each crank angle. On the other hand, though the difference between the values of γ_{xy} under the unsteady condition and those under the quasi-steady condition is small at the time of the maximum valve lift ($\Theta=104°$), the difference is large at other crank angles; at $\Theta=60°$, the values of γ_{xy} under the unsteady condition are smaller than those under the quasi-steady condition, and at $\Theta=148°$ the inequality between them is reversed. This means that the swirl generation in the helical port delays at high engine speed. Therefore, it is concluded that the difference between the unsteady swirl ratios and the quasi-steady swirl ratios in the range of high engine speed is due to the delay of the development of flow in the intake port.

6 CONCLUSIONS

The swirl ratio and volumetric efficiency of a 6-cylinder production engine have been estimated under firing conditions by means of a numerical method. Furthermore, using an experimental apparatus to generate intermittent flows and also employing a single-cylinder test engine, the effect of the transitional behavior of intake flow on the swirl ratio has been investigated. On the basis of these experimental and calculated results, the factors influencing the swirl ratio have been examined. The main conclusions are as follows:

(1) The present hybrid calculation method which employs the one-dimensional and three-dimensional approaches has made it possible to estimate the swirl characteristics, along with the volumetric efficiency, in each cylinder of a multi-cylinder production engine with a complicated intake system under firing conditions.
(2) The swirl ratio is not constant but varies considerably with engine speed under operating conditions in both cases of single-cylinder and multi-cylinder engines with long intake passage. The transitional behavior of flow in the intake system is a cause for the variation of swirl ratio with engine speed.
(3) In the 6-cylinder engine investigated, no small differences exist in swirl ratios between cylinders, though the configurations of all intake ports are the same. This is because the transitional behaviors of flows in the intake ports are different between cylinders due to the difference in pipe element length of the intake manifold.
(4) The important factors influencing the swirl ratio to be generated by each intake port under actual operating conditions in a multi-cylinder engine are not only the volumetric efficiency, but also the pattern of the temporal variation of mass flow rate through each intake port.

ACKNOWLEDGMENTS

This work has been partially supported by a Grant-in-Aid for Scientific Research from the Ministry of Education, Science and Culture of Japan. Calculations were carried out on FACOM M780/30 at Kyoto University Data Processing Center.

REFERENCES

(1) FITZGEORGE, D. and ALLISON, J. L. Air Swirl in a Road-Vehicle Diesel Engine. Proceedings of IMechE(A.D.), 1962–63, Vol. 177, 151–177.
(2) PISCHINGER, F. Entwicklungsarbeiten an einem Verbrennungssystem für Fahrzeugdieselmotoren. *ATZ*, 1963, Vol. 65, No. 1, 11–16.
(3) TIPPELMANN, G. A New Method of Investigation of Swirl Ports. SAE Paper No. 770404, 1977.
(4) TANABE, S., IWATA, H. and KASHIWADA, Y. On Characteristics of Impulse Swirl Meter. Proceedings of International Symposium on Diagnostics and Modeling of Combustion in Reciprocating Engines(COMODIA 85), Tokyo, 1985, 267–272.
(5) WILLIS, D. A., MEYER, W. E. and BIRNIE, Jr., C. Mapping of Airflow Patterns in Engines with Induction Swirl. SAE Paper No. 660093, 1966.
(6) OHIGASHI, S., HAMAMOTO, Y. and TANABE, S. Swirl — Its Measurement and Effect on Combustion in a Diesel Engine. Proceedings of IMechE, 1971, C134/71, 129–136.
(7) DENT, J. C. and DERHAM, J. A. Air Motion in a Four-Stroke Direct Injection Diesel Engine. Proceedings of IMechE, 1974, Vol. 188, 269–280.
(8) KATOH, J., OHKUBO, Y., OHTSUKA, M. and SUGIYAMA, K. LDV Measurement of Swirl Flow in Internal Combustion Engines. Proceedings of International Symposium on Diagnostics and Modeling of Combustion in Reciprocating Engines(COMODIA 85), Tokyo, 1985, 193–202.
(9) ARCOUMANIS, C., VAFIDIS, C. and WHITELAW, J. H. Valve and In-Cylinder Flow Generated by a Helical Port in a Production Diesel Engine. *J. Fluids Engineering*, 1987, Vol. 109, 368–375.
(10) BRANDSTÄTTER, W., JOHNS, R. J. R. and WIGLEY, G. The Effect of Inlet Port Geometry on In-Cylinder Flow Structure. SAE Paper No. 850499, 1985.
(11) WAKISAKA, T., SHIMAMOTO, Y. and ISSHIKI, Y. Induction Swirl in a Multiple Intake Valve Engine — Three-Dimensional Numerical Analysis. Proceedings of IMechE, 1988, C40/88, 73-84.
(12) MARGARY, R., NINO, E. and VAFIDIS, C. The Effect of Intake Duct Length on the In-Cylinder Air Motion in a Motored Diesel Engine. SAE Paper No. 900057, 1990.
(13) ISSHIKI, Y., SHIMAMOTO, Y. and WAKISAKA, T. Numerical Prediction of Effect of Intake Port Configurations on the Induction Swirl Intensity by Three-Dimensional Gas Flow Analysis. Proceedings of International Symposium on Diagnostics and Modeling of Combustion in Reciprocating Engines (COMODIA 85), Tokyo, 1985, 295–304.
(14) SHIMAMOTO, Y., ISSHIKI, Y., WAKISAKA, T. and UEDERA, M. The Numerical Prediction of Gas Flow in Intake Ports of Four-Cycle Internal Combustion Engines (3rd Report, Improvement of Accuracy by Applying a Porosity Approach). (in Japanese) *Trans. JSME(B)*, 1989, Vol. 55, No. 518, 3246–3250.
(15) SHIMAMOTO, Y., ISSHIKI, Y. and WAKISAKA, T. The Numerical Prediction of Gas Flow in Intake Ports of Four-Cycle Internal Combustion Engines (4th Report, Improving the Treatment of the Valve Outlet Boundary). (in Japanese) *Trans. JSME(B)*, 1990, Vol. 56, No. 523, 864–868.
(16) BRANDL, F., REVERENCIC, I., CARTELLIERI, W. and DENT, J. C. Turbulent Air Flow in the Combustion Bowl of a D.I. Diesel Engine and Its Effect on Engine Performance. SAE Paper No. 790040, 1979.

(17) MONAGHAN, M. L. and PETTIFER, H. F. Air Motion and Its Effect on Diesel Performance and Emissions. SAE Paper No. 810255, 1981.
(18) PARTINGTON, G. D. Development and Application of a Fully Machined Helical Inlet Port for High Speed DI Engines. Proceedings of IMechE, 1982, C121/82, 277–283.
(19) SHIGEMORI, M., TSURUOKA S., and SHIMODA M. Development of a Combustion System for a Light Duty D.I. Diesel Engine. SAE Paper No. 831296, 1983.
(20) SHIMAMOTO, Y., OKA, M. and TANAKA, Y. A Research on Inertia Charging Effect of Intake System in Multi-Cylinder Engines. Bull. JSME, 1978, Vol. 21, No. 153, 502–510.
(21) SHIMAMOTO, Y., NISHIWAKI, K., OKADA, C. and TANAKA, Y. Gas Flow Analysis in Intake and Exhaust Pipe Systems of a Multi-Cylinder Engine. (in Japanese) Proceedings of the Joint Symposium on Internal Combustion Engines, 1979, 175–180.
(22) SHIMAMOTO, Y., ISSHIKI, Y., WAKISAKA, T. and FUJIMOTO, T. Investigation of Swirl Generating Characteristics of Helical Ports by Numerical Simulation. Proceedings of International Symposium on Diagnostics and Modeling of Combustion in Reciprocating Engines (COMODIA 90), Kyoto, 1990, to be published.

APPENDICES

Appendix 1 <u>Swirl ratio</u>

When the velocity vector at the intake valve curtain area is projected on the plane perpendicular to the valve axis, the vector is resolved into the radial and tangential components with respect to the valve axis. The angular momentum flux Ω_C (with respect to the cylinder axis) through the valve curtain area is expressed as

$$\Omega_C = \int_s \rho v_r v_{xy} r_c ds$$

where s: area of the valve curtain area, ρ: density, v_r: radial velocity component, v_{xy}: resultant velocity of radial and tangential velocity components, r_c: length of a perpendicular from the cylinder axis to the velocity vector v_{xy}.

When the wall friction in the cylinder is neglected, the swirl ratio S_C can be expressed as

$$S_C = \int_{\Theta_{IO}}^{\Theta_{IC}} \frac{\pi}{180\omega_e} \Omega_C d\Theta \Big/ \frac{\omega_e}{2} R_C^2 M_{IC}$$

where Θ: crank angle, Θ_{IO}: crank angle of intake valve opening, Θ_{IC}: crank angle of intake valve closing, ω_e: engine angular velocity, R_C: cylinder radius, M_{IC}: mass of the cylinder contents at Θ_{IC}.

Appendix 2 <u>Average vane speed</u>

When the wall friction in the cylinder and the decay of swirl in the period of intake valve closing are neglected, the average vane speed $\bar{N}_v$ can be expressed as

$$\bar{N}_v = \int_{\Theta_{IO}}^{\Theta_{IC}} \Omega_C d\Theta \Big/ \pi R_C^2 \int_{\Theta_{IO}}^{\Theta_{IC}} m d\Theta$$

where m: mass flow rate through the valve curtain area.

Appendix 3 <u>Quasi-steady swirl ratio</u>

The angular momentum flux (with respect to the cylinder axis) through the valve curtain area was calculated under quasi-steady conditions at the valve lifts of 1/5, 2/5, 3/5, 4/5 and 5/5 of the maximum valve lift. As the number of calculated values was not sufficient, the values of angular momentum flux under the conditions where they were not calculated, were determined by a linear interpolation. The swirl ratio at the time of intake valve closing was calculated by integrating the obtained angular momentum flux in the period of intake valve opening.

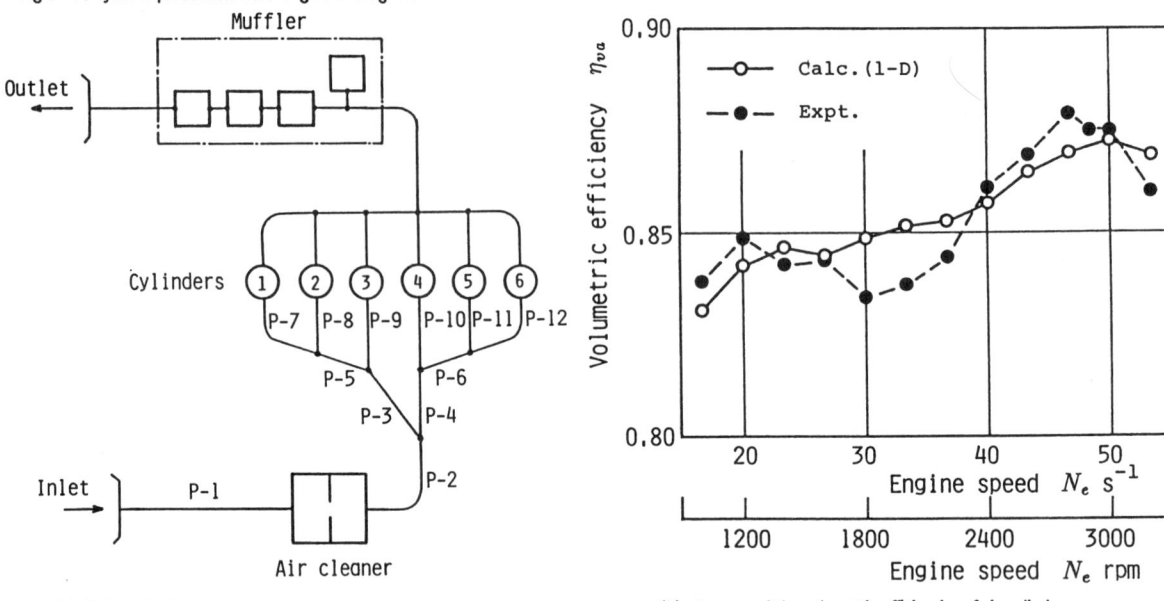

Fig 1 A 6-cylinder production diesel engine investigated

(a) Schematic diagram of intake and exhaust systems

(b) Average of the volumetric efficiencies of six cylinders

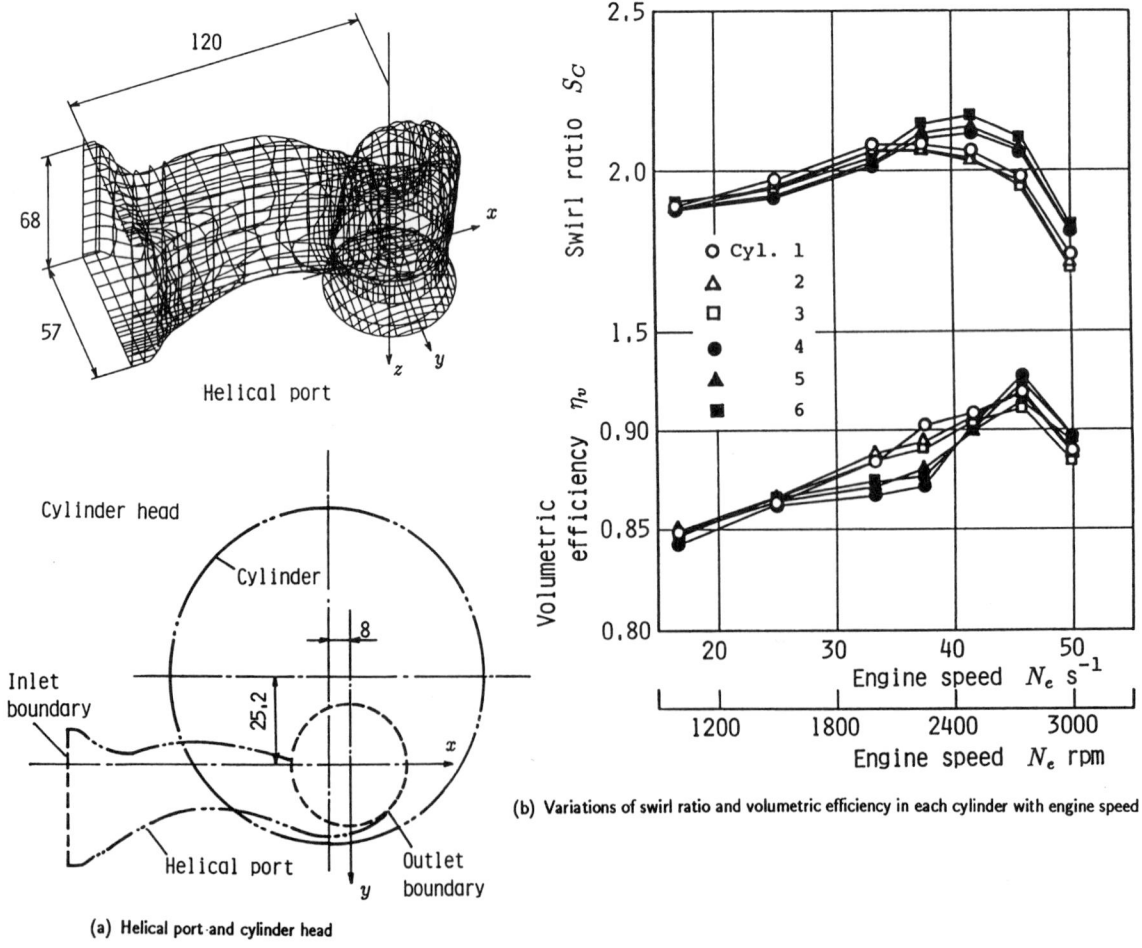

Fig 2 Estimation of swirl ratio and volumetric efficiency in the 6-cylinder engine

(a) Helical port and cylinder head

(b) Variations of swirl ratio and volumetric efficiency in each cylinder with engine speed

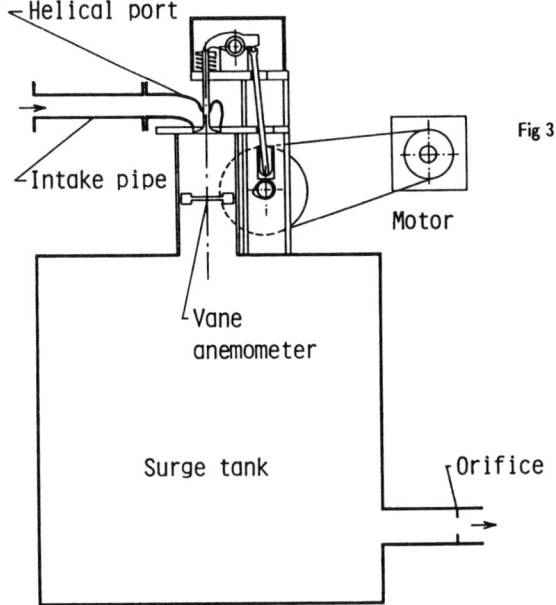

(a) Schematic diagram of an experimental apparatus

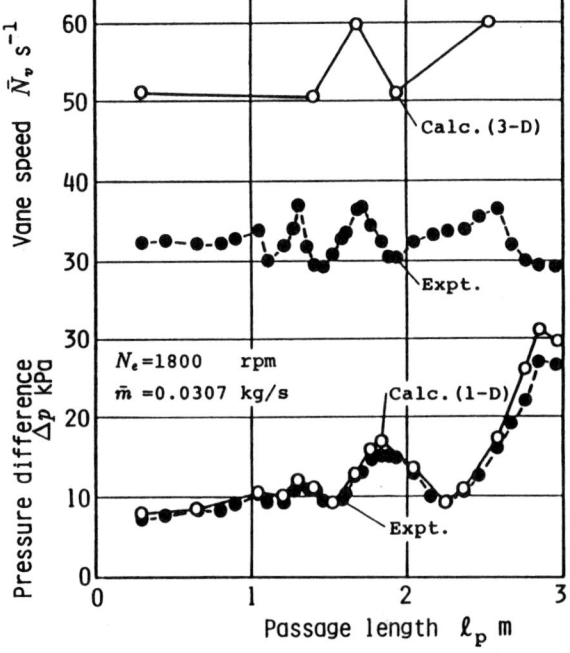

(b) Variation of average vane speed and that of average mass flow rate with passage length under the condition of a given average pressure difference

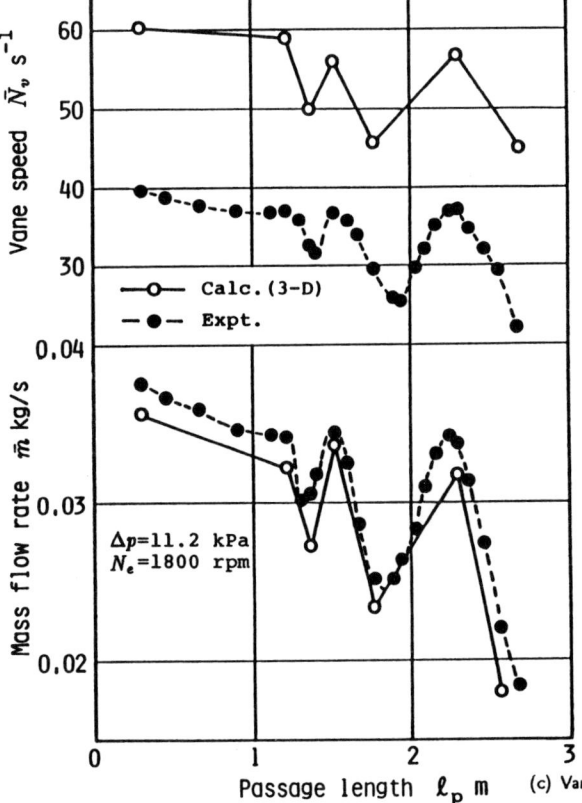

(c) Variation of average vane speed and that of average pressure difference with passage length under the condition of a given average mass flow rate

Fig 3 Experimental investigation under the condition of intermittent flow

Fig. 4 Investigation using a single-cylinder test engine

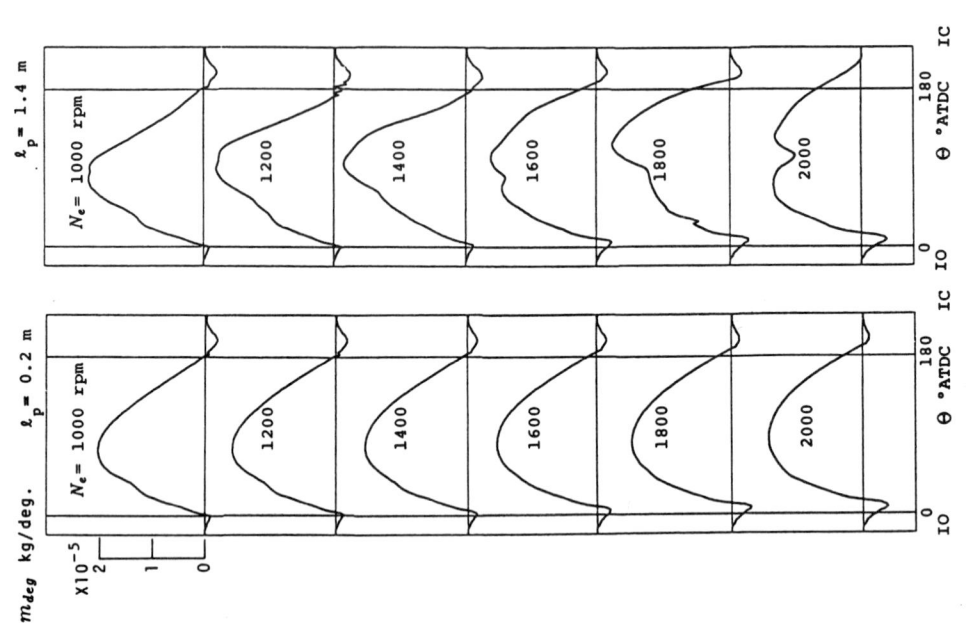

(a) Variations of swirl ratio and volumetric efficiency with engine speed

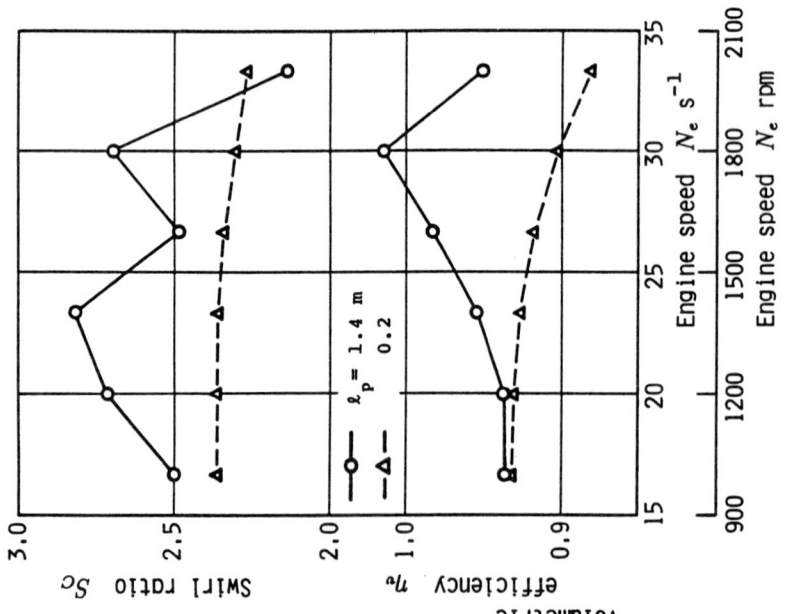

(b) Variation of mass flow rate through the valve curtain area with crank angle

Fig. 5 Investigation using the 6-cylinder engine

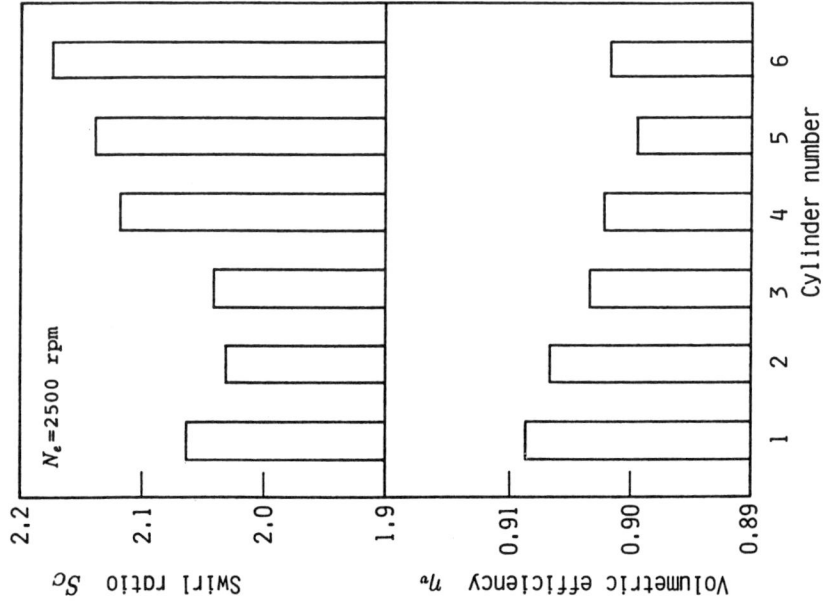

(b) Differences in swirl ratios and volumetric efficiencies between cylinders

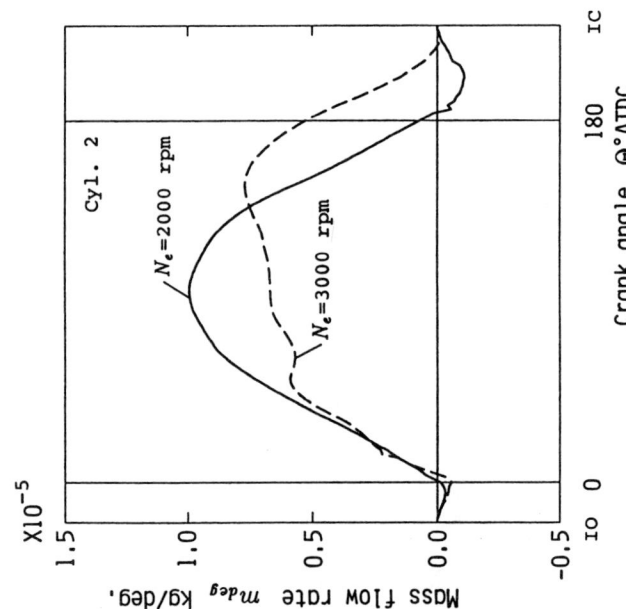

(a) Variations of mass flow rates through the valve curtain area with crank angle at 2000 rpm and 3000 rpm (No. 2 cylinder)

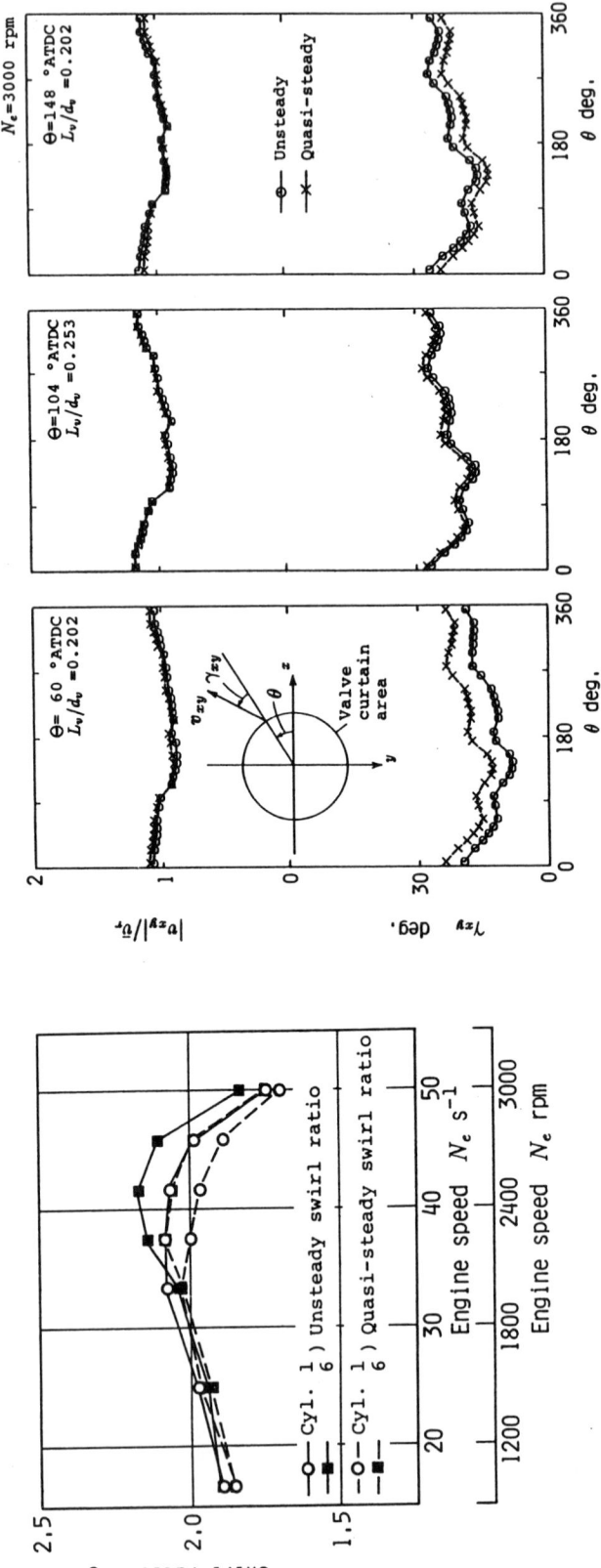

Fig. 6 Investigation of the development of flow in the intake port

(a) Variations of quasi-steady and unsteady swirl ratios with engine speed

(b) Velocity distributions at the valve curtain area under the quasi-steady and unsteady conditions; L_v = intake valve lift; d_v = intake port diameter

C413/079

Numerical simulation of fluid flow in poppet valves

N D VAUGHAN, BSc(Eng), PhD, CEng, MIMechE,
D N JOHNSTON, BSc, PhD and K A EDGE, BSc, PhD, CEng, FIMechE
Fluid Power Centre, School of Mechanical Engineering, University of Bath

SYNOPSIS: Simulations of flow through poppet valves were performed using a finite volume Computational Fluid Dynamics program. A range of valve geometries was simulated, and the flow was turbulent, incompressible and steady. Simulations were compared with experimental measurements and visualised flow patterns. Overall correlation between simulated and visualised flow patterns was good. Jet separation and re-attachment proved difficult to model accurately. This was due to inaccuracies caused by the upwind differencing scheme employed, and in the representation of turbulence by the k-ε model which is known to be inaccurate when applied to re-circulating flow.

NOTATION

a	orifice area
A_0	face area of valve = $\pi d_0^2/4$
C_Q	flow coefficient of valve = $\dfrac{Q}{a\sqrt{(2\Delta P)/\rho}}$
C_F	force coefficient of valve = $\dfrac{Thrust}{\Delta P\, A_0}$
d_0	inlet diameter
f	geometric weighting factor for cell face
F_ϕ	flux of ϕ through cell face
k	turbulent kinetic energy
$\dot{m}$	mass flux through cell face
Q	flow rate
R	Reynolds number
u	axial component of fluid velocity
v	radial component of fluid velocity
z	axial opening of valve
Z	non-dimensional opening z/d_0
ΔP	pressure drop across valve
γ	blending factor for upwind differencing scheme
ε	turbulent dissipation rate
μ_t	effective turbulent viscosity
ρ	fluid density
ϕ	dependent variable (u, v, k or ε)

Subscripts

in	conditions at inlet
out	conditions at outlet
U	property at node immediately upstream of face
UU	property two nodes upstream of face

1. INTRODUCTION

A great deal of research has been published on the flow characteristics of hydraulic valves. This serves to indicate that valve flow can be a highly complex process which is strongly dependent on the details of the valve geometry, the fluid properties and the operating conditions. Separation and re-attachment of jets can have a profound effect on the flow, pressure and force characteristics as well as influencing the susceptibility to cavitation.

Experimental work on valve flow stretches back over many years (1) and the extent is quite comprehensive. The complementary analytical study is less well developed, however. Analysis is complicated by the occurrence of flow re-circulation and jet re-attachment which make it difficult to formulate accurate models. A number of different approaches have been taken by various researchers for modelling of valve flow. Von Mises (2), for example, predicted the contraction coefficient for flow through an orifice using potential flow analysis. However this assumes inviscid flow and does not take into account re-attachment or pressure recovery occurring downstream of the vena contracta, which can have a considerable effect on the flow coefficient C_Q. It is common to predict the flow force by estimation of the fluid momentum change through the valve (3), but this relies on similar assumptions to those made by von Mises. Nevertheless, von Mises' predictions and the momentum analysis of flow force are very commonly used, though sometimes erroneously.

The above analyses are best suited to sharp-edged orifices in which little or no pressure recovery takes place. A poppet valve with a sharp-edged seat is an example of such an orifice, although discrepancies can still occur. Long orifices, in which the flow restriction extends for a finite distance along the direction of flow, can have quite different flow characteristics, as the jet downstream of the vena contracta can expand and re-attach to the walls causing pressure recovery to occur. Examples of this type of orifice are disc, plate or flapper valves and poppet valves with chamfered seats of similar angle to the poppet face. The characteristics of such valves have been examined by several researchers (1,4) and have been found to deviate considerably from those predicted by von Mises or by momentum analysis.

Several attempts have been made to model the flow through disc or flapper valves analytically. Takenaka *et al* (4) and Hagiwara (5) attempted to solve the Navier-Stokes equations for the flow between the disc and valve seat, but neglected the effects of flow separation. Thus the predictions did not correlate well with their experimental measurements, particularly at larger openings where flow separation has a greater effect. Lichtarowicz and Markland (6) and Jones (7) used a conformal transformation technique to solve the Navier-Stokes equations for separated potential flow in a disc valve. Duggins (8) extended this technique to model flow re-attachment, and obtained good agreement with experimental tests. However, these analyses require certain assumptions, including that of inviscid, irrotational flow, and are difficult to apply in all but the simplest of geometries.

The more recent development of computational fluid dynamics (CFD) techniques provides great scope for study of valve flow. The effects of complex geometry, turbulence, flow separation and reattachment can be incorporated. Early CFD codes were limited in their geometric flexibility and generally required a flow domain which could be described by a set of rectangular or wedge-shaped blocks. Thus the early attempts were aimed at simulating flow in spool valves or in disc valves which could be simply defined. Hayashi *et al* (9) simulated the flow in nozzle-flapper valves. Their results showed jet separation and re-attachment and compared well with experimental results. However, they were limited to laminar flow. Jet re-attachment in turbulent flow is a much more difficult phenomenon to predict. Pountney *et al* (10) performed a study of the turbulent flow through spool valve orifices. Some favourable comparisons were made with von Mises' results, but no comparison was made with experimental results in their paper.

Considerable research has been carried out recently on the simulation of flow in the cylinders of internal combustion engines. This is a particularly difficult problem because of the geometric complexity, time dependence of flow and geometry, and the combustion process. For these reasons, most researchers, such as Ramos & Sirignano (11) or Gosman *et al* (12), simplified the modelling of the flow through the valves and concentrated on the in-cylinder flow. However, this ignores the strong effect which the detailed design of the valve profile can have. With advances in computer speed and storage, attempts have been made to model the valve flow more accurately, but it has generally been necessary to simplify the problem, such as by assuming axi-symmetric flow (11) or by ignoring combustion.

1.1 Inaccuracies due to turbulence modelling and numerical effects

Fleming *et al* (13) published some simulations of the flow through disc valves as used in reciprocating compressors, using proprietary CFD software. They found considerable differences between the measured and simulated valve forces and suggested that these were due to the difficulty in accurately predicting the form of the separation region and the inadequacy of the turbulence models used; indeed different turbulence models were compared and found to give different results. These results can be taken as an indication that the commonly available turbulence models, such as the k-ε model, may give unreliable predictions where flow separates and reattaches. It is generally accepted that those turbulence models which employ the eddy-viscosity concept, including the k-ε formulation and most other practical formulations, do not perform very well in regions of re-circulating flow or curved streamlines, as described by Leschziner (14). More accurate models based on a Reynolds stress formulation are being developed but are complex and not generally available for incorporation into proprietary CFD software. Thus care must be taken in interpreting turbulent flow simulations, and comparisons with experimental results made wherever possible.

Further numerical inaccuracy can be caused by the form of the mesh discretisation and the differencing scheme used. Finite volume methods generally use some form of upwind differencing scheme for their approximation of convection, but there are several formulations of this each with their own weaknesses, as described by Leschziner (14). Some formulations are prone to 'numerical diffusion' or artificial smearing of the results, while some are less stable and prone to producing oscillatory results. A recent development is the finite analytic technique in which local analytical solutions to the fluid flow equations are developed for each computational cell. This technique appears to avoid the difficulties normally encountered in simulating the effects of convection. However, it is not as yet in general use. Chen *et al* (15) used this technique for simulating the unsteady turbulent flow through prosthetic heart valves. Unfortunately no experimental measurements were presented for comparison.

1.2 Outline of the present study

The work presented in this paper is part of a project to study the performance characteristics of self-acting poppet valves as used in reciprocating hydraulic pumps working with water-based fluids. The speed of such pumps is often limited by the dynamic response and pressure drop of the valves. Broadly speaking, for best operation the valves should have a low flow resistance, i.e. a low pressure drop for a particular flow rate. This is particularly important for the inlet valve, in which cavitation may also be a problem.

In this paper, the authors have described a study of incompressible, steady state, turbulent flow through a range of poppet valves. The STAR-CD program, from Computational Dynamics Ltd, was employed for the simulations. This program allows a high degree of geometric flexibility enabling complex valve geometries to be modelled. In common with most commercially available CFD software, the k-ε turbulence model is employed and this would be expected to introduce some inaccuracy, as described above.

For the numerical approximation of convection the STAR-CD code employs two alternative upwind differencing formulations, these being first and second order linear upwind differencing. The flux F_ϕ of the scalar variable ϕ through a cell face is described by the equation

$$F_\phi = \dot{m}[\phi_U + \gamma(\phi_U - \phi_{UU})f]$$

First order upwind differencing ($\gamma=0$) is recognised to be relatively stable but prone to errors due to numerical diffusion. Second order upwind differencing ($\gamma=1$) is more prone to instability or convergence problems. Oscillations in the solution may result because of unboundedness due to the negative coefficient of ϕ_{UU}. However it is less prone to numerical diffusion. STAR-CD permits blending of the two schemes, in that γ can be set between 0 and 1. The value of γ is constant over the whole flow domain.

A programme of experimental tests was carried out concurrently with the CFD simulations, in order that the simulations could be compared with measurements. The experimental tests were performed by passing a steady flow of water through an axi-symmetric model of a poppet or disc valve. A variety of different valve designs could be tested. Dimensions of the valve chambers and some of the poppets tested are shown in Fig 1. Flowrate, upstream and downstream pressures, valve lift and valve thrust were measured, from which C_Q and C_F could be calculated. The upstream and downstream chambers were constructed from transparent perspex in order that the flow could be visualised. A d.c. voltage was applied to two electrodes upstream of the valve in order to produce bubbles of hydrogen and oxygen by electrolysis. Using suitable slit illumination, the paths of these bubbles could be captured photographically, by which means the flow could be visualised. The test rig and experimental work are described in more detail by Johnston et al (16).

The valve types for which results are described in this paper are shown in Fig 1. Valve 'A' is a conical poppet with a sharp-edged seat. Valve 'B' is a similar poppet with a lip on the rear edge of the poppet. Valve 'C' is a conical poppet with a chamfered seat. Valve 'D' is identical to valve 'C' except that the inner edge of the chamfer is curved. In all cases flow passes through the valves in an outward direction.

2 SIMULATION OF A CONICAL POPPET VALVE WITH A SHARP EDGED SEAT

A series of simulations was performed on a poppet valve with a sharp-edged seat (valve 'A'). The results were compared with experimental measurements in the form of the flow coefficient C_Q and force coefficient C_F, and experimentally visualised flow patterns.

Axi-symmetric flow was assumed in the simulations, which employed a narrow wedge-shaped mesh composed of a single layer of cells. Several computational mesh arrangements were investigated in order to ascertain the sensitivity of the results to the geometric discretisation. The meshes were designed to give maximum concentration of cells in the crucial region around the orifice, whilst avoiding excessive distortion of individual cells. Upstream and downstream of the orifice, the density of the mesh was reduced as the gradients of the flow variables are generally less and the flow in these regions is of lesser interest. This arrangement ensures maximum accuracy with good computational efficiency. One such arrangement is shown in Fig 2. It was found that increasing the mesh density did not affect the results significantly; hence this density was considered adequate.

Simulations were non-dimensionalised by setting the inlet diameter d_0, inlet velocity u_{in} and density ρ to unity. By consideration of the simulated pressure drop across the valve, the value of C_Q was calculated. The force acting on the poppet was calculated by integrating the pressure distribution over its faces, from which C_F was calculated. Shear forces on the poppet were ignored in the computation of the force as they were negligible.

Table 1 lists some details of the simulation conditions; these details also apply to simulations described later.

Table 1: Simulation conditions

no. of nodes	typically 60×25 (i direction × j direction)
variables solved for	$u, v, p, k, \varepsilon, \mu_t$
density	$\rho = 1.0$
laminar viscosity	$\mu = 10^{-4}$
Reynolds number	$R = \rho u_{in} d_0/\mu = 10\,000$
inlet boundary	$u_{in} = 1.0, v_{in} = 0.0$
outlet boundary	constant pressure: $p_{out}=0.0$
solution algorithm	SIMPLE (17)
under-relaxation factors	$u, v, k, \varepsilon, \mu_t$: 0.7 p: 0.1
empirical coefficients of turbulence equations	default values as described in STAR-CD manual (18)

A simulated flow pattern is shown in Fig 3(a), in which the velocities at alternate nodes are shown. The equivalent visualised flow is shown in Fig 3(b). Correlation between simulation and experiment is good. The simulated jet contracts downstream of the orifice, then travels at the angle of the poppet face. It is deflected by the side wall and two re-circulation zones are formed on either side of the jet. A smaller, less intense re-circulation zone occurs behind the rear face of the poppet. Because of turbulence it was not possible to obtain numerical measurements of the experimental velocity field using this flow visualisation technique; laser-Doppler anemometry would be more suited for that purpose.

The predicted value of C_Q, obtained from a series of simulations with different mesh configurations and values of γ, ranged from 0.75 to 0.81 as compared to the experimental value of 0.79. The predicted value of C_F varied from 0.75 to 0.83 compared to the experimental value of 0.83. Thus good correlation between experiment and prediction was observed.

These results show good agreement with von Mises, who predicted $C_Q = 0.77$, and with momentum theory, which predicts $C_F = 0.79$. This agreement is due to the valve behaving as a sharp-edged orifice and little pressure recovery taking place. The experimental tests were performed with water and the Reynolds number was between 10^4 and 10^5. Little change in C_Q and C_F was observed experimentally over this range of Reynolds number. Correspondingly, simulations were performed with $R=10^4$ and $R=10^5$, and the differences were negligible.

For more detailed comparison it would be desirable to measure the pressure experimentally at various points within the valve. However with the available apparatus it was not practical to provide the necessary pressure tappings. Velocity measurements are currently being taken using laser-Doppler anemometry, and these should provide a useful comparison with the simulated velocity and turbulence energy fields.

3 FLOW FORCE COMPENSATED POPPET VALVE

The flow force acting on a poppet valve with a sharp-edged seat is generally accepted to be related to the change in momentum flux in the axial direction as the fluid passes through the valve, as described in section 1 (3). Stone (19) reasoned that this could be reduced to a minimum by directing the jet radially using a rim on the rear edge of the poppet. Tests performed by Stone on such a valve indicated that this was indeed the case and the flow force was greatly reduced. Tests performed by Johnston et al (16) on a similar valve (valve 'B') drew the same conclusion provided that the opening was small; however at a larger opening a discontinuity occurred in the values of C_Q and C_F as shown in Fig 4. A flow discontinuity was observed (16) when the opening Z increased past 0.1. For $Z < 0.1$, the jet re-attached to the valve seat, and for $Z > 0.1$ the jet did not re-attach to the seat but travelled in a radial direction until it struck the side wall. Such an effect was observed in the simulations, as shown in Fig 5, and the nature of the predicted flow patterns agrees well with the observed ones. It was difficult to ascertain the exact opening at which transition between the two simulated flow patterns occurred as convergent simulations could not be obtained near to this point. However the transition in the simulated flow was observed to occur at approximately the same opening as in the experimental tests.

The predicted values of C_Q and C_F are shown with the experimental measurements in Fig 4. The values of C_Q are in good agreement, whereas the values of C_F show some deviation but similar trends are exhibited.

4 POPPET VALVE WITH CHAMFERED SEAT

A series of experimental tests was performed on a range of different poppet valves with chamfered seats. The aim was to design a valve which would provide low flow resistance (i.e. high C_Q) over a wide range of openings and which would not be overly prone to cavitation. For comparison, CFD simulations were performed on two of the valve geometries. These were a 60 degree conical poppet with a chamfered seat (valve 'C'), and a similar valve with a radius on the upstream edge of the chamfer (valve 'D'). The details of these valves are shown on Fig 1.

Experimental measurements of C_Q and C_F for valve 'C' are shown in Fig 6 (theoretical results are also shown and are discussed later). For $Z < 0.04$, C_Q is approximately 1.2. This contrasts with a value of approximately 0.7 for the equivalent valve with a sharp-edged seat, indicating a reduction in pressure drop of about 3:1 for a given flowrate and opening. At larger openings, C_Q drops to a steady value of about 0.8, which is close to the measured value for the valve with a sharp-edged seat. C_Q and C_F exhibit a discontinuity at $Z = 0.04 - 0.05$, and this was observed to correspond to a change in the flow pattern; for $Z < 0.04$, the jet attached to the seat, and for $Z > 0.05$, the jet detached and travelled at the angle of the poppet face. C_F shows a minimum at $Z = 0.06$ and rises with increasing opening. Clearly this valve has the desired low resistance to flow at small openings, but is less effective at larger openings.

The equivalent experimental measurements for valve 'D' are shown in Fig 7. It is apparent that large values of C_Q are maintained for a wider range of openings, with C_Q rising to a maximum of 1.5 for $0.125 < Z < 0.145$. For $Z > 0.145$, C_Q drops very sharply. Both C_Q and C_F exhibit discontinuities at $Z = 0.1$. This was observed to be due to a similar change in flow pattern to that of valve 'C'. It is clear then that the curved edge of the chamfer causes significant improvements in the flow performance.

A typical simulated flow pattern for valve 'C' is shown in Fig 8. The simulated flow pattern downstream of the valve corresponds well to the observed flow pattern.

Fig 9(a) shows the flow within the restriction of valve 'C' with $Z = 0.05$. It can be seen that a small separation region occurs on the chamfer on the valve seat but the flow subsequently re-attaches. Due to this re-attachment, pressure recovery takes place, thus explaining the large value of C_Q. Note that the jet, on leaving the restriction, curves and attaches to the seat, an effect observed experimentally at small openings. A similar flow pattern occurs for valve 'D' as shown in Fig 9(b), but in this case no separation occurs in the restriction because the flow is able to follow the smooth profile. At larger openings, Figs 9(c-d), the jet can be seen to detach completely from the chamfer on the valve seat. Thus little or no pressure recovery takes place and C_Q is reduced. As with the experimental observations, with small openings the jet issuing from the orifice curves and attaches itself to the valve seat, but at larger openings it travels in a straight line at the angle of the poppet face. However, the opening at which the transition occurred varied and was not always consistent with the experimentally observed transition.

Values of C_Q and C_F calculated from simulation results are shown on Figs 6 and 7. Some degree of scatter is apparent between the experimental and simulated results. Particularly for valve 'D' (Fig 7), it can be seen that the value of γ employed has a considerable effect on the results, and the best correlation is generally achieved using second order upwind differencing ($\gamma = 1$). However, correlation is less satisfactory in the cases where the simulated jet attaches to the valve seat after leaving the orifice. It is thought that errors in the prediction of the small recirculation zone on the valve seat may affect the pressure distribution within the orifice and hence C_F and to a lesser extent C_Q.

The errors incurred from use of first order upwind differencing can be attributed to numerical diffusion. Some idea of the scale of this effect can be obtained by examining the influence of turbulence on the results. Simulations were performed with a Reynolds number of 10^4 but assuming laminar flow. With $\gamma = 0$, the results were very similar to those for turbulent flow. Conversely, with $\gamma = 1$, realistic results could not be obtained. This indicates that, for $\gamma = 0$, numerical diffusion masks the effect of turbulent dissipation.

With turbulent flow, the value of γ was found to have a very strong effect on the turbulence parameters k, ε and μ_t. In the case of $\gamma = 1$, and to a lesser extent $\gamma = 0.5$, the numerical scheme predicts negative values of k and ε in a small region upstream of the orifice. These values are adjusted to a small positive value by the numerical solver as they have no physical meaning, but this itself prevents complete convergence, and causes unrealistically high values of μ_t. The negative values are a result of the oscillatory behaviour caused by the unboundedness of this scheme and tend to manifest themselves at the lower edges of high-gradient regions, as reported by Leschziner (14). Relatively long simulation times were necessary to reduce the residual errors to acceptable values.

The most reliable results overall were generally obtained using a hybrid differencing scheme with $\gamma = 0.5$; this is essentially a compromise between accuracy, simulation time and convergence. With the hybrid scheme the numerical oscillations seem to be largely suppressed, resulting in a reasonable rate of convergence, and better agreement with experimental results is achieved than with $\gamma = 0$. Perhaps improved accuracy would be obtained with the finite analytic technique (15), but this technique is complex and not at present implemented in any commercially available CFD program.

5 CONCLUSIONS

A series of simulations has been performed using a finite volume technique to predict flow through poppet valves. A range of valve types was studied, and results were compared with experimentally visualised flow patterns and with measured flow and force characteristics. The overall form of the predicted flow patterns compares well with the experimentally visualised flow, and the path of the jet and the form of the recirculation zones correlate well.

Predicted results for a poppet valve with a sharp-edged seat compare well with experimental results. A similar poppet valve with a lip on its outer edge has the effect of minimising the flow force at small openings, but a change in the flow pattern occurs at larger openings, accompanied by an increase in the magnitude of the flow force. This change was observed both experimentally and in simulated results.

Flow in long orifices with re-attachment appears to be rather more difficult to simulate precisely. However the simulations agree qualitatively with experimental results.

Discrepancies can be attributed to the following factors:

1) Shortcomings in the turbulence model. The common k-ε model is known to give rise to significant errors in situations with strong body forces due to streamline curvature and recirculation. Thus the complex flow in a valve is a severe test for the turbulence model. Furthermore the flow-pressure characteristics of valves tend to be very sensitive to flow separation and re-attachment. A more precise turbulence model based on the Reynolds stresses would be expected to give more accurate results.

2) Errors due to discretisation and numerical differencing. The first order upwind differencing scheme is well known to cause errors when high shear rates occur with streamlines skewed relative to grid lines and with insufficiently fine meshes. These conditions are difficult to avoid in valve flow because of the complex geometry and re-circulatory flow and the high velocity, pressure and turbulence gradients. The second order linear upwind differencing scheme is less diffusive but prone to instability and oscillatory behaviour. A hybrid scheme was found to give best results.

A better understanding of the results would be obtained if detailed flow field measurements were available for comparison. A valve model is currently being instrumented for velocity and turbulence energy measurements using laser-Doppler anemometry, and the results will be published at a later date.

It is clear from the experimental and simulation results that the small-scale features of the geometry of a valve can have a disproportionate effect on the overall flow field, and the pressure distribution may be particularly strongly affected. Re-circulation is an important feature of flow through valves, but it appears that the flow characteristics are less sensitive to the large-scale re-circulation which occurs downstream of the valve than to the small-scale but intense re-circulation which may occur in the orifice region. In this region the high velocities and streamline curvatures cause high body forces on the fluid. Account of this sensitivity to the valve geometry should thus be taken when one attempts to model flow in the cylinder of an internal combustion engine, for example, in which errors may be incurred by an overly coarse representation of the valve region.

6 ACKNOWLEDGEMENTS

The authors would like to acknowledge the support of the Science and Engineering Research Council for providing the necessary funding for this project. Thanks are also due to the Ford Motor Company Ltd for provision of the STAR-CD program, and David Foulkes of Ford for his liaison work with Computational Dynamics Ltd.

7 REFERENCES

(1) SCHRENK, E. Disc valves, flow patterns, resistance and loading (Translation from German), BHRA publication T547, 1957

(2) VON MISES, R. The calculation of flow coefficient for nozzle and orifice, *V.D.A.*, 61, no. 21,22,23, 1917

(3) McCLOY, D., MARTIN, H.R. *Control of fluid power - analysis and design*, Ellis Horwood, Wiley, 1980

(4) TAKENAKA, T., YAMANE, R., IWAMIZU, T. Thrust of the disc valves, *Bulletin of JSME*, Vol. 7, No. 27, 1964, 559

(5) HAGIWARA, T. Studies on the characteristics of radial-flow nozzles (1st report, theoretical analysis of outward flow), *Bulletin of JSME*, Vol 5, No. 20, 1962, 656

(6) LICHTAROWICZ, A., MARKLAND, E. Calculation of potential flow with separation in a right-angled elbow with unequal branches, *Journal of Fluid Mechanics.*, Vol 17, part 4, 1963, 596

(7) JONES, J.E. Potential flow from a tapering nozzle impinging on a flat plate, *Trans. ASME, Journal of Fluids Engineering*, Vol. 106, March 1984, 54

(8) DUGGINS, R.K. Further studies of flow in a flapper valve, 3rd International Fluid Power Symposium, Turin, Italy, 1973, B2-25

(9) HAYASHI, S., MATSUI, T., ITO, T. Study of flow and thrust in nozzle-flapper valves, *Trans. ASME, Journal of Fluids Engineering*, March 1975, 39

(10) POUNTNEY, D.C., WESTON, W., BANIEGHBAL, M.R. A numerical study of turbulent flow characteristics of servo-valve orifices, *Proc. Instn. Mech. Engrs*, Vol. 203, Part A, 1989, 139

(11) RAMOS, J.I., SIRIGNANO, W.A. Axisymmetric flow model in a piston-cylinder arrangement with detailed analysis of the valve region, SAE paper 800286, 1980

(12) GOSMAN, A.D., MELLING, A., WHITELAW, J.H., WATKINS, P. Axisymmetric flow in a motored reciprocating engine, *Proc. Instn. Mech. Engrs*, Vol. 192, 1978, 213

(13) FLEMING, J.S., TRAMSCHEK, A.B., ABDUL HASSAN, J.M.H. A theoretical and experimental investigation of the flow of gas through reciprocating compressor valves, Instn Mech. Engrs, C390/011, 1989, 117

(14) LESCHZINER, M.A. Modeling turbulent recirculating flows by finite-volume methods - current status and future directions, *Int J. Heat and Fluid Flow*, vol 10, no 3, Sept. 1989, 186

(15) CHEN, C.J., YU, C.H., CHANDRAN, K.B. Finite analytic numerical method and fluid dynamics of disc type valves, ASME Winter Annual Meeting, Chicago, 1988, 347

(16) JOHNSTON, D.N., EDGE, K.A., VAUGHAN, N.D Experimental investigation of flow and force characteristics of hydraulic poppet and disc valves, Submitted for publication, *Proc. Instn Mech. Engrs*

(17) PATANKAR, S.V. *Numerical heat transfer and fluid flow*, Hemisphere, McGraw-Hill, New York, 1980

(18) STAR-CD manual, Computational Dynamics Ltd, 1989

(19) STONE, J.A. *Discharge coefficients and steady state flow forces for hydraulic poppet valves*, Trans. ASME, Journal of Basic Eng., March 1960, 144

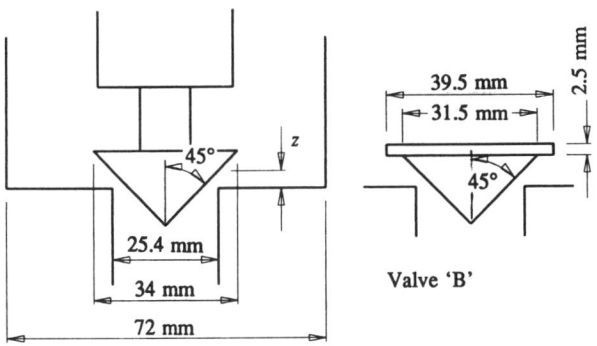

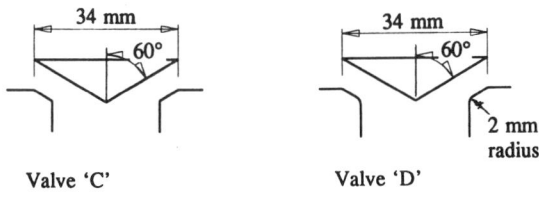

Fig 1 Shapes of poppet valves

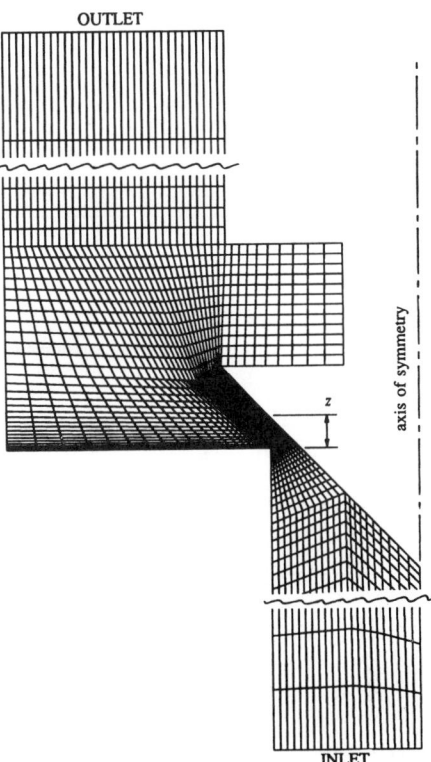

Fig 2 Mesh geometry, valve 'A'

Fig 3 Flow pattern, valve 'A'

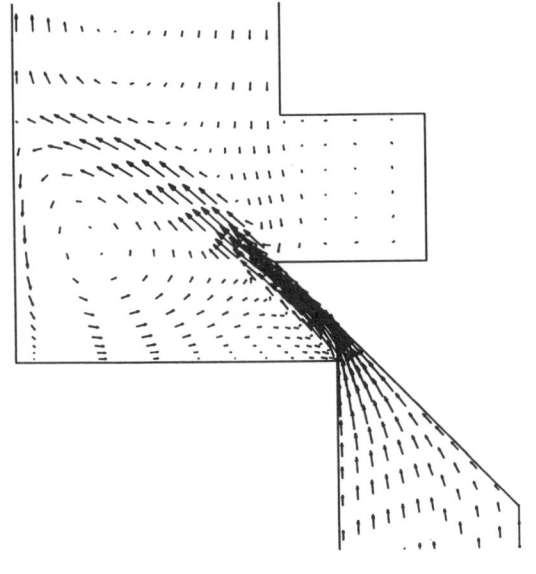

(a) simulated velocity vectors

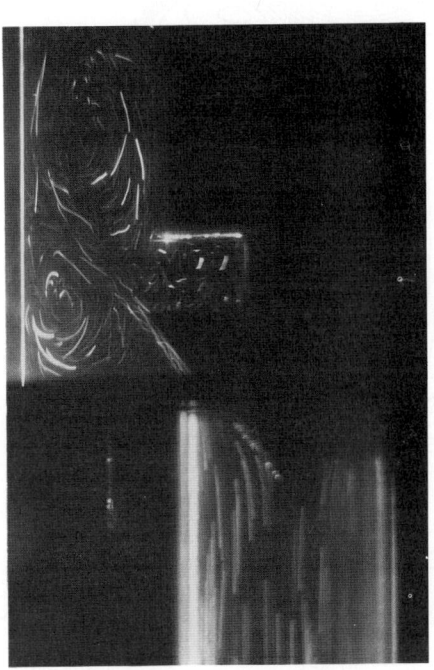

(b) experimentally visualised

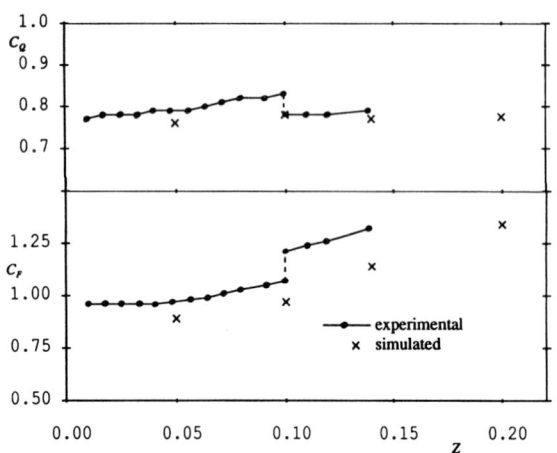

Fig 4 Flow and force coefficients, valve 'B'

Fig 5 Simulated streamlines, valve 'B'

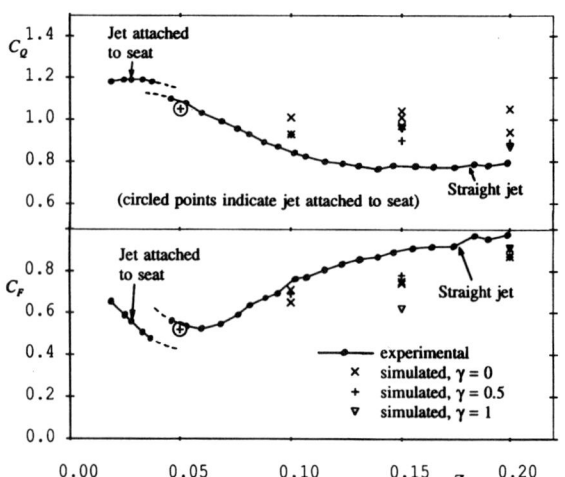

Fig 6 Flow and force coefficients, valve 'C'

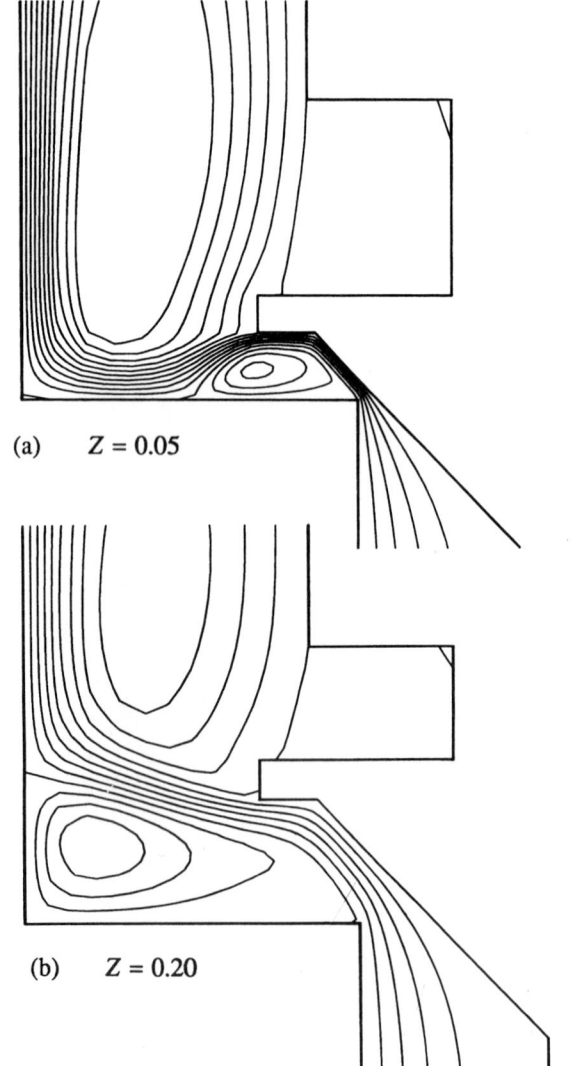

(a) Z = 0.05

(b) Z = 0.20

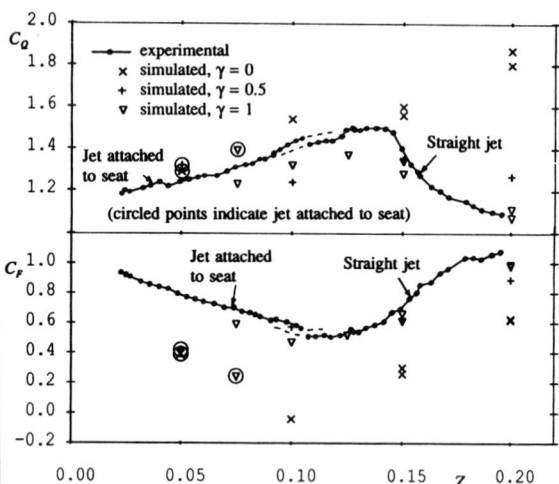

Fig 7 Flow and force coefficients, valve 'D'

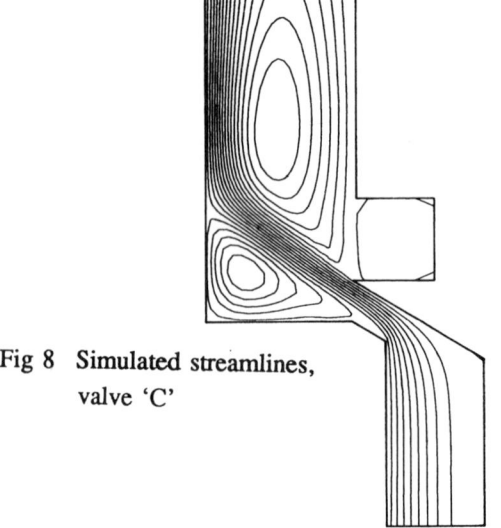

Fig 8 Simulated streamlines, valve 'C'

126

© IMechE 1991 C413/079

Fig 9 Simulated velocity vectors in orifice

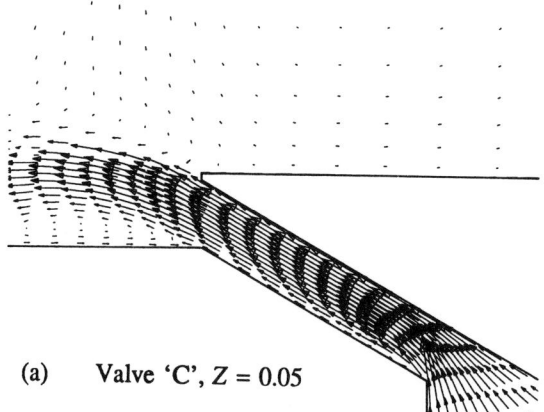

(a)   Valve 'C', $Z = 0.05$

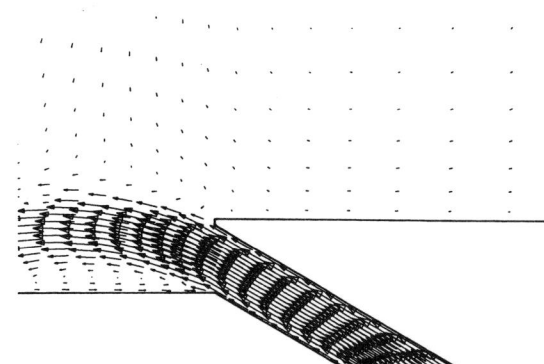

(b)   Valve 'D', $Z = 0.05$

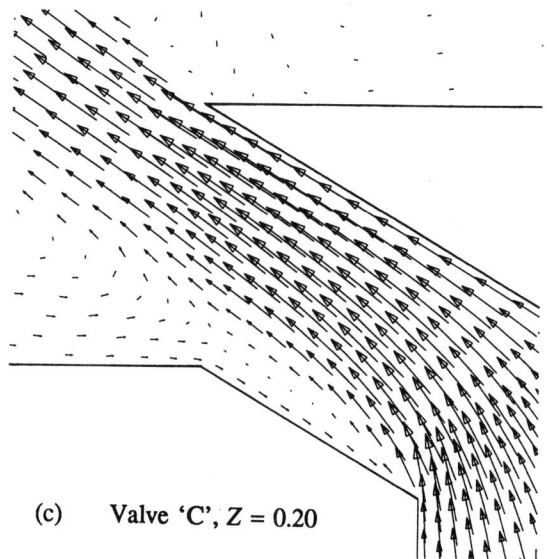

(c)   Valve 'C', $Z = 0.20$

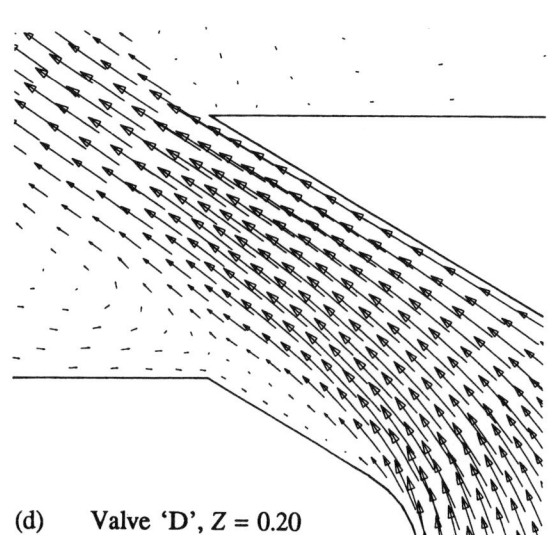

(d)   Valve 'D', $Z = 0.20$

C413/023
Some developments in the application of computational fluid dynamics to flows in internal combustion engine cylinders

A P WATKINS, BA, MSc, PhD, DIC, MSAE, C DIOMATARIS, BSc and M Z GÜL, BSc, MSc,
Department of Mechanical Engineering,
UMIST, Manchester
H KHALEGHI, BSc, MSc, PhD, MAIAA, Tarbiat Modarres University,
Tehran, Iran
C J LEA, BSc, MSc, AMIMechE and D M WANG, BSc, MSc
Department of Mechanical Engineering, UMIST, Manchester

SYNOPSIS A number of sub-models for certain aspects of the calculation of fluid flows in ic engines are described and their performances are assessed against experimental data and earlier models. The investigations involve two models for the turbulence, namely a differential stress model, and a three equation eddy-viscosity model, a second order accurate spatial discretisation scheme and a diesel spray wall impaction model with gaseous crossflow. It is concluded that all these sub-models are capable of predicting flows better than those models currently employed in computer codes for ic engine applications. A code for calculating flows in dual-inlet-valve engines is also reported on.

1 INTRODUCTION

Fluid flow within ic engine cylinders is extremely difficult to analyse numerically or experimentally. The flow is transient, due to the periodic motions of piston and valves, and is highly three-dimensional in space. This latter is usually because of a combination of geometric constraints, such as the off-centre positions of inlet and exhaust valves, piston bowl, spark plug or fuel injector.

The nature of the flow varies markedly through an engine cycle. Thus high values of Reynolds and Mach numbers pertain during inlet flow, squish flow and around direct injection fuel sprays. Whereas during compression and expansion those numbers will be many orders of magnitude less. The fluid pressure varies from around one bar during intake to perhaps 70 bar during combustion. Likewise the fluid temperature may vary between 300k and 2000k. For most of the engine cycle the flow is definitely turbulent in nature, making analysis very much more difficult, than if it were laminar flow. These difficulties apply not only to numerical analysis but also to experimental investigations. For example, cycle to cycle variations make extraction of an experimental value of turbulence intensity open to many interpretations and poses a problem for both the experimenter analysing his results and the numerical modeller using the results to assess model performances.

The combination of complex geometry, with transient, turbulent flow undergoing a variety of physical processes makes the task of accurate analysis probably beyond the capabilities of today's supercomputers. To achieve accurate results in space, numerical grids are required which divide the engine chamber into hundreds of thousands of control volumes. Storage requirements are therefore extremely high. For fast calculations these stored numbers need to be readily accessible. For temporal accuracy, calculations must be performed over small crank angle intervals. This interval depends both on the nature of the solution method employed, ie whether explicit or implicit, and also on the time scales of the flow phenomena being calculated. For example much smaller time scales are required to accurately calculate spray flows or combustion flame fronts than are required to calculate flow during compression. The central processing times of computer codes calculating three-dimensional ic engine flows are therefore extremely long.

The constructors of some computer codes have sought to build codes which are of such a geometric capability that virtually any engine shape can be accommodated, and all the relevant physical phenomena solved for. Examples are the KIVA (1) code and the GM-FIRE (2) code.

The philosophy of the ic engine computational fluid dynamics (CFD) group at UMIST has been somewhat different. We seek to improve the various building blocks or sub-models within computer codes, by addressing the physical and numerical bases of the models. In order to calculate complex flows in reasonable computing times, models are almost invariably introduced, at either the physical level (in particular, the turbulence model) and/or at the numerical level (eg low-order-accurate spatial discretisation schemes), which stabilise the calculations. The penalty paid is that accuracy is lost. We are attempting to remove the necessity for these artifices. The progress made is outlined in section four of the paper. Before that the mathematical model is outlined so that the work on turbulence modelling can be placed in context. This is followed by an outline of the finite-volume procedure adopted to solve these equations, and it is here that the work on spatial discretisation schemes is rooted. In section four, as well as the work cited above, the progress made in developing a new sub-model to account for the impaction of fuel spray on the chamber wall is discussed. This is a problem which is now receiving more attention because of the development of small high-speed direct-injection diesel engines for passenger car use. Due to the small size of the engine chambers, the fuel spray will usually impact on the chamber wall before evaporating. This is despite the fact that the air flow in such engines is highly swirled to enhance air/fuel vapour mixing. Efforts to model the combined effects of swirl and impaction form the major part of the work described here.

Not all our work is concerned with the basic sub-models as described above. Some work is directed more towards the design of engines. Section five therefore contains a brief account of progress made in developing a computer code which allows some aspects of multiple-valve intake flows to be analysed.

2 MATHEMATICAL MODELS

Computational fluid dynamics attempts to solve the partial differential conservation equations which govern the fluid motion. At the very minimum, these consist of equations for the conservation of mass and momentum. In ic engine flows, additional equations are required for the conservation of energy, and, because the flow is turbulent, some form of model to account for the effects of turbulence. Additional equations for the conservation of liquid mass, momentum and energy, and fuel vapour mass, are also required if the motion and evaporation of liquid fuel sprays are to be calculated. For the calculation of combustion, at the very least, two additional conservation equations for oxidant and fuel vapour are needed. To calculate emissions a further set of conservation equations for the species are required.

Thus for a full three-dimensional calculation of all the processes in a direct injection diesel engine, for example, up to 20 such partial differential conservation equations must be solved. In this paper, combustion will not be considered, but all the other processes will be covered.

The forms taken by the Eulerian conservation equations depend on the particular coordinate system employed. Here a variety of such systems are used. Thus for the turbulence model and spatial discretization studies a simple two-dimensional axisymmetric cylindrical polar coordinate system is used. For the spray wall impaction/swirl studies, which use experimental data obtained from a rectangular test rig rather than an operating engine, a three-dimensional Cartesian grid is employed. Finally, for the multiple-inlet valve studies an orthogonal curvilinear coordinate system is used. This will be described in more detail later. A very full description of this system and the resulting forms of the conservation equations are given by Ahmadi-Befrui et al (3).

The conservation equations are most compactly expressed in Cartesian tensor notation. A general form of the equations is then

$$\frac{\partial(\rho\theta\Phi)}{\partial t} + \frac{\partial(\rho\theta U_j\Phi)}{\partial x_j} = \frac{\partial}{\partial x_j}\left(\theta\Gamma\frac{\partial\Phi}{\partial x_j}\right) + S, \qquad (1)$$

where Φ is the dependent variable and can stand for 1 (mass), U_i (momentum), e_s (stagnation energy), f (fuel vapour mass fraction), etc. The gas density is denoted by ρ. If a second, liquid phase is present, the fraction of a volume occupied by the gas is termed the void fraction and is denoted by θ. Thus in a volume V, $\rho\theta V$ is the mass of gas present. The conservation equations vary between each other mainly through the diffusivities Γ, and the source terms S. Reference (3) gives full details of these. We will discuss here only those which pertain to the present studies.

2.1 Turbulence models

For steady flows the instantaneous flow equations are averaged over time, to obtain the time average or mean value $\bar{U}_i$ of the velocity components. For reciprocating flows, an ensemble average is taken over many cycles. In the momentum equations there results a source term in the form

$$-\frac{\partial}{\partial x_j}(\rho\overline{u_i u_j}) \quad \text{(the Reynolds stresses)}$$

here, u_i and u_j are turbulent components of the instantaneous velocities, ie $U_i = \bar{U}_i + u_i$. These turbulent correlations are unknown, so to provide a closed set of equations, they must be modelled in terms of known or calculable quantities.

The majority of ic engine codes which use any turbulence model at all (many rely on some false viscosity method which bears little or no relation to the physical phenomena involved), employ the $k - \epsilon$ turbulence model (4). Here k is the turbulence kinetic energy and ϵ is its rate of dissipation. This model is based on the eddy viscosity concept,

$$\mu_t = \rho c_\mu \frac{k^2}{\epsilon}, \qquad (2)$$

where μ_t is a turbulent viscosity, c_μ is a constant and k and ϵ are calculated from semi-empirical transport equations having forms similar to equation 1. The Reynolds stresses are then found by application of the Boussinesq hypothesis

$$-\rho\overline{u_i u_j} = \mu_t\left(\frac{\partial\bar{U}_i}{\partial x_j} + \frac{\partial\bar{U}_j}{\partial x_i}\right) - \frac{2}{3}\rho\delta_{ij}k \qquad (3)$$

The $k - \epsilon$ model was originally developed for the calculation of thin shear flows. It does not perform well when attempting to calculate recirculating flows, as in ic engine cylinders. Nor does it allow anisotropy of turbulence, ie the value of μ_t is the same for all combinations of u_i and u_j. The Boussinesq hypothesis does not give the correct behaviour of turbulence to secondary strains, eg streamline curvature.

The differential stress model (DSM) (5) seeks to model the exact transport equations for the Reynolds stresses. There are then up to six partial differential equations of the form of equation 1 for the Reynolds stresses, plus one for ϵ. The Reynolds stress source terms in the mean flow momentum equations are then given.

Lying between the $k - \epsilon$ model and the DSM in complexity is the algebraic stress model (ASM) (6). The convective and diffusive transport of the Reynolds stresses are related algebraically to that of k. The DSM transport equations then reduce to algebraic relations and only transport equations for k and ϵ are therefore required to be solved. It has been demonstrated eg Fu et al (7) that the DSM has the most comprehensive applicability to different flows.

In section four of this paper, the application of these various models to ic engine flows will be discussed and assessed.

A further turbulence model will also be assessed in section four. This is a new model based on the $k - \epsilon$ model, but which employs a further transport equation for a second turbulence time scale τ. The $k - \epsilon - \tau$ model was first developed by Wu et al (8) who recognised that in highly compressed flow, when the compression is fast, the normal turbulence dissipation time scale, given by k/ϵ, is not adequate to describe the time scale of the energy-bearing eddies. This model gave much superior predictive performance in compressed flow situations than variants of the $k - \epsilon$ model in which the ϵ equation is adjusted to account for compression. (see, Ahmadi-Befrui and Gosman (9) for descriptions of these models).

Unfortunately, Wu et al only tested their model for homogeneous turbulence. At UMIST, we have been attempting to extend the model to accurately predict shear flows, as well as compression flows. This must be done to allow the model to be used for the calculation of flows throughout all strokes of ic engine cycles. The progress made in this direction is described in section four.

2.2 Liquid phase equations

The fuel spray in a direct injection diesel engine can be solved analytically in a number of ways. Traditionally the spray has been assumed to vaporise rapidly, thus allowing the assumption of a gas jet. However this is not a tenable assumption, particularly in small diesels in which the fuel is still liquid when it reaches the chamber walls. Alternative approaches are required. One is to assume the spray is another continuum, like the gas phase. Eulerian conservation equations can then be written to describe the spray motion and evaporation. However it is very apparent that sprays in engines are not continuous, for much of the travel. Rather they consist of many millions of discrete droplets, once atomisation has taken place. Near the injector however the spray consists of a liquid core surrounded by a cloud of ligaments and droplets. The discrete droplet model (DDM) is adopted here, and is assumed to apply throughout the spray.

To reduce computer storage, only representative drops are calculated for. Each drop then represents many thousands of drops each having the same size, temperature, velocity components, etc. For each drop, Lagrangian ordinary differential equations can be written to describe the droplet position, velocity components, energy, and mass. These equations will not be given in detail here. As with the gas phase, their actual forms depend on the coordinate system employed. Watkins (10) describes the equations in great detail in a general orthogonal curvilinear coordinate system. In this study second phase equations are used only in a three-dimensional Cartesian grid computer code.

Full interactions between the phases are accounted for in terms of mass, momentum and energy exchanges. Within the droplet phase, sub-models account for collision and coalescence, and secondary droplet break-up.

Recently a new sub-model has been developed by Watkins and Wang (11) to describe the impaction of spray on a wall. The model uses experimental evidence to determine the fate of droplets as they approach the wall. Depending on the Weber number, the drops either rebound with a loss of energy, or they attach to the surface whilst shattering into a number of smaller drops. In either case the drops are then free to move along the surface of the wall. This basic model was found not to accord with experimental evidence of diesel sprays impacting on walls. In particular the dispersion of the drops into the gas after impact was not adequately modelled. To overcome this the collision model of O'Rourke and Bracco (12), which already formed part of the code, was extended to allow preferential movement of drops away from areas of high droplet concentrations. This had the desired effect of dispersing the drops away from the wall after impact.

In section four of this paper this sub-model is further assessed by examining its behaviour when the spray is subjected to a gaseous cross-flow as well as impaction on a wall.

3 SOLUTION METHODS

The gas phase conservation equations are solved by finite-volume means. A computational grid is generated in two or three dimensions to cover the cylinder or rig geometry. The volumes formed by the intersections of the grids are the control volumes or cells over which the conservation equations are solved. Just as for the coordinate systems, the grids can be of varying complexity, ranging from straight lines in the case of Cartesian grids, to orthogonal curvilinear. They can be generated analytically, as in the Cartesian case, or numerically. Watkins and Diomataris (13) describe the problems encountered and solutions found, when generating numerically the orthogonal curvilinear grid used here in section five for the multi-inlet-valve flow calculations.

The general transport equ 1 is integrated over each of the control volumes and a time (or crank angle) step to form sets of algebraic equations. In this process a number of assumptions and approximations must be made, due to the fact that values of the dependent variables, and auxiliary variables such as density and viscosity, are only available at finite locations within the grid, and at finite intervals in time.

In ic engine codes, for the most part, the temporal differencing used is either Euler explicit or implicit. The former has the advantage that extremely simple algebraic equations result, so that no specific solution algorithm need be devised. It has the major drawback however that the time step must be very small to avoid the algebraic equations becoming unstable. The Euler implicit method, on the other hand, has no such stability restriction, but the equation sets now become coupled both within a set for a given dependent variable and also between sets for different dependent variables. A solution algorithm is then required for the coupled sets. The choice of a good algorithm can greatly influence the computational time taken by the code to achieve convergent solutions of the algebraic equations.

The choice of the Euler implicit method appears to offer a reasonable compromise between the desires for using larger time steps on the one hand, and not introducing very complicated solution schemes on the other. It is the choice made here for the time dependent problems. The Euler implicit method is first order accurate in time step. A second order accurate method such as the alternating direction implicit (ADI) scheme could be implemented relatively simply within the current framework, and would allow larger time steps to be used. Higher order methods become much too complicated, would require large scale changes in the computer code, and run the risk of introducing spurious solutions.

For the spatial discretisation of the diffusion terms in the transport equations, a second order accurate central-differencing formulation is invariably used. This is stable and accurate when coupled with Euler implicit methods. The discretisation of the convective terms is more difficult. A number of schemes have been implemented in ic engine flow codes. The most popular is the HYBRID scheme of Spalding (14). Another scheme which has also been used is the power law differencing scheme (PLDS) of Patankar (15). Both these schemes employ second order accurate central differencing at low cell Peclet numbers (Pe $\equiv \rho U \delta x / \Gamma$, where δx is a typical grid spacing), and the first order accurate upwind differencing scheme at higher values. The essential difference lies in the Pe value at which the change-over takes place. For HYBRID this value is 2, for PLDS it is 10. For the turbulent flows considered here, Peclet numbers are almost always larger than ten, and so the two schemes become effectively the same in these applications, reducing to upwind differencing.

The major problem with both these schemes is that because the upwind scheme is only first order accurate, the errors in the solutions are second order. This is the same order as diffusion. Thus the errors act like false or numerical diffusion. In certain circumstances, particularly when the flow is crossing grid lines at large angles (up to 45°), the false diffusion can be many times larger than the physical diffusion, even when this is the turbulent value. As a result, for example, sharp gradients are smoothed out much faster than they should be.

To combat false diffusion, a number of higher order accurate discretisation schemes have been developed. One such is the QUICK scheme of Leonard (16). This is based on the use of an upwind-shifted parabolic interpolation for each control volume face. It is free from false diffusion, but is unbounded, ie values may be generated which lie outside known physical bounds. This can be alleviated by, for example, blending QUICK interpolation with, say, PLDS. Because the true diffusivity is used and because of the unboundedness problem, the algebraic equation sets obtained using the QUICK scheme can be difficult to converge. This leads to instability problems and greater use of computer time than for the same flow configuration solved using either HYBRID or PLDS schemes. Currently then, in most uses of the QUICK scheme, it is restricted to the mean flow momentum equations only. All other equations are solved using either the HYBRID or PLDS scheme.

The QUICK scheme has hardly been used in ic engine flow calculations. Where it has been used, by El Tahry (17) and Vayonitis (18), no clear presumption could be made in its favour over the other schemes. For a steady flow past a valve into an open-ended cylinder however, Watkins et al (19) found a great improvement when using QUICK in place of either HYBRID or PLDS. We shall return to this matter in section four.

The Lagrangian ordinary differential droplet equations can also be integrated and written in Euler explicit or implicit algebraic forms. For the same reasons as cited above for the gas phase equations we choose here to write these equations in an Euler implicit formulation. The same time step is used as for the gas phase equations, although this is not an essential requirement. However the two phases are inextricably linked through the drag on the droplets, so that the time scales of the two phases become similar. Thus the use of the same time step for both phases seems reasonable, and it may also reduce the complexity of the computer code.

The solution of the gas phase equations is hampered by the fact that, although the pressure gradients appear as source terms in the mean flow momentum equations, there is no explicit equation for pressure. Instead the mass conservation equation can be used, along with the momentum equations, to generate pressure correction equations as used in the SIMPLE method of Patankar (15). There was a major breakthrough in this area when Issa developed the non-iterative but implicit solution scheme called PISO (20). This stands for pressure implicit by splitting of operators. For a first order temporal accurate scheme such as Euler implicit, Issa showed that a predictor step for the momentum equations need only be followed by two pressure corrector steps. This method has subsequently been extended to ic engine flow calculations (3), and to include droplet phase equations (10). This method is used for most of the calculations shown here, because of the greatly reduced computing times required compared with iterative implicit solution (3). The exceptions are the calculations of steady flow past a valve, which were made by iterative solutions.

4 RESULTS WITH NEW SUB-MODELS

4.1 Turbulence models

The only previous application of a differential stress model of turbulence to engine-like flows is due to EL Tahry (21). He investigated the flow in an axisymmetric model reciprocating engine with a permanently open central valve. Comparison was made with data obtained by means of Laser Doppler Anemometry (LDA). These were of profiles, at various distances from the cylinder head, of the mean axial velocity and the rms of the turbulence, giving the axial turbulence intensity.

From this paper alone, it is not possible to elucidate the improvements made by the DSM over results using the $k-\epsilon$ turbulence model. This is because the discretisation scheme was altered also, to the Skew Upwind Differencing Scheme (SUDS) of Raithby (22). In this scheme the numerical diffusion caused by the upwind nature of the scheme is alleviated by aligning the direction from which the upwind information is drawn, with the local streamline. However, some further results for the same case are given by El-Tahry (17) in which a combination of the $k-\epsilon$ turbulence model with the SUDS is employed, and by Gosman et al (23), who used the $k-\epsilon$ model with the HYBRID scheme. An examination of all these predictions reveals that as far as the mean flow quantities are concerned the turbulence model used is much less important than the discretisation scheme. However the turbulence levels are much improved by using a DSM in place of the $k-\epsilon$ model.

As part of the development of our DSM of turbulence, the model has been tested on a number of engine and engine-like flows. In particular the axisymmetric steady flow past a stationary valve into an open ended cylinder is used as a test case. LDA measurements are available to allow refinement of the model to be made. An initial comparison of results obtained using either $k-\epsilon$, the algebraic stress model (ASM) and the DSM was presented by Watkins et al (19).

The ASM results were found to be very similar to $k-\epsilon$ results. These predictions were made using a 63 x 41 line grid. The rms velocity results were in general improved by the use of the DSM, particularly in the important jet flow region. The mean flow predictions however were better for the ASM, than for the DSM. There were some improvements made to the mean flow predictions if a fine 101 x 61 line grid was employed. Much more significant improvements were made to the turbulence predictions.

In (19) it was speculated that the poor resolution of the mean flow at some measuring stations, when using the DSM, was due to the numerically diffusive nature of the HYBRID scheme. In the DSM the cell Peclet number was composed with the laminar viscosity, resulting in high values. This in turn results in the use of the upwind scheme over most of the computational volumes. In the $k-\epsilon$ turbulence model the effective turbulent viscosity, as given by equ 2, is used to form the cell Peclet numbers. In the ASM, the apparent viscosity concept of Huang and Leschziner (24) is employed, wherein turbulent diffusion is introduced into the mean flow momentum equations, via equ 2, and an equivalent amount is deducted from the source terms. This again allows the Peclet numbers to be formed using the turbulent viscosity.

The apparent viscosity concept has now been incorporated into the DSM, and the test case rerun on the fine grid. The predictions are summarised in Fig 1, where comparison is made with the earlier predictions in which the Peclet number employed the laminar viscosity. Clearly, near the valve, the mean flow velocities are greatly improved by the new practice. This must be because of reduced numerical diffusion. The jet is predicted as being thinner and stronger than hitherto. Even so, by 15mm downstream from the head, the jet is predicted as being slightly too wide, and hence the peak velocities are still underpredicted.

Changes have been made in the rms velocities also. On the valve face plane, Fig 1(c) and (d), the peaks in the profiles in the jet region are generally better captured. However lower values than hitherto are obtained in the recirculation region nearer the cylinder wall, with a decrease in accuracy. The same is true for the axial rms velocities at the 15mm plane, however the radial rms values have clearly been degraded by the new practice over much of the profile.

It is not generally recommended that the apparent viscosity concept be utilised in the HYBRID scheme with a DSM of turbulence. However it serves to illustrate the importance of numerical diffusion.

The next step is to remove false diffusion everywhere by incorporating the QUICK scheme in place of the HYBRID scheme, at least for the mean momentum equations. Section 4.2 is concerned with progress made towards this objective.

The $k-\epsilon-\bar{\tau}$ turbulence model of Wu et al (8) has been extended and developed to accommodate shear flows, as well as compression flows. The k equation is exactly the same as for the standard $k-\epsilon$ model, however the ϵ equation is different and of course there is a transport equation for the turbulence time scale $\bar{\tau}$. These run:-

$$D(\epsilon) = -\frac{\epsilon}{\bar{\tau}} + \frac{C_{\epsilon 1}}{C_{\epsilon 2}} \frac{P}{\bar{\tau}} - \frac{2C_{\epsilon 4}}{3} \epsilon \nabla \cdot \bar{u} \quad (4)$$

$$D(\bar{\tau}) = C_3 + C_5 z + C_7 \frac{P}{\epsilon} z + C_6 \bar{\tau} \nabla \cdot \bar{u} \quad (5)$$

The operator D contains the time change terms, and convective and diffusive fluxes, P is the incompressible part of the production of k, and $z = \epsilon \bar{\tau}/k$. The constants are given by Wu et al as $C_{\epsilon 4} = 1.0$, $C_5 = -1.1$, $C_3 = 1 - z_0(C_5 + 1)$, and C_6 depends on the form of compression (1D or 3D, etc). In an equilibrium flow with zero strain the two time scales, k/ϵ and $\bar{\tau}$, should be identical, hence z is then a constant, $z_0 = 6/11$. $C_{\epsilon 1}$ and $C_{\epsilon 2}$ have the standard values of 1.44 and 1.92 respectively. There are two major changes to the equations as given by Wu et al. One is the replacement of k/ϵ by $C_{\epsilon 2}\bar{\tau}$ in the second term on the RHS of the ϵ equation. In calculating both axisymmetric and plane jets using the original $k-\epsilon-\bar{\tau}$ model it was found that in the entrainment zone the value of $\epsilon/\bar{\tau}$, the destruction term of the ϵ equation, reduced, and this was not balanced by the production of dissipation. As a result the spreading rate was far too low. When the above replacement was made, good predictions of spreading rates were obtained. Previous to that, the extra term $C_7 z P/\epsilon$ had been introduced into the $\bar{\tau}$ equation. Differentiation of z with respect to time and introduction of the equations for $\partial k/\partial t$, $\partial \epsilon/\partial t$ and $\partial \bar{\tau}/\partial t$ leads to the conclusion that such a term is necessary, and yields a value for C_7 of less than -0.37.

An optimum value of $C_7 = -0.59$ has been found for plane jets. Figure 2 shows comparison with the data of Bradbury (25) and Gutmark and Wygnanski (26) for k and the Reynolds stress $\overline{-uv}$. Also included in the figures are $k-\epsilon$ and ASM predictions. Table 1 shows the spreading rate obtained for the jet compared with the experimental value and values calculated using other turbulence models. It is clear that the new model produces excellent results for this case.

Similar results, using the same value of C_7, are shown in Fig. 3 and Table 1 for the round jet. Here the values of k and $\overline{-uv}$ are underpredicted compared with the experimental data of Rodi (28) and Hussain and George (29), however the agreement is at least as good as for the $k-\epsilon$ and ASM turbulence model results of Huang (27). Indeed Table 1 shows that the spreading rate is predicted much more closely by the new model.

Agreement with the data for k and $\overline{-uv}$ for the round jet can be increased if the value of C_7 is altered. Very close agreement is obtained if $C_7 = -0.66$. However the spreading rate becomes 0.103 which is in worse agreement with the data.

Table 1 Spreading rates for free jets

	Measured	$k-\epsilon$	ASM	$k-\epsilon-\bar{\tau}$
Plane	0.110	0.108	0.114	0.109
Round	0.094	0.114	0.149	0.096

It would appear from these, and other, calculations that C_7 is not a universal constant. Work is underway to ascertain if it is a function of P/ϵ, the production to dissipation ratio of k.

The $k-\epsilon-\bar{\tau}$ model, as it now is constituted, has been inserted into a computer code for the calculation of axisymmetric flows in i.c. engine cylinders. Calculations have been made for a model engine operating at 200rpm with a single centrally-located valve with a 3.5:1 compression ratio. A number of runs have been made in which the values of C_6 and C_7 have been varied. It is found that the former 'constant' does not affect the mean flow or turbulent intensity values, at least for the relatively low value of strain imposed in this case during compression. However the turbulence length and time scales are affected by the value of C_6. Smaller values of scale are produced if larger values of $|C_6|$ are used.

The effects of altering C_3 are more profound, indicating the influence of shear on in-cylinder flows. Figure 4 shows the predicted mean and rms axial velocity profiles at two crank angles. These are compared with the LDA measurements of Ahmadi-Befrui et al (30), and $k-\epsilon$ model predictions. For this run the standard value of $C_3=1.05$ has been used. At 90° crank angle during intake, the mean flow predictions made by the two models are virtually identical. This might be expected since the two models differ mainly in handling the compression effects. However the evidence provided by the jet test cases indicates that differences could also be expected arising from shearing effects. These do not appear in the mean flow for this case. The turbulence intensities do differ between the models. In the jet region higher values are predicted by the $k-\epsilon-\bar{\tau}$ model in agreement with the results for the plane jet in Fig. 2. However the peak value is still greatly underpredicted, and in the region behind the valve the intensities are more underpredicted by the $k-\epsilon-\bar{\tau}$ model than the $k-\epsilon$ model.

During compression the superiority of the $k-\epsilon-\bar{\tau}$ model is evident, at least for the mean flow calculations. The $k-\epsilon$ model predicts the persistance of a recirculating flow, up to TDC of compression. The $k-\epsilon-\bar{\tau}$ model damps this circulating motion, giving much better agreement with experimental data. The mean flow energy is presumably converted into turbulence energy, hence the $k-\epsilon-\bar{\tau}$ model predicts higher turbulence intensity levels at TDC, giving generally over-predicted levels.

The effect of changing C_3 is illustrated by comparing Fig. 4 with Fig. 5, in which are shown results obtained using $C_3=0.9$. The mean flow during intake is hardly effected by the change, however the turbulence levels are substantially increased, whilst maintaining a similar profile to that in Fig. 4. During compression the mean flow is even more damped than previously, leading to superior comparison with data. However the turbulence intensity levels are again increased, yielding poorer agreement with data.

The mean flow and turbulence intensities during compression can be manipulated by changing C_3. However the effects of such a change on the jet flows previously examined, need to be quantified.

The evidence from these flows is therefore that the values of C_3, C_6 and C_7 are crucial in obtaining good predictions.

4.2 Spatial discretisation

As was outlined in the introduction, almost all calculations of flows in ic engine cylinders use the numerically diffusive PLDS or HYBRID schemes to approximate convective and diffusive fluxes. The exceptions are El-Tahry (17), Vayonitis (18), and Gosman et al (31) who were calculating flows in axisymmetric model engines. El-Tahry used both the SUDS and QUICK scheme, Vayonitis used the QUICK scheme, and Gosman et al the SUDS.

Both users of the SUDS found considerable improvement in predictive accuracy over the use of the HYBRID or upwind differencing scheme. Theoretically the QUICK scheme should be more accurate than SUDS as it removes false diffusion entirely, rather than reducing its effect as in the SUDS. However its unbounded nature may cause problems. In (17), El-Tahry found that the QUICK results were far inferior to those using the SUDS. Indeed, a comparison of his results with those of Gosman et al (23) for the same flow, shows that the QUICK results are inferior even to HYBRID results, when the same $k-\epsilon$ turbulence model is employed. Vayonitis (18) found that at some crank angles the QUICK results were superior to HYBRID results, at others they were inferior. This apparent failure of the QUICK scheme is puzzling, given that in all other applications found in the literature the QUICK scheme gives much superior predictive ability than the HYBRID or PLDS schemes. In particular Leschziner (32) examined the behaviour of a plane jet issuing from a wall at an angle to the computational grid. He found that the QUICK and SUDS results were virtually identical, and much superior to HYBRID results.

Currently at UMIST the QUICK scheme has been incorporated for the mean flow equations only into ic engine computer codes also containing the $k-\epsilon$ turbulence model, and into steady flow codes containing the $k-\epsilon$ model and the algebraic stress model. Work is on hand to do the same for an engine code containing the DSM of turbulence. Attention has been focused initially on the test case of steady flow past a stationary valve. An example of results obtained using the $k-\epsilon$ turbulence model on a 101 x 61 line grid is shown in Fig 6. For the mean flow axial velocity, the QUICK results are clearly superior to either HYBRID or PLDS results which are almost identical. As for the turbulence intensities, the use of QUICK in the mean flow equations produces some contradictory results. The peaks in the jet flow region are captured better than when HYBRID or PLDS is used, however the intensities in the backflow region below the valve are substantially over predicted.

The mean flow results obtained here using QUICK are certainly much better than El-Tahry's (17) near-valve results for a reciprocating engine. It may be that the QUICK scheme does not work well in time-dependent situations as also found by Vayonitis (18), or this may be a feature of the Euler implicit scheme which was used in both cases. Nasser (33) shows that use of QUICK with a second order accurate ADI scheme in place of the first order accurate Euler scheme results in considerable improvements in accuracy. Such a scheme could be implemented fairly simply in the current codes.

4.3 Fuel spray wall impaction with gaseous cross-flow

In (11), Watkins and Wang tested their new model for fuel spray impaction on a wall by comparing with experimental data obtained in test rigs. Some cases were for normal impaction, some were at oblique angles. In all cases the gas pressure was elevated above atmospheric but the gas temperature was at room level. The shapes of the wall sprays were well simulated, although the extent of the dispersion into the gas was usually underpredicted.

In none of the cases in (11) was there cross-flow, as there would be in an operating diesel engine, in the form of swirl. To simulate the effects of swirl on diesel sprays, Mirza (34) ran a series of tests in a wind tunnel in which the non-burning pulsed sprays were subjected to a uniform cross-flow. The gas pressures were elevated but the gas was at room temperature. A floor, upon which the spray impacted, was placed in the wind tunnel 32mm from the nozzle.

Table 2 Test cases for spray wall impaction and cross-flow modelling

Case No	Gas Pressure (bar)	Mean Cross-flow Velocity (m/s)	Turbulence Level(%)	Expt. Data(34)
1	20.7	18.5	1.0	Yes
2	20.7	18.5	16.0	No
3	20.7	14.0	16.0	Yes
4	13.8	14.0	16.0	Yes
5	6.9	14.0	16.0	Yes

Table 2 gives details of a number of computer simulations which have been made, employing the spray wall impaction model. In some cases experimental data is available for comparison purposes. The higher turbulence levels were obtained by means of a mesh set upstream of the spray location. In all the cases shown here the spray was injected normally to the cross-flow, although Mirza performed other runs in which the spray was injected at an oblique angle to the cross-flow.

Reference (35) compared the simulated spray results of test case 1, with photographs taken from Mirza (34). A number of points were apparent. Firstly the free spray was predicted as penetrating slightly more slowly than in the experiment, so that the wall is reached 0.1 msec later. This is probably due to the relative coarseness of the computational grid. Khaleghi (36) has shown the effects grid density has on spray penetrations. Secondly the simulated spray was deflected more by the cross flow than in the experiment. This would seem to indicate that there is too large a coupling between the phases. It is likely that the single-droplet drag coefficient used in these calculations should be modified when there are a large number of droplets present. This may be done by including the void fraction in the drag coefficient. Thirdly there was an upstream wall spray in the experiment which was not reproduced by the calculations. However the development of the downstream wall spray appeared to be reasonably well predicted.

Figure 7 compares the results for cases 1 and 2. This allows the effects of increased turbulence in the cross-flow to be analysed. Firstly increased dispersion of the droplets in the spray results in the free spray penetrating more slowly. Secondly the cross-flow has an even greater effect of bending the spray, probably because of the slower motion of the droplets. Thirdly the spray on the wall is more dispersed with a greater height to the wall spray. However there appears to be little overall effect on the speed with which the spray is swept downstream, this again quickly approaching the mean cross-flow velocity.

The effects of reducing the cross-flow velocity in case 3, whilst still retaining the equivalent percentage turbulence in the flow, are illustrated in Fig. 8. The spray is deflected less by the cross-flow and penetrates further. The sizes of the droplets are generally increased, denoting an increase in collisions between droplets. Despite the increased penetration, the lower cross-flow velocity results in the wall spray (not shown) travelling more slowly and hence later lagging behind the higher cross-flow velocity case.

The effects of the gas pressure on free sprays has been analysed in detail by Khaleghi (36). Here we look at the effects on the wall sprays. These are illustrated in Fig. 9, which shows results of cases 3, 4 and 5. The downstream wall spray penetrates further for a lower gas pressure. This is partly due to reduced drag on the droplets in the wall spray itself, so that droplet velocities stay above the cross-flow velocity for a longer time, but is also due to the increased penetration of the free spray resulting in an earlier formation of the wall spray. These effects are somewhat reduced because the height of the wall spray is also increased for a reduced gas pressure, ie the wall spray is more dispersed, resulting in a reduced ability to penetrate. The dispersion also leads in general to smaller drop sizes. The reduced gas pressure also allows the formation of an upstream wall spray.

Figure 10 shows both the free spray and downstream wall spray penetrations for cases 3 and 5. In both cases the free spray penetration is underpredicted, resulting in late arrival at the wall. Therefore the wall spray predictions are shifted to the right in this figure, compared with the experimental values. However the slopes are in good agreement, indicating that the leading edge velocities are well predicted. These velocities are significantly greater than the cross-flow velocity of 14.0 m/s, particularly for the low pressure case 5, but appear to be decreasing towards that value, as time progresses.

The spray wall impaction model seems to reproduce experimental evidence and expected trends well when combined with a cross-flow. It remains now to test the model in more engine-like circumstances. In particular attention is now focused on high temperatures cases. In the longer term, the model needs to be introduced into three dimensional engine codes in which swirl provides the cross-flow.

5 MULTI-INLET-VALVE FLOWS

This section is concerned with describing the progress made in developing a computer code intended to allow some aspects of multi-inlet-valve flows to be investigated. As mentioned in the introduction, the code used here does not have the geometric applicability of some ic engine codes. In particular the grid arrangement in the axial direction between the cylinder head and piston crown is rectilinear. This is because orthogonal curvilinear grid systems are employed in two dimensions only. As a consequence the pentroof shape of the cylinder head of most multi-inlet-valved engines cannot be matched. Instead a flat cylinder head is assumed. Despite this restriction it is believed that some useful information can be obtained with the restricted geometry. For example, the major differences in flow pattern between single and dual valved engines can be exhibited, as can differences between different combinations of valve types. A further area of considerable interest, which only arises with dual valved engines, is the question of whether differential valve timing can play a significant role in influencing the fluid flow and turbulence structures, and hence, ultimately, the combustion.

On the numerical side, it would be useful to assess the ability of the type of boundary fitted grid used here to provide reasonably accurate predictions of the flow and turbulence structure. The grid is fitted to both the inlet valves and the cylinder wall in order to accurately simulate both the inlet flow and the in-cylinder flow. An example is shown in Fig. 11. It is clear that there are a number of areas in the grid where potentially the numerical errors could be large. In particular the non-orthogonal grids at both ends of the symmetry diagonal will give rise to errors in equations based on an orthogonal system of coordinates. Further errors may well arise because of the large aspect ratios of some of the control volumes, near, for example, the symmetry diagonal between the two valves, and between the valves and the cylinder wall.

When the project was initiated there appeared to be no experimental data available of flows in dual-valve engines. Some data has now been published eg. Le Coz et al (37) and Khalighi (38). As a consequence the first calculations made were a repeat of a previous numerical simulation, due to Wakisaka et al (39).

5.1 Comparison with previous calculations

Wakisaka et al used a cylindrical polar grid system in order to calculate flows in a dual-valve engine with a flat cylinder head. Thus the inlet flow was not well simulated, but the grid was fitted to the cylinder wall. For their simulations they used a combination of directed and swirl ports, as illustrated in Fig. 12. We have calculated their 'model FP/D-H', that is with valve number one in the directed port and valve number two in the helical or swirl port.

Figure 13 compares the two sets of predictions at 90° ATDC. The flows in the elevation planes are in quite reasonable agreement. Thus on the $\theta = 0-\Pi$ plane the downward flow beneath the directed port valve, and the flow across the cylinder head and down the farther wall, are shown in both sets of calculations. The flow in the central part of the cylinder is however predicted differently. A similar situation exists on the $\theta = \Pi/4 - 5\Pi/4$ plane. On the $\theta = 3\Pi/4 - 7\Pi/4$ plane, the predictions are in agreement between the ports, but differ on the other side of the cylinder, particularly of the flow down the cylinder wall.

Such differences in detail are not unexpected given the radically different grid structures used. It should also be noticed that for our calculations a 26 x 34 x 30 line grid is employed, whereas Wakisaka et al used an 11 x 27 x 14 line grid. In terms of possible temporal errors, here we use 0.25° crank angle intervals whereas in (39) a 2.5° interval was employed.

However the flows predicted in the plan views are very different. Clearly in (39) a solid body type of swirling flow is set up throughout the cylinder volume. For the present calculations this organised flow does not appear. Instead there are a number of isolated recirculation zones, and the flow velocities appear to be considerable less than predicted in (39). Calculations at later crank angles confirm these observations.

It may be that the wall function approach coupled to the $k-\epsilon$ turbulence model employed in the present calculations, is not adequate to cope with the widely varying grid spacings and shapes near the cylinder wall. For it is here that any swirling flow must decay. It was originally thought that too high levels of turbulent viscosity were being generated near the cylinder wall leading to excessive damping of the swirl velocities. Later predictions for a laminar flow showed this to be only part of the cause.

The swirl is maintained for a longer period of crank angle, but still begins to decay too early. Attention is therefore turning to the grid, and in particular the singular cells at the ends of the symmetry diagonal, as the main cause of this decay.

5.2 Differential valve timing

Here we examine the effects on the flow in an ic engine cylinder if the two inlet valves are not synchronous. As an example we have re-simulated the above case from Wakisaka et al (39), but with the swirl port valve timing delayed by 20° crank angle behind the directed port valve.

This simulation requires that the computer code be altered to accommodate the two valves having different lifts. In the example above both valve faces are on the same grid line in the axial direction. Here they are on different grid lines. A major difficulty arises when the first valve starts to close whilst the second valve is still opening. For at some crank angle the valve faces pass each other. The grid line numbers of the valve faces must then be swapped. Around this point in time there will also be the possibility of having very thin control volumes between the valve face grid lines. This could lead to high aspect ratios for these cells, with associated numerical problems.

The overall swirl levels in both the synchronous and asynchronous valve cases have been evaluated, and then normalised by the moment of inertia of the mass of gas in the cylinder, and by the piston rotation rate. The same procedure may be done for flow about other axes. Here x and y axes are set, as shown in Fig. 11, with their origin 5 mm from the cylinder head. Angular flow about these axes are termed tumble motions. Fig. 14 illustrates the effects of asynchrony on these properties of the flow. The production of swirl is delayed, presumably because of the later opening of the swirl valve, and hence lower values of swirl from this source up to and beyond 90° ATDC. However given the comments made previously concerning the prediction of swirl, these results should be treated with caution. The y-tumble ratio is also reduced, indicating more disorganised flow in the asynchronous valve case. The x-tumble ratio values remain relatively small, but have changed sign, the tumbling motion now occuring in the opposite sense.

6 CONCLUSIONS

This paper has presented some of the work currently underway at UMIST on the multidimensional calculation of fluid flows in ic engine cylinders.

The work on discretisation schemes and turbulence models indicates that it is not possible to properly distinguish between turbulence models if a numerically-diffusive discretisation scheme is employed. If a non-diffusive scheme such as QUICK is used, then the mean flow properties are substantially better predicted whatever turbulence model is employed. The differential stress model appears to be capable of producing a much better prediction of the turbulence intensities, and presumably of time and length scales, too, than the eddy-viscosity based k-ϵ model. However a final judgement on the DSM in ic engine applications must await the combining of the model with the QUICK scheme.

The k-ϵ-$\bar{\tau}$ turbulence model is still being developed. It would appear that at least one, if not more, of its current 'constants' is not constant, since different flows require a different value to obtain good agreement with experimental data. A probable way forward is to make the 'constant' a function of the ratio of production and dissipation of turbulence kinetic energy.

The latest development in the area of computing diesel engine sprays is a wall impaction sub-model. This has been tested here in conjunction with cross-flowing gas, to simulate swirl. Comparison with experimental data is encouraging.

Finally some work on calculating flows in engines equipped with dual intake ports has indicated a number of areas where the current code, in conjunction with the particular grid employed, may be inadequate. Nevertheless some calculations do agree with earlier predictions. The use of asynchronous valve timings has been investigated, and seem to have substantial effects on the overall swirl and tumble ratios.

ACKNOWLEDGEMENTS

The work reported here has been supported financially, and by the provision of computing resources, by the SERC under grants GR/D/90512, GR/E/30652 and GR/D/78237.

REFERENCES

1 Amsden, A.A., Butler, T.D., O'Rourke, P.J. and Ramshaw, J.D., KIVA-A comprehensive model for 2-D and 3-D engine simulations. SAE 850554, 1985.

2 Haworth, D.C., El Tahry, S.H., Huebler, M.S. and Chang, S., Multidimensional port-and-cylinder flow calculations for two-and four-valve-per-cylinder engines: influence of intake configuration on flow structure. SAE 900257, 1990.

3 Ahmadi-Befrui, B., Gosman, A.D., Issa, R.I. and Watkins, A.P., EPISO - an implicit non-iterative solution procedure for the calculation of flows in reciprocating engine chambers. Comp. Meth. Appld. Mech. Engrg, 1990, 79 , 249-279.

4 Launder, B.E. and Spalding, D.B., <u>Mathematical models of turbulence</u>. Academic Press, London, 1972.

5 Launder B.E., Second moment closure and its use in modelling turbulent industrial flows. Int. J. Num. Methods in Fluids, 1989, 9, 963-985.

6 Rodi, W., A new algebraic relation for calculating the Reynolds stresses. ZAMM, 1976, 56, T219-T221.

7 Fu, S., Huang, P.G., Launder, B.E. and Leschziner, M.A., A comparison of algebraic and differential second-moment closures for axisymmetric turbulent shear flows with and without swirl. Trans. ASME, J. Fluids Eng., 1988, 110, 216-221.

8 Wu, C.T., Ferziger, J.H. and Chapman, D.R., Simulation and modelling of homogeneous, compressed turbulence. Fifth symp. of turbulent Shear Flows, Cornell Univ., 1985, 17.13-17.20.

9 Ahmadi-Befrui, B. and Gosman, A.D., Assessment of variants of the k-ϵ turbulence model for engine flow applications. Int. J. Num. Methods in. Fluids, 1989, 9, 1073-1086.

10 Watkins, A.P., Three-dimensional modelling of gas flow and sprays in diesel engines. <u>Computer Simulation of Fluid Flow, Heat and Mass Transfer and Combustion in Reciprocating Engines</u> (Markatos, N.C. ed.) Hemisphere, 193-237, 1989.

11 Watkins, A.P. and Wang, D.M., A new model for diesel spray impaction on walls and comparison with experiment. Int. Symp. on Diagnostics and Modelling of Combustion in Internal Combustion Engines, Kyoto, 1990.

12 O'Rourke, P.J. and Bracco, F.V., Modelling of drop interactions in thick sprays and a comparison with experiment. Proc. I.Mech.E. Conf. on Stratified Charge Automotive Engines, 1980.

13 Watkins, A.P. and Diomataris, C.M., Case study of numerical grid generation by means of transformed Laplace equations. Int. J.Mech.Eng. Education, 1989, 17, No. 2, 139-148.

14 Spalding, D.B., A novel finite-difference formulation for differential equations involving both first and second derivatives. Int. J.Num. Meth. in Engg., 1972, 4.

15 Patankar, S.V. Numerical heat transfer and fluid flow. Hemisphere, Washington, 1980.

16 Leonard, B.P., A stable and accurate convective modelling procedure based on quadratic upstream interpolation. Comput. Meths, in Appl. Mech. Engrg, 1979, 19, 59-98.

17 El-Tahry, S.H., A comparison of three turbulence models in engine-like geometries. Int. Symp. on Diagnostics and Modelling of Combustion in Reciprocating Engines. JSME, SAEJ and MESJ, Tokyo, 1985.

18 Vayonitis, I., Evaluation of a second order discretisation scheme for flows in internal combustion engines. M.Sc. Dissertation, University of Manchester, 1987.

19 Watkins, A.P., Kanellakopoulos, P. and Lea, C.J., An assessment of discretisation schemes and turbulence models for in-cylinder flows. Int. Symp. on Diagnostics and Modelling of Combustion in Internal Combustion Engines, Kyoto, 1990.

20 Issa, R.I., Solution of the implicitly discretised fluid flow equations by operator splitting. J. Comput. Phys., 1986, 62, No. 1, 40-65.

21 El-Tahry S.H., Application of a Reynolds stress model to engine-like flow calculations. Trans. ASME, 1985, 107, 444-450.

22 Raithby, G.D., Skew upstream differencing schemes for problems involving fluid flows. Comput. Meths. in Appl. Mech. Engrg., 1975, 9.

23 Gosman, A.D., Johns, R.J.R., and Watkins, A.P., Assessment of a prediction method for in-cylinder processes in reciprocating engines. Combustion Modelling in Reciprocating Engines, eds. J.N. Mattavi and C.A. Amann, Plenum Press, New York, 1980.

24 Huang, P.G. and Leschziner, M.A., Stabilization of recirculating flow computations performed with second-moment closures and third-order discretization. Fifth symp. on Turbulent Shear Flows, Cornell Univ., 1985.

25 Bradbury, L.J.S., The structure of a self-preserving turbulent plane jet. J. Fluid Mech., 1965, 23, pt.1, 31-64.

26 Gutmark, E. and Wygnanski, I., The planar turbulent jet. J. Fluid Mech., 73, pt.3, 465.

27 Huang, P.G., The computation of turbulent elliptic flows with second moment closure models. Ph.D. Thesis, University of Manchester, 1985.

28 Rodi, W., The prediction of free boundary layers by use of a two-equation model of turbulence. Ph.D. Thesis, University of London, 1972.

29 Hussain, H.J. and George, W.K., Measurement of small scale turbulence in an axisymmetric jet using moving hot-wires. Seventh symp. on Turbulent Shear Flows, Stanford Univ, 1989.

30 Ahmadi-Befrui, B., Arcoumanis., C., Bicen, A.F., Gosman, A.D., Jahanbakhsh, A. and Whitelaw, J.H., Calculations and measurements of the flow in a motored engine and implications for open-chamber, direct-injection engines, Three Dimensional Turbulent Shear Flows, Proc. AIAA/ASME Conf., pp 1-9, 1982.

31 Gosman, A.D., Jahanbakhsh, A. and Watkins, A.P. Evaluation of multidimensional predictions of flow in piston-bowl configurations. Flow in internal combustion engines -III, ASME FED - Vol. 28, 1985, 115-124.

32 Leschziner, M.A., Practical evaluation of three finite-difference schemes for the computation of steady-state recirculating flows. Comput. Meths. Appl. Mech. Engrg, 1980, 23, 293-312.

33 Nasser, A.G., Compact finite-difference/finite-volume schemes for unsteady recirculating flows. Ph.D. Thesis, University of Manchester, Faculty of Technology, 1990.

34 Mirza, R., Studies of diesel spray interacting with cross-flow and solid boundaries. Ph.D. Thesis, University of Manchester, Faculty of Technology, 1989.

35 Watkins, A.P., Khaleghi, H. and Wang, D.M., Modelling Spray Phenomena in Direct-Injection Diesel Engines, I. Mech. E. Conf. on Internal Combustion Engine Research in Universities, Polytechnics and Colleges, Jan. 1991.

36 Khaleghi, H., Three-dimensional modelling and comparison with experiment of sprays and gas flow in test rigs and diesel engines. Ph.D. Thesis, University of Manchester, Faculty of Technology, 1990.

37 Le Coz, J.-F., Henriot, S. and Pinchon, P., An experimental and computational analysis of the flow field in a four-valve spark ignition engine-focus on cycle-resolved turbulence. SAE 900056, 1990.

38 Khalighi, B., Intake-generated swirl and tumble motions in a 4-valve engine with various intake configurations-flow visualisation and particle tracking velocimetry. SAE 900059, 1990.

39 Wakisaka, T., Shimamoto, Y. and Isshiki, Y., Induction swirl in a multiple intake valve engine-three-dimensional numerical analysis. I. Mech. E. Conf. on Combustion in engines-technology and applications, London, 1988, C54/88.

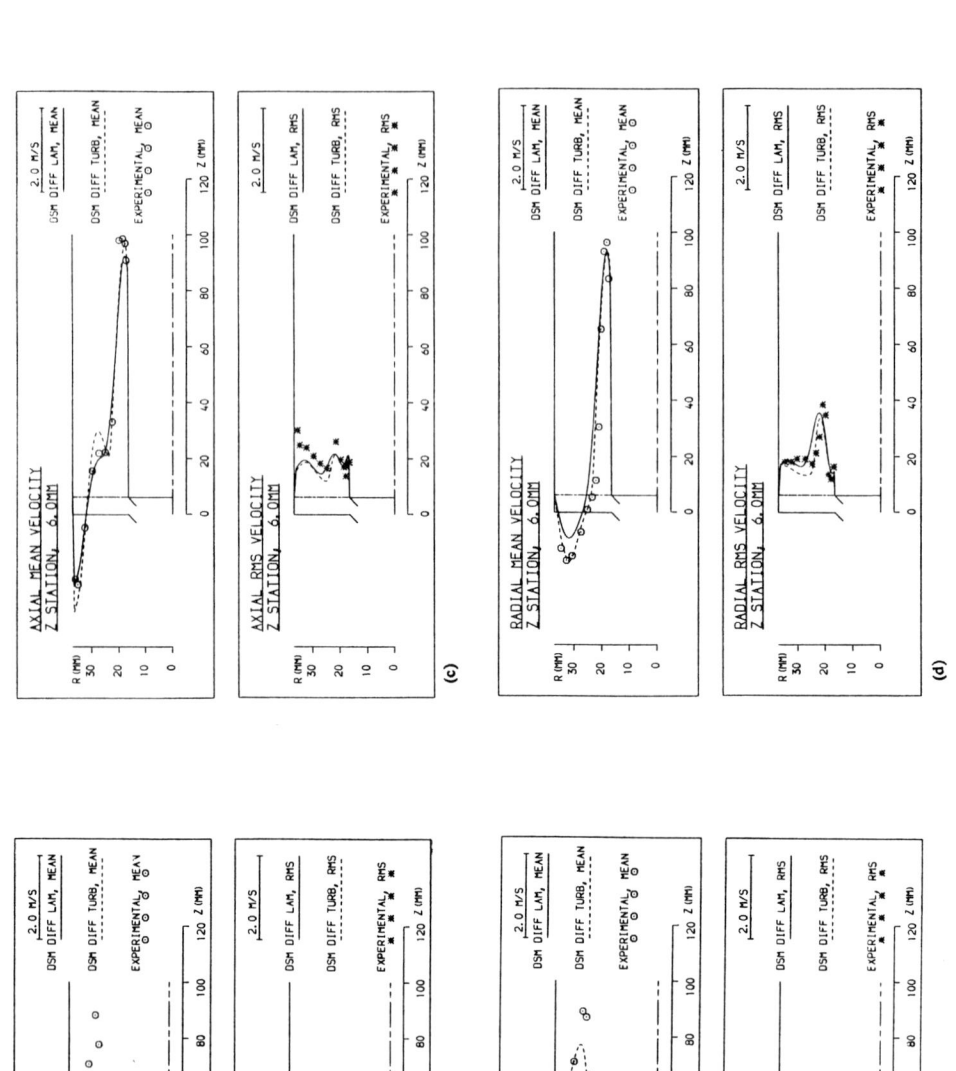

Fig 1 Effects of viscosity used in HYBRID scheme, fine grid DSM.

Fig 2 Calculations of plane jet

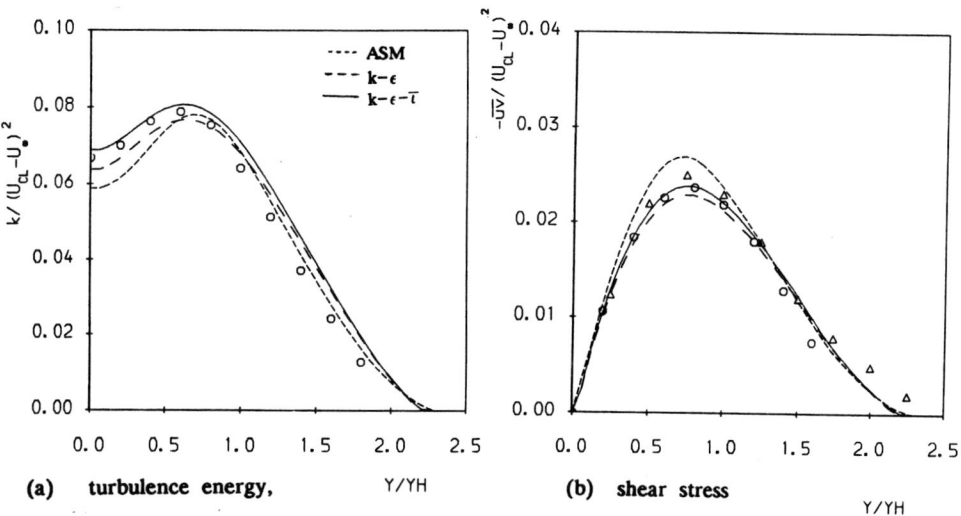

(a) turbulence energy,   (b) shear stress

Fig 3 Calculations of round jet

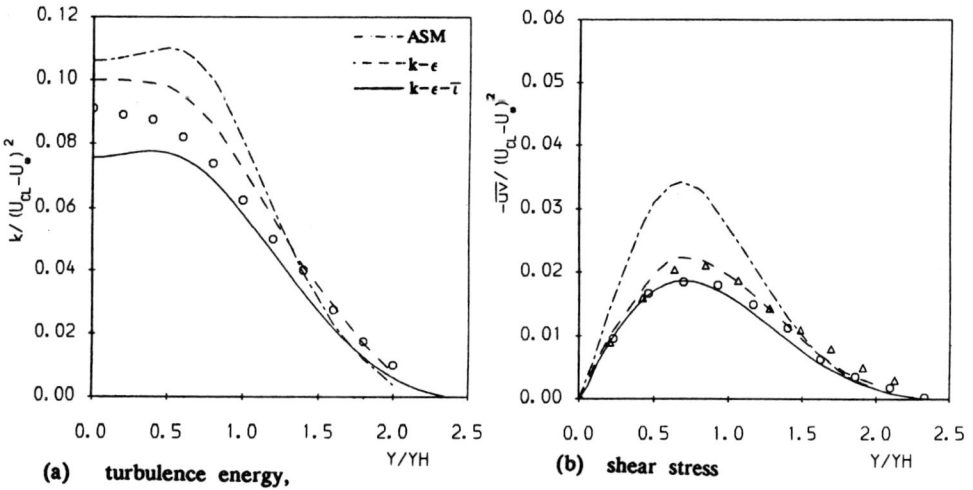

(a) turbulence energy,   (b) shear stress

Fig 4 Predictions of reciprocating engine flow, $C_3 = 1.05$

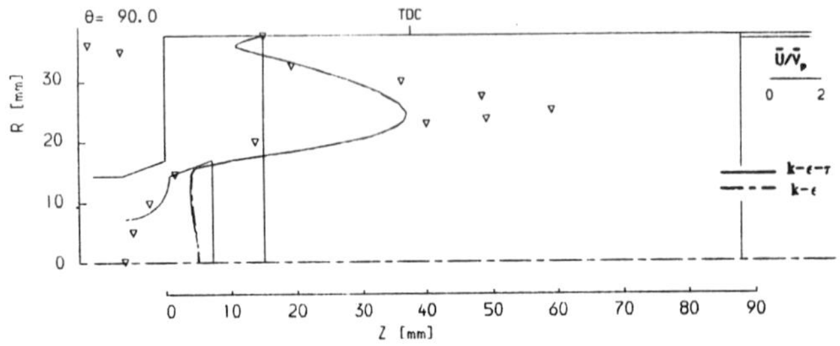

(a) mean flow, 90°

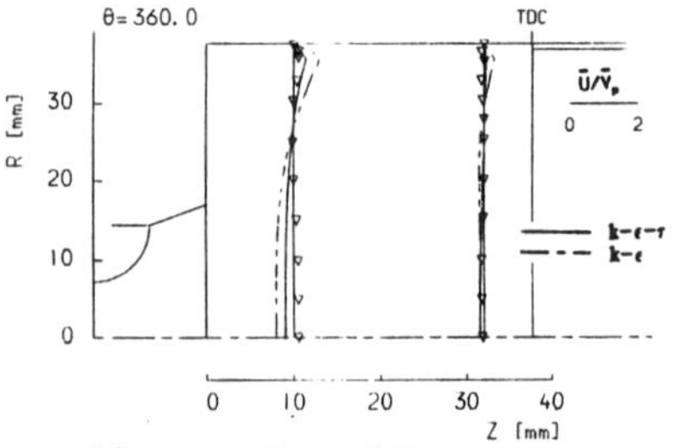

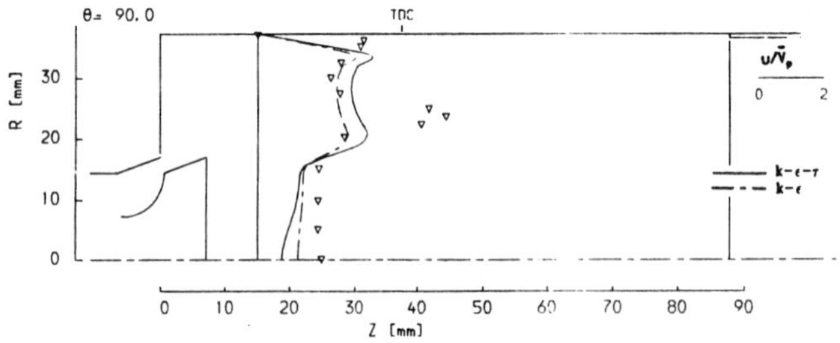

(b) turbulence intensity, 90°

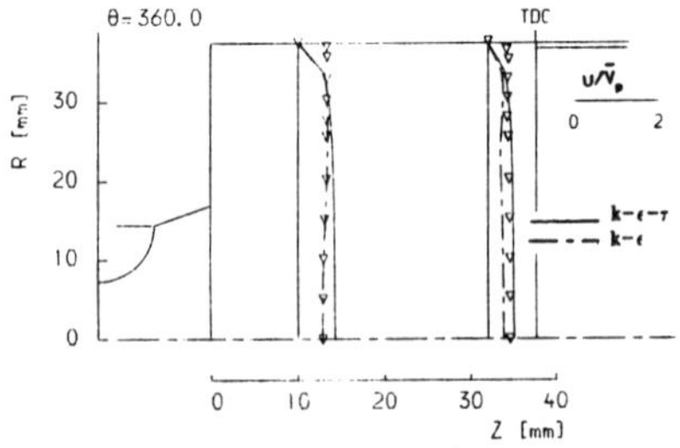

(c) mean flow, 360°

(d) turbulence intensity, 360°

Fig 5 Predictions of reciprocating engine flow, $C_3 = 0.9$.

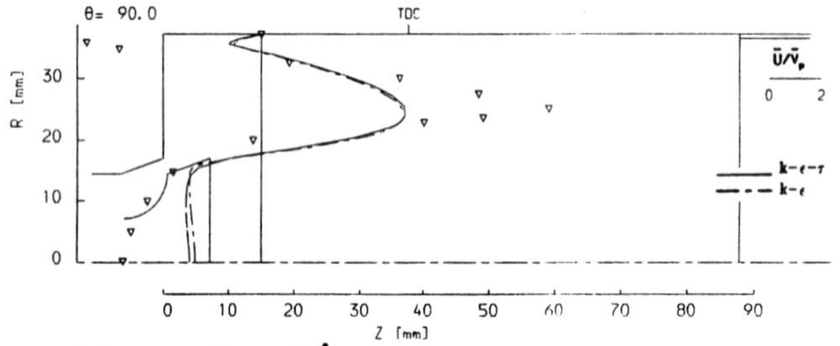

(a) mean flow, 90°

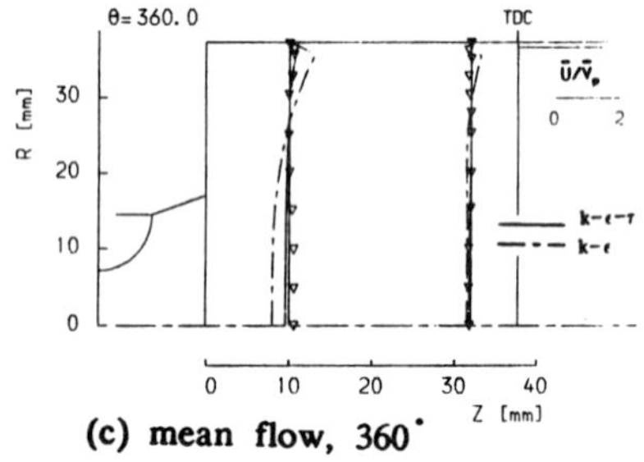

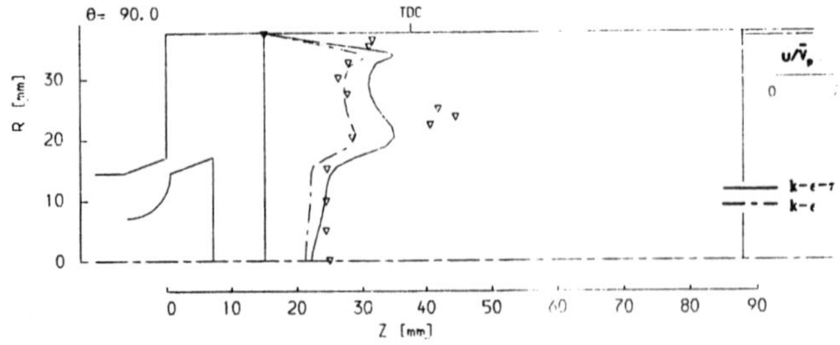

(b) turbulence intensity, 90°

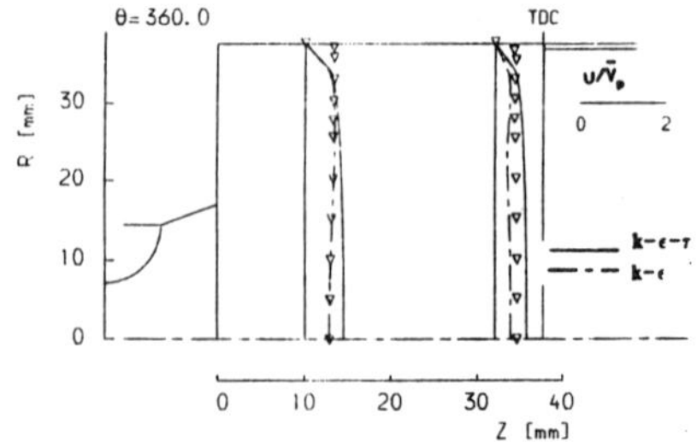

(c) mean flow, 360°

(d) turbulence intensity, 360°

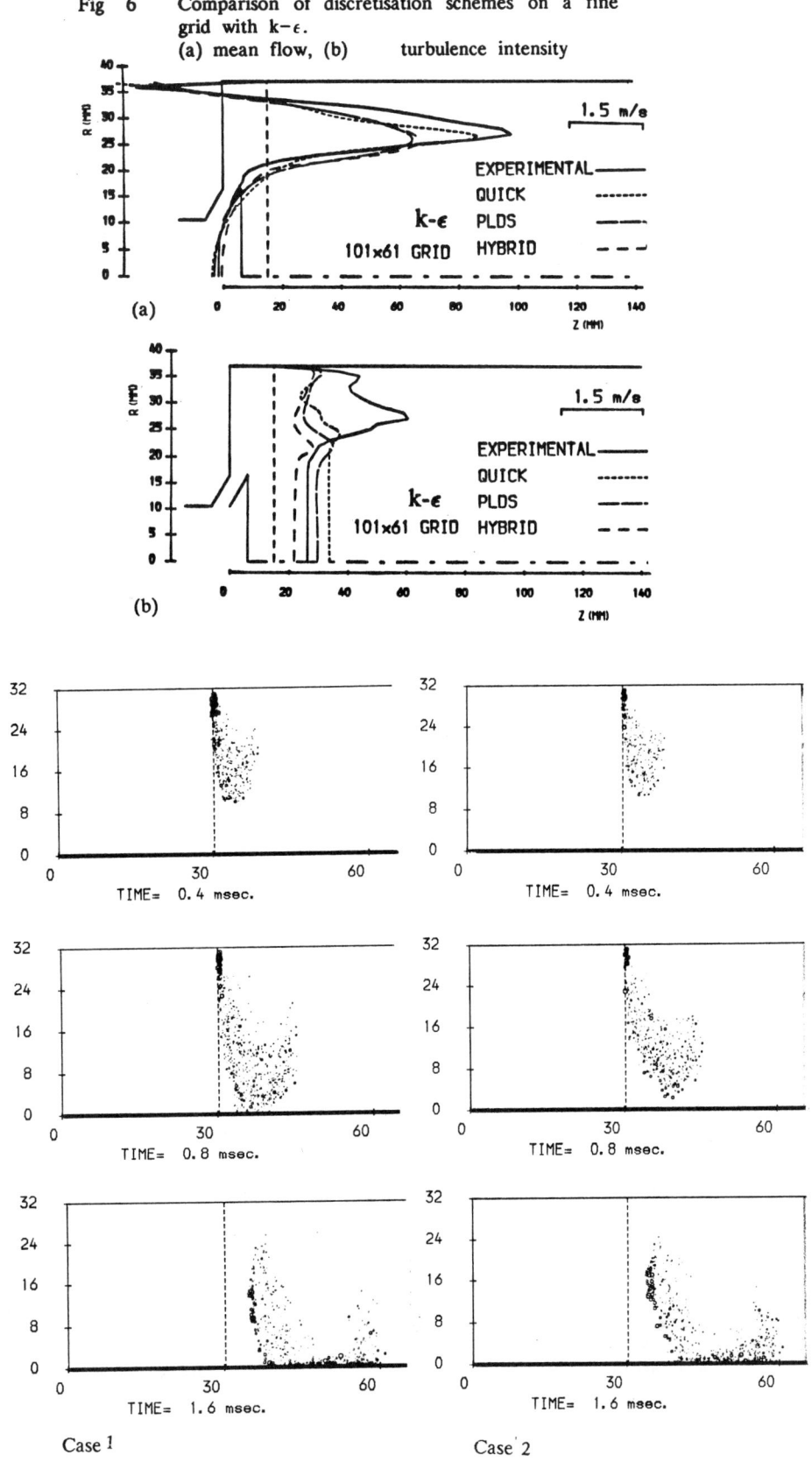

Fig 6 Comparison of discretisation schemes on a fine grid with k-ε.
(a) mean flow, (b) turbulence intensity

Fig 7 Spray developments for Cases 1 and 2.

Fig 8   Effect of cross-flow velocity on free spray

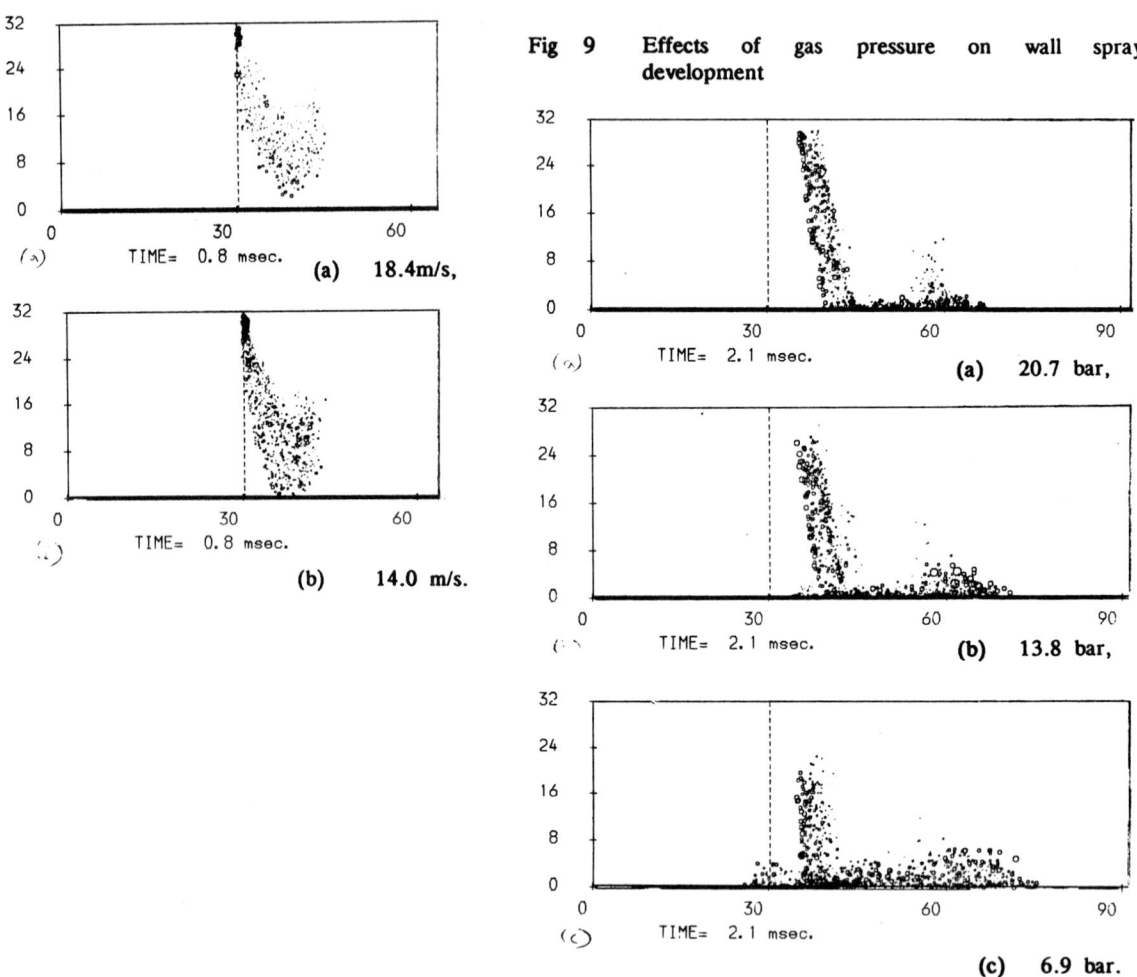

(a) 18.4m/s,

(b) 14.0 m/s.

Fig 9   Effects of gas pressure on wall spray development

(a) 20.7 bar,

(b) 13.8 bar,

(c) 6.9 bar.

Fig 10   Free and wall spray developments

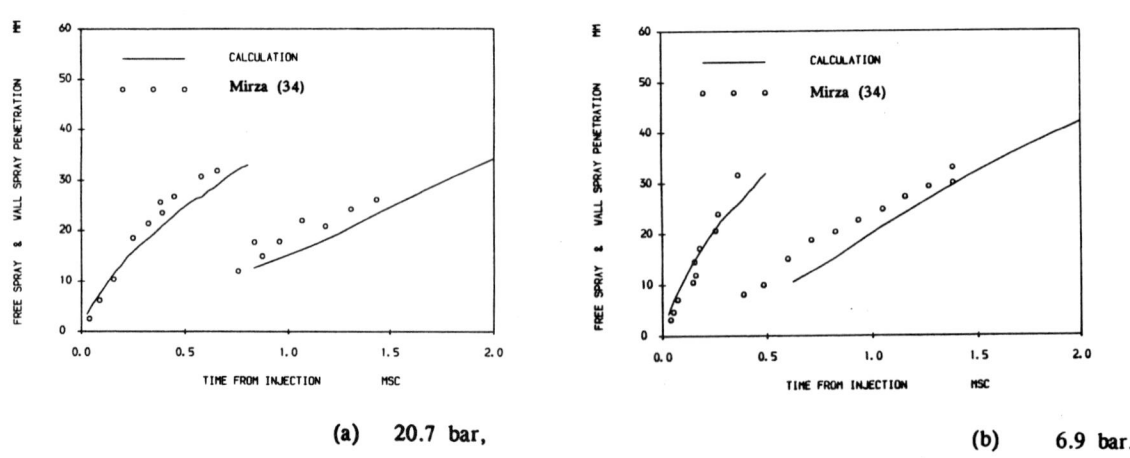

(a) 20.7 bar,

(b) 6.9 bar.

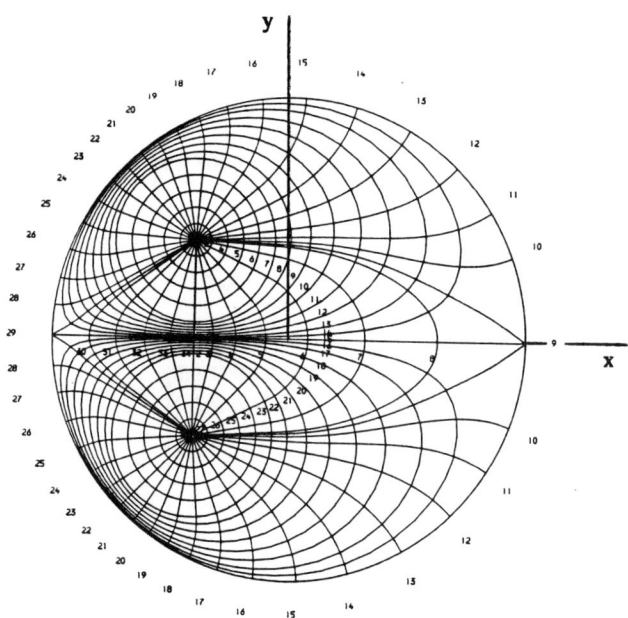

Fig 11 Typical grid for dual-inlet-valve calculations.

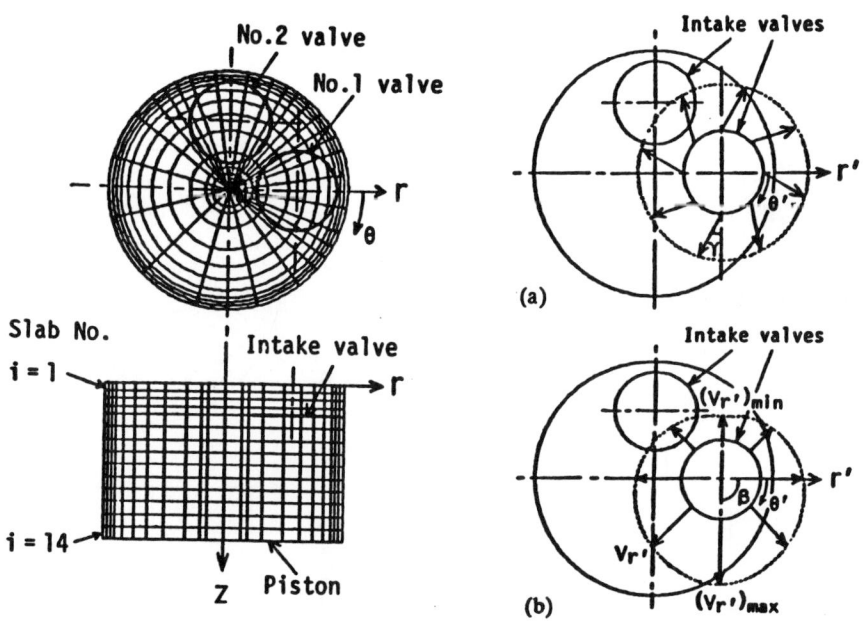

Fig 12 Grids and (a) swirl port and (b) directed port geometries of Wakisaka et al (39).

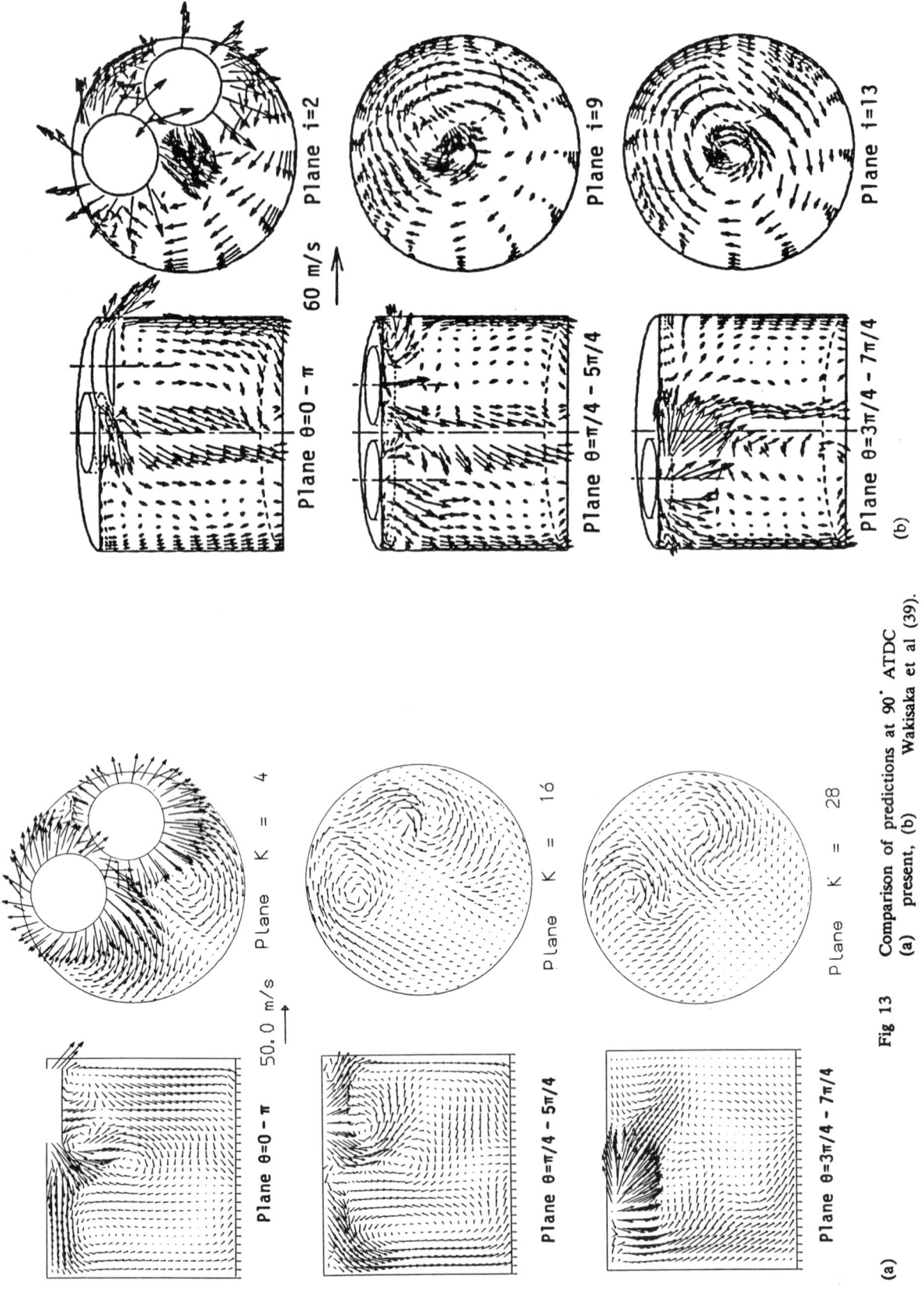

Fig 13 Comparison of predictions at 90° ATDC (a) present, (b) Wakisaka et al (39).

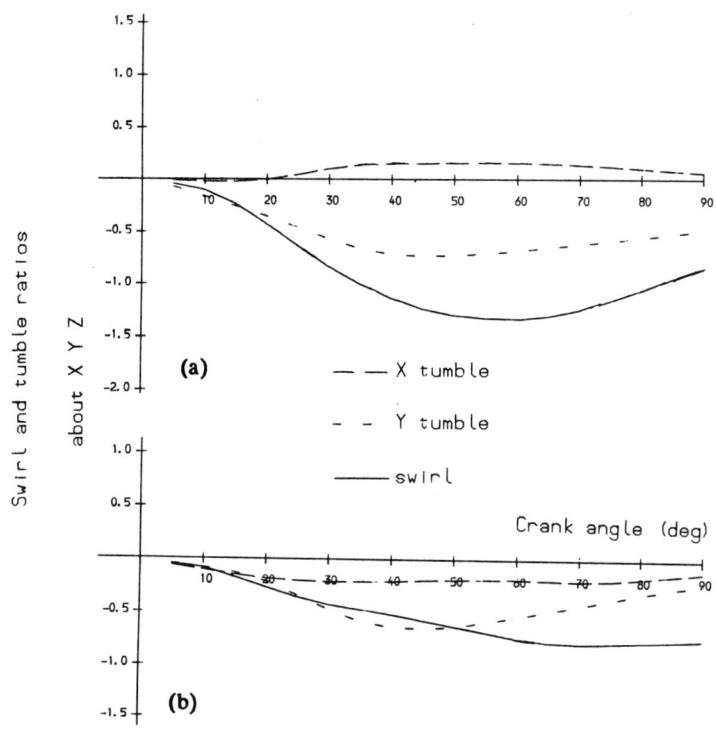

Fig 14 Effects of valve asynchrony on overall swirl and tumble ratios
(a) Synchronous, (b) 20° delay.

C413/049

Local void fraction measurement in flow channels of irregular shape

I K SMITH, BSc(Eng), PhD, DIC, CEng, FIMechE, Z G XU, BSc and C A ALDIS, BSc, MSc, PhD
Department of Mechanical Engineering, City University, London

SYNOPSIS A computer based local void fraction measurement system developed primarily for use in flow channels of irregular cross section is described. It has been designed to operate in sealed systems which contain fluids such as refrigerants. A single fibre optic probe is employed which is positioned by a specially designed mechanical traversing unit driven and controlled by a stepper motor system. The void fraction is obtained by means of a high speed signal processing module installed in a PC/AT microcomputer which alternately controls the traversing unit. The dwell time of the probe at any point can be varied from 1 to 100 seconds. Test results are presented with Refrigerant 113 as the working fluid.

1 INTRODUCTION

Two-phase flow and heat transfer phenomena occur frequently in a wide range of equipment in oil, chemical, nuclear and power industries. A knowledge of the void fraction profile across passages in which it occurs is of significance to the design and safe operation of such equipment. This is because of its direct relevance to pressure drop calculation and flow regime detection as well as detailed flow and heat transfer modelling.

Although there has been much research in the field of two-phase flow over the last three decades, information on void fraction distribution is generally scarce, especially for fluids other than air-water or water-steam mixtures. Previous investigations of this have been limited to flow in circular ducts or, in more complex cross sections, to the minimum of test points [1].

The work presented in this paper was the experimental part of a research programme to investigate two-phase flashing flow in the inlet port of a Lysholm screw machine used to develop power from the expansion of hot saturated liquids. Such an expander has been shown to have considerable potential for the recovery of power from low grade heat resources [2]. The inlet port basically functions as a convergent nozzle. However its exit cross section, which matches the male and female rotor meshing area, is irregular in shape, as shown in Fig.1. In order to validate the theoretical modelling, it was important to have a detailed knowledge of void fraction distribution in this plane.

It is normal practice to measure the local void fraction with manually positioned probes. However it was recognised from the very outset that to scan the entire port cross section by this means would be both complex and time consuming. Moreover the fluid of interest was a refrigerant which had to be retained in a completely sealed containment. A remotely controlled automatic traversing system was considered to be the best means of fast scanning under these conditions. The system finally developed operated very well and can easily be used in any channel of irregular cross section.

2 EXPERIMENTAL APPARATUS AND OPERATION

2.1 Test Rig Description

The layout of the closed loop test facility developed for the experimental programme is shown in Fig.2. It consists of a three stage mixed flow feed pump driven by a 37 kW electric motor which delivers pressurised refrigerant 113 to the tube side of a shell and tube heat exchanger. This can supply up to 700 kW of heat to the refrigerant from steam generated in a gas fired boiler which is passed through the shell side. The hot pressurised liquid leaves the heater and returns to a water cooled condenser either through a by-pass valve, for start up or emergency shut down, or through a pneumatically operated throttle control valve to enter the test section at the required temperature and dryness fraction

Test section details are given in Fig.3. The overall length is 445 mm along its axis which is at an angle of 15° to the horizontal. The inlet cross section is circular with a diameter of 101.6 mm. This is reduced to 76.2 mm in an axial displacement of 50 mm from the inlet. From here onward the cross section undergoes a smooth transition from circular to the irregular cross section at the exit without any further change of area.

The temperatures around the loop are measured with grade A resistance thermometers. The mass flow rate is measured with a turbine flow meter in the subcooled liquid region between the pump and the heater. Pressure tapping points are located at intervals of approximately 60 mm along the test section. All pressure measurements are made with strain gauge type transducers. The void fraction probe tip is located approximately 5 mm downstream of the test section exit. The probe is supported by a traversing mechanism, located approximately 50 mm further downstream to minimise any flow disturbance.

2.2 Local Void Fraction Measurement

Various techniques have been developed over the last decade for local void fraction measurement. These include hot wire, electrical and fibre optic probes. A fibre optic probe was chosen for the current work because it is suitable for both non-conductive and conductive fluids and therefore may be used for both refrigerants and air-water mixtures.

A fibre optic probe is either a single filament or loop of fibre optic glass contained in a sheath with an exposed tip. Incident light transmitted from a source at one end is either reflected internally at its tip to return in the reverse direction or is refracted to the surrounding medium. The amount of incident light transmitted from entry to exit depends on the refractive index of the medium surrounding the tip. If the medium is gas then most of the light is reflected internally to give a high signal. If it is liquid then most of the light is refracted externally at the tip to give a low exit signal. The local value of the void fraction in a gas-liquid flow is defined as

$$\alpha = \frac{t_g}{t_g + t_l} = \frac{t_g}{T}$$

where t_g is the time gas or vapour is present at the probe tip and t_l is the time there is liquid. T is the total measurement time.

The probe used in the investigation is a modified version of a standard unit made by RBI of Grenoble, France. It uses a single fibre with a tip diameter of 50 μm and has a response of less than 0.1 μs. The main dimensions are given in Fig.4. The rigid stainless steel casing which retains the fibre is 250 mm long to enable the probe to be attached to its traversing system well downstream of the plane of measurement.

The signal developed by the probe is processed on line by a high speed signal processing module installed in one of the free slots of a PC/AT type computer. The program for this operation was written in Pascal and Assembly language. The integrated void fraction over various time intervals ranging from 1 to 100 seconds can be obtained.

2.3 Probe Traversing System

Fig.5 is a front view of the mechanical traversing system used to move the probe across the test section. Two horizontal drive shafts control the movement. The upper is a ball screw shaft which moves the probe retaining housing horizontally. The lower is a spline shaft, which rotates a vertical ball screw shaft via a pair of bevel gears, to move the housing up or down along a pair of guide rods on which it slides. Each drive shaft is rotated by a separate externally mounted stepper motor and leakage between it and the casing is prevented by a pair of lip seals. The range of area over which the traversing mechanism can operate is determined by the diameter of the opening at the front and rear which is 203.2 mm.

The stepper motor control unit consists of a single industrial controller for the two stepper motors. This is connected to a PC/AT microcomputer via an RS232 interface card using a serial link. All the instructions to the stepper motors from the computer are handled and translated by the controller. Both motors have a 0.9o step angle with tolerance within 5.0% and are fitted with a 5:1 reduction gearbox to provide sufficient torque.

2.4 Programming

All programming operations were carried out on the same microcomputer with an Intel 80286 main processor and 80287 co-processor. The computer alternately controlled the traversing unit and high speed signal processing module. These were coordinated by a program written in Pascal. Fig.6 shows a flow chart of this program. The input for this program defines the cross section to be traversed and the coordinates of the points at which void fraction measurements are taken. The output comprises the measured void fractions and their corresponding coordinates. This is recorded and stored in a file which is then used in the post processing to construct isovoid diagrams and also to calculate the cross sectional average void fraction.

3 RESULTS AND DISCUSSION

For a complete traverse of the test section exit plane for one flow condition, void fractions were measured at 73 points at the positions shown in Fig 7a. This took approximately 45 minutes. The void fraction distribution over the cross section thus measured was presented graphically in an isovoid diagram. A total of twelve tests was made at inlet nominal two-phase mixture temperatures of 90, 100 and 110°C. However, due to space limitations, only three results are shown. These are given in Fig 7. The others are given in reference [3].

The traverses show clearly the effects of gravity on the phase distribution under all flow conditions. The highest void fraction regions are at the top and, as can be seen, these extend further downwards as the dryness fraction increases. It should be noted that due to the low vapour/liquid density ratio of the refrigerant, the average void fraction is much higher for a given dryness fraction than in air-water or water-steam mixtures. Also the rate of change of void fraction is greatest at regions where the change of the port cross sectional profile is the sharpest.

4 CONCLUDING REMARKS

The performance of the probe and associated traversing system described in this report was very satisfactory despite the highly irregular cross sectional form of the passage examined. It follows that such a system could be used for a wide range of applications in which void fraction measurements are required in other unusual shaped channels.

ACKNOWLEDGMENTS

The work reported here was sponsored jointly by the British Council and China State Education Commission and their support is gratefully acknowledged. The authors also wish to thank TFC Power Systems Ltd for the use of their equipment and financial support.

REFERENCES

(1) SALCUDEAN, M. and CHUN, J.H. Effect of flow obstruction on void distribution in horizontal air-water flow. Int.J.Multiphase Flow, Vol.9, No.1, 1983.

(2) Smith, I. K. and Martin, P.R. Power from low grade heat: Project SPHERE - an evaluation of the Trilateral Wet Vapour Cycle. C.M.E., January 1985.

(3) Xu, Z. G. Theoretical and experimental investigation of two-phase flow in the inlet port of a Lysholm screw expander. Ph.D thesis, 1991, Department of Mechanical Engineering, City University.

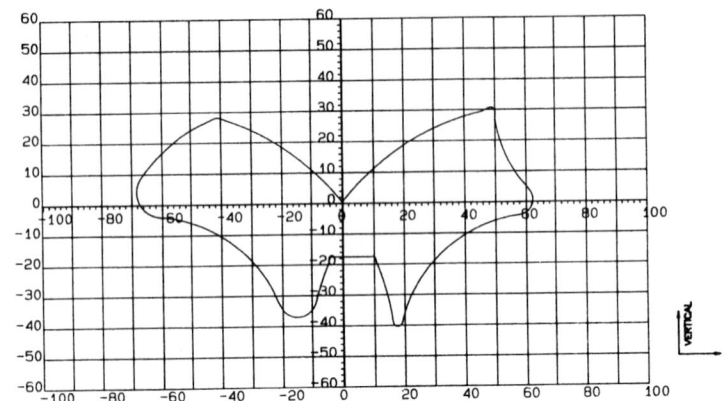

Fig.1 Outlet cross section of the inlet port
Scale (1:1), Unit (mm)

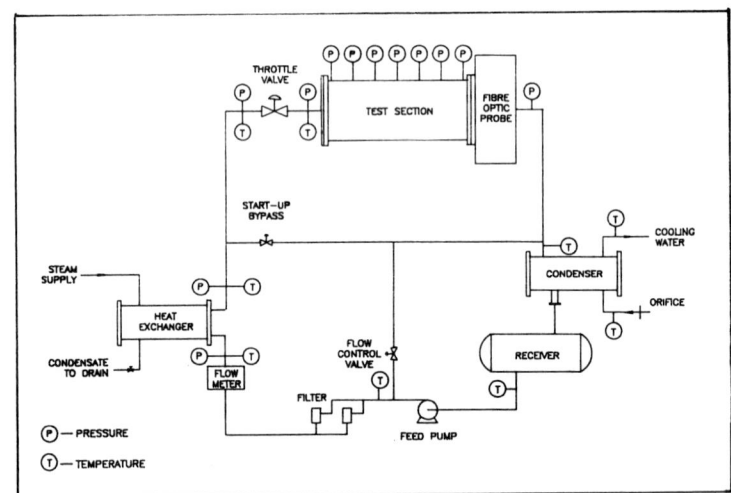

Fig.2 Closed loop test facility layout

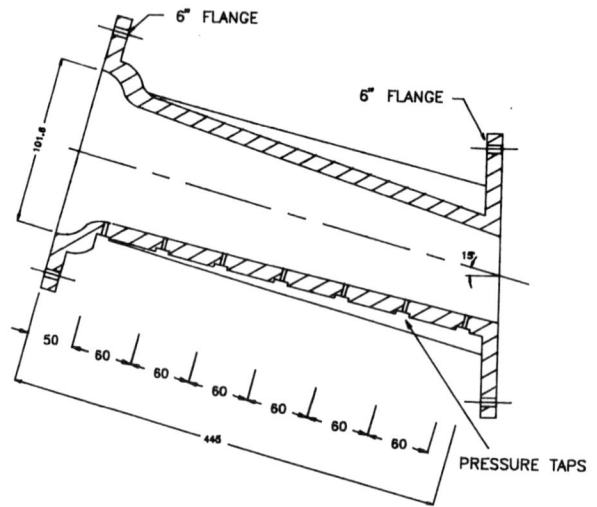

Fig.3 Test section details (not to scale)

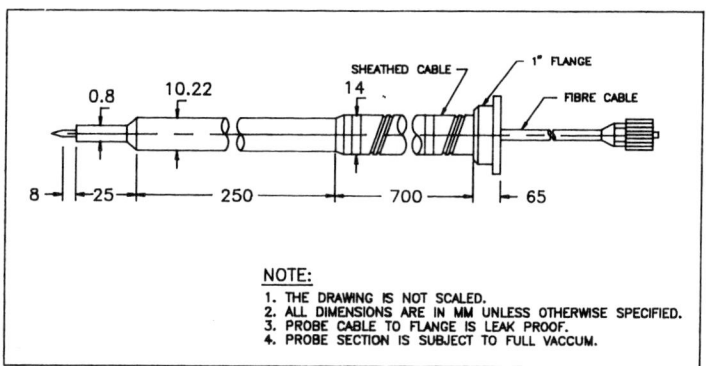

Fig.4 Fibre optic probe arrangement

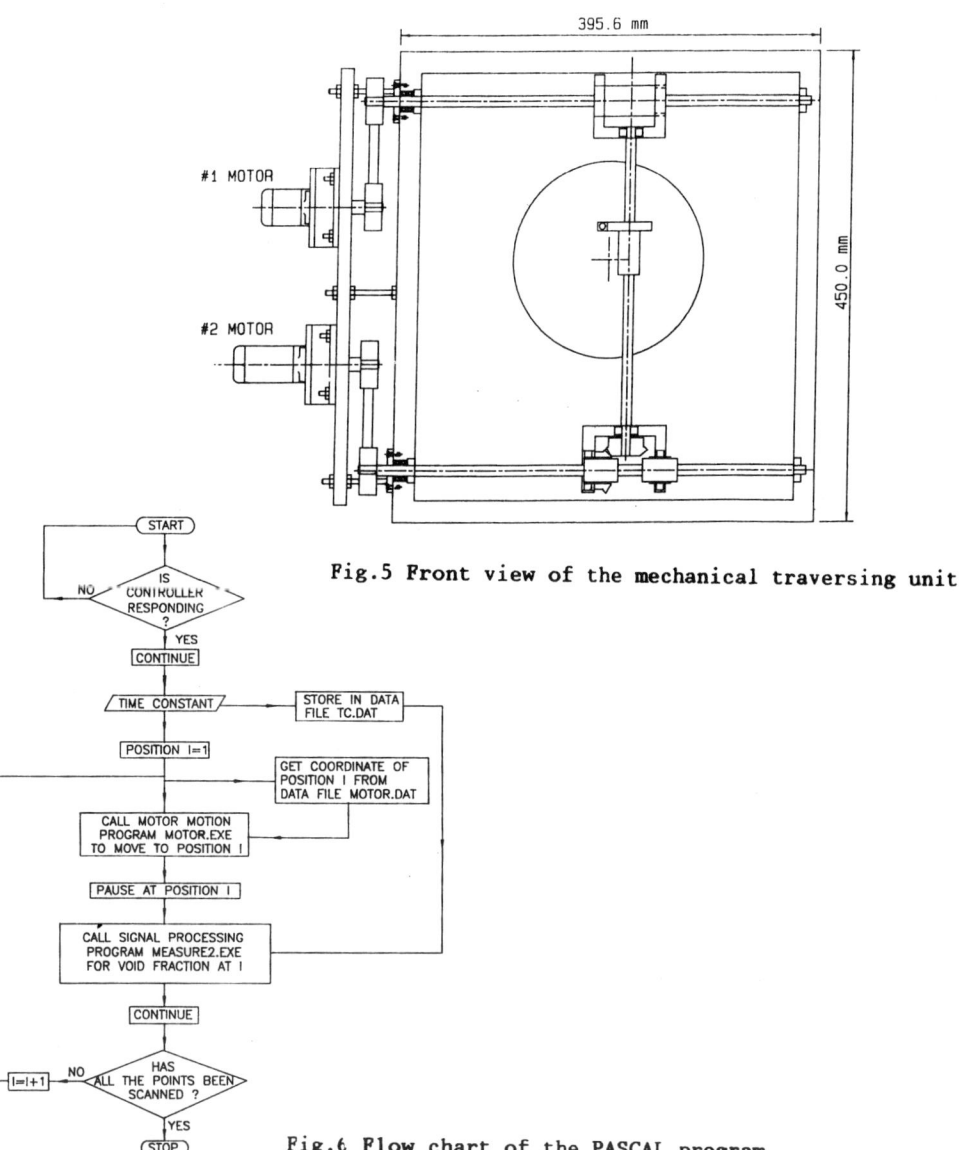

Fig.5 Front view of the mechanical traversing unit

Fig.6 Flow chart of the PASCAL program

Fig.7 Isovoid diagrams

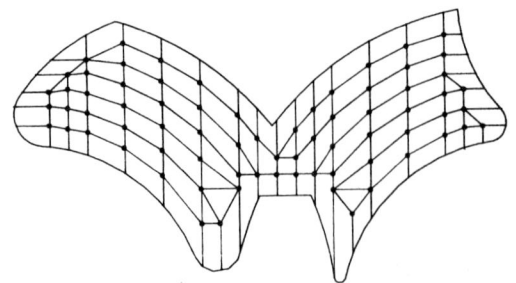

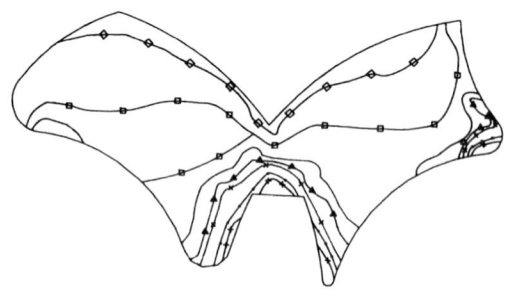

FLOW CONDITION AT INLET:

PRESSURE	:	3.43679 bar
TEMPERATURE	:	90.0000 °C
DRYNESS FRACTION	:	0.0778
MASS FLOW RATE	:	11.3330 kg/s

PRESSURE AT EXIT: 3.29734 bar

AVERAGE VOID FRACTION: 72.65 %

ISO-VOID LINES (%)
- 90
- 80
- 70
- 60
- 50
- 30
- 20

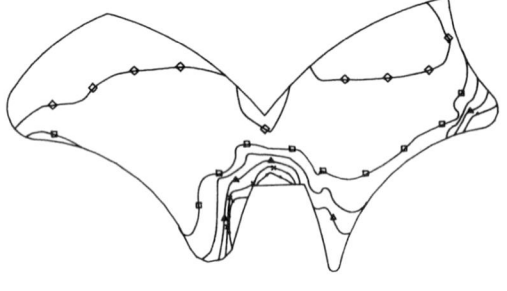

FLOW CONDITION AT INLET:

PRESSURE	:	3.43679 bar
TEMPERATURE	:	90.0000 °C
DRYNESS FRACTION	:	0.1318
MASS FLOW RATE	:	10.2559 kg/s

PRESSURE AT EXIT: 3.26529 bar

AVERAGE VOID FRACTION: 80.46 %

ISO-VOID LINES (%):
- 90
- 80
- 70
- 60
- 50
- 40

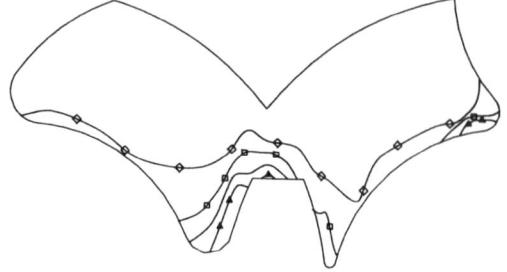

FLOW CONDITION AT INLET:

PRESSURE	:	3.43679 bar
TEMPERATURE	:	90.0000 °C
DRYNESS FRACTION	:	0.2378
MASS FLOW RATE	:	7.5818 kg/s

PRESSURE AT EXIT: 3.25536 bar

AVERAGE VOID FRACTION: 88.72 %

ISO-VOID LINES (%):
- 90
- 80
- 70
- 60

C413/044

Pressure drop in foam flow

J R CALVERT, BA, PhD
Department of Mechanical Engineering, University of Southampton

SUMMARY

A liquid-gas foam flowing through a pipe behaves more like a homogeneous non-Newtonian fluid than a two-phase flow. This means that conventional two-phase flow calculation methods produce results very seriously in error if applied to foams. The flow of the foam is dominated by an effective slip arising out of bubble migration away from the wall leaving a liquid-rich layer.

Experimental results for pressure drop in a variety of conditions are presented, together with a relatively simple model which can predict the pressure drops for many flows to within a factor of 2 in most cases. The effects of pressure variation on the parameters of the model and the flow of foam are discussed.

Notation

D	Pipe diameter
E	Foam expansion ratio
Q	Foam volumetric flow rate
Q_f	Flow rate due to shear
Q_s	Flow rate due to slip
d	Mean bubble diameter
k	Foam consistency
n	Foam flow index
γ	Shear rate
δ	Slip layer thickness
μ_g	Gas viscosity
μ_l	Liquid viscosity
τ	Shear stress
τ_y	Foam yield stress
τ_z	Wall shear stress
χ^2	Lockhart-Martinelli parameter

1 INTRODUCTION

A liquid-gas foam consists of a liquid matrix enclosing a large number of gas bubbles. Because of the excess surface energy of the interfaces, foams are inherently unstable and tend to decay with time. There are several mechanisms for this decay: one involves the liquid migrating from the lamellae between bubbles to the nodes (known as Plateau borders). The lamellae become thinner and eventually rupture thus uniting two bubbles or releasing the contents of one to the surroundings. A similar one involves liquid migration downwards under gravity. A rather different one occurs when gas diffuses through the lamellae from small bubbles to larger ones due to the excess pressure in the smaller bubbles. Foams can also decay through mechanical damage.

On the other hand, mechanical processes often put energy into a mixture to create the foam in the first place. Shaking, beating, forcing through a gauze or orifice and general turbulence can all lead to the creation of a foam.

The interaction of these two effects - the general tendency of a foam to decay, and the creation of a foam through mechanical activity - leads to the possibility of a dynamic equilibrium state where the foam, although continuously changing in detail, remains stable on some statistical basis. This is what appears to happen when foam flows as a fluid.

2 FOAM GEOMETRY

The structure of a foam is described in terms of the expansion ratio (the ratio of foam volume to liquid volume) or equivalently the quality (the ratio of gas volume to foam volume, the same as the void fraction). Typical values of expansion ratios for industrially important foams range from about 2 (food products, cosmetics) to 1000 or more (high expansion fire fighting foam). Some authorities describe low expansion foams (below about 10) as gas emulsions, others describe high expansion foams as froths; there is no clear consensus on the nomenclature.

Foam of any given expansion ratio can be made up of bubbles of almost any average size, although there does seem to be a trend to larger bubble sizes at higher expansion ratios, presumably linked via some stability requirements. Also, any given average bubble size can be obtained from many different bubble size distributions. What few results have been published (e.g. Calvert and Nezhati[1]) suggest a log-normal distribution.

Typical average bubble sizes are of the order of 100 μm. On the assumption of spherical bubbles, it is possible to calculate the average lamella thickness as a function of expansion ratio. For example, with an expansion ratio of 10 and 100 μm bubbles the thickness is about 1.9 μm.

To fully describe the foam, therefore, we need to be able to specify its expansion ratio and bubble size distribution, and also know how these quantities will vary from place to place and with time. This is impossible with the current state of the art, and we usually have to make do with bulk average values.

3 FLOW OF FOAM

The actual mechanism of foam flow is not fully understood. At low expansion ratios it probably involves shear in the (relatively thick) lamellae, while at high expansion ratios, a lamella migration process is probably involved. A two dimensional model of this type of flow has been published by Kraynik and Hansen[2].

Many reported experimental results show changes in flow behaviour at expansion ratios around 5. This is probably where the spherical bubbles at low expansion ratios start to interfere with each other and become polyhedral, and the flow mechanism changes from one of the above types to the other.

In either of these cases, the gas component remains in relatively discrete bubbles and is convected with the liquid component. This means that the phase velocities are almost the same, and the foam flows as a single phase non-Newtonian fluid, rather than a two phase fluid.

3.1 Pipe flow of foam

In pipe flow, the shear stresses are a maximum at the pipe wall, and zero on the centre line. The high stresses at the wall lead to bubble break-up. The layers nearer the wall will have a smaller average bubble size than elsewhere, and be liquid rich. In contact with the wall there will generally be a pure liquid layer. Towards the centre of the pipe, the bubble size would be expected to be larger.

In general, the liquid-rich boundary layer will have a lower viscosity than the bulk foam, and so a significant proportion of the velocity change will appear across this layer. Since one can argue on dimensional grounds that the layer will be of the same order of magnitude thickness as the (very thin) inter-bubble lamellae, this will give the appearance of boundary slip.

Away from the boundary, the behaviour will depend on the stress/strain properties of the bulk foam. Various models have been proposed for this (Hella and Kuntumukkula[3]), but most of them feature a yield stress below which no flow occurs, and a non-linear flow curve above the yield stress.

The velocity profile will thus consist of three segments: near the wall a thin layer of shearing liquid, which may be considered Newtonian and at constant shear stress equal to the wall shear stress, near the centre a plug of foam which is below its yield stress and not shearing, and in between these, if the wall shear stress is above the yield stress, a layer of shearing foam (Figure 1).

3.2 Prediction of flow behaviour

The flow behaviour described above is very different from that found in 'conventional' two-phase situations, and it is likely that methods useful for such flows may be inappropriate in foam flows. (Unfortunately, it is not always obvious whether a particular flow is a foam or not. It is quite possible for the same mixture of gas and liquid to flow as two separate phases in one set of conditions, but as a foam in another not very different set. Methods of determining this are beyond the scope of this paper.)

As an example, we may consider the Lockhart-Martinelli correlation (see Kay and Nedderman[4]). This correlation, and extensions and variants of it, are widely used and generally quoted as giving answers correct to around 20%.

We will consider an experiment in which foam of expansion ratio 6.1 flowed at a volumetric rate of 3.1 litres/min (both measured at ambient pressure) through a pipe 10 m long and 12.7 mm diameter. The superficial velocities are directly related to the foam velocity via the expansion ratio, and assuming both flows are laminar, the Lockhart-Martinelli parameter χ^2 is given by

$$\chi^2 = \frac{\mu_l}{\mu_g(E-1)} \quad \ldots\ldots(1)$$

where μ_l and μ_g are the dynamic viscosities of the liquid and gas respectively, and E is the expansion ratio.

In this case, χ^2 comes out to 10.8. The predicted single-phase liquid pressure drop is 13.3 Pa giving a predicted two-phase pressure drop of about 35 Pa. The actual measured foam pressure drop in this experiment was 58 kPa, a factor of 1660 higher.

There are other two-phase flow pressure drop calculation methods; some are discussed in Whalley[5] (p.53 to 58). Two of these have also been applied to the above foam flow case. The results were a predicted pressure drop of 757 Pa for Homogeneous Frictional approach, and 571 Pa for Friedel's correlation. While these are much larger than the Lockhart-Martinelli predictions, they are still too low by a factor of 100.

It is clear that two-phase correlations are unsatisfactory in this case; in fact in about 400 experimental runs, the pressure drop was always at least 10 times the value predicted by the Lockhart-Martinelli method, and often very much more. A different pipe flow model is clearly needed for pressure drop estimation.

3.3 Flow Model

The general structure of foam flow in pipes is described above. With a suitable equation to represent the flow curve, and empirically determined coefficients, we can calculate pressure drop treating the foam as a homogeneous non-Newtonian fluid.

A suitable model on which to base the calculation is the Calvert-Nezhati (C-N) model[6]. This has two components: a slip layer of thickness δ with the viscous properties of the base liquid, and a bulk layer characterised by a yield stress τ_y and a power law term with a consistency k and an index n:

$$\tau = \tau_y + k\gamma^n \quad \ldots\ldots(2)$$
$$(\tau > \tau_y)$$

where τ is the shear stress and γ is the shear rate.

Calvert and Nezhati note that the consistency and index are relatively constant across their data, with values of about 2.5 (SI units) and 0.4, respectively. Predictions are not very sensitive to the precise value of τ_y, since at low shear stresses, most of the flow arises from the slip layer, while at high shear stresses errors in τ_y will be unimportant. A typical value is 1 N/m^2.

Predictions are, however, sensitive to the value of slip layer thickness δ. It has been shown (Calvert[7]) that a reasonable correlation exists which enables the slip layer thickness to be calculated from the average bubble size and expansion ratio:

$$\frac{\delta}{d} = \frac{2}{3(E-1)} \quad \ldots\ldots(3)$$

where d is the mean bubble diameter and E is the expansion ratio.

Provided the bubble size can be estimated, this allows the slip layer thickness to be calculated in any given case. The bubble size is in general a function mainly of the method of foam generation.

Using this model, it is possible to write down equations giving the flow rate as a function of pressure drop, which may be solved numerically to give the pressure drop for a given flow rate:

$$Q = Q_s + Q_f$$

where:

$$Q_s = \frac{\pi \delta D^2 \tau_w}{4\mu_l} \quad \ldots\ldots(4)$$

and

$$Q_f = 0$$

$$\frac{8Q_f}{\pi D^3} = \left(\frac{\tau_w}{k}Y\right)^{\frac{1}{n}} \left(\frac{n}{n+1}Y\right)\left[1 - \frac{2n}{2n+1}Y\left(1 - \frac{n}{3n+1}Y\right)\right]$$

$$\ldots\ldots(5)$$

where $Y = (1 - \tau_y/\tau_w)$.

4 EXPERIMENTS

Measurements were made of the pressure drop of low expansion foam flowing through straight circular section polyethylene pipes. The pipes ranged from 4 m to 10 m long, and from 3.175 mm to 25.4 mm nominal diameter. The foam compound was an 3% aqueous solution of BP Byprox detergent expanded with air to expansion ratios ranging from 3.3 to 18.8 (measured at ambient pressure). The foam generator was a packed bed of glass ballotini in the size range from 0.87 mm to 1.275 mm diameter. Foam flow rates ranged from 0.1 litres/min to 13.4 litres/min (at ambient pressure). Approximately 400 sets of experimental data were collected.

4.1 Results

Figure 2 shows the results for one particular geometry, in the form of values of pressure drop as a function of expansion ratio and flow rate. (The lines of constant pressure drop ratio are sketched in by eye.) It may be seen that pressure drop increases with both flow rate and expansion ratio; this is also true for all the other geometries studied. The results, although not presented here, are all included in the calculations below.

5 CALCULATIONS

Calculations of pressure drop were made using the C-N model, extended by the relation between bubble size and slip layer thickness given above. The parameter values chosen were the typical values, $k=2.5$, $n=0.4$, $\tau_y=1$ N/m^2, as stated above.

The average bubble size generated was unknown, but was assumed to be the same for all runs (since it will primarily be determined by the geometry of the packed bed). A value was obtained by trial and error to make the average of all 400 ratios of experimental pressure drop to calculated pressure drop equal to 1. The value obtained by this method was 94 µm, which seems physically reasonable.

The results of this were that the calculated results were within a factor of 3 of the experimental results in more than 99% of the cases, and within a factor of 2 in more than 95%.

To test the sensitivity of this approach to the value of τ_y, the same procedure was carried out for other values. The results are shown in Table 1, which shows the number of experimental points for which the calculated pressure drop was a factor of more than 2 or 3 in error. The first 4 rows have the bubble size selected to give an average ratio of 1.0, and the last four use a convenient round figure for bubble size.

As can be seen, the results are not very sensitive to the value of τ_y, and it does not seem particularly worthwhile selecting bubble size precisely.

6 OPTIMISATION

Attempts were made to optimise the values of the parameters for the minimum residual on pressure drop. These were only partially successful, for reasons which will be discussed.

The first attempt used commercially available optimisation routines on mainframe computers (e.g. the NAG routine E04CCF). These routines are extremely general, and their operation is somewhat obscure to the non-expert. No acceptably stable results were obtained; output was very dependent on initial values of parameters, and problems were encountered with variables going outside permitted ranges during the optimisation process.

The next attempt was to do a manual optimisation on a desk-top computer. The least-squares residual was displayed and each parameter was adjusted in turn to minimise this. The whole process was iterated until a reasonable stable set of results was obtained. It proved possible to get a minimum residual for the flowrate (treating the pressure drop as independent) in about 200 iterations. However, when the values obtained from this process (k=4.33, n=0.41, τ_y=0.83 and d=152 µm) were used, the distribution was significantly inferior to those for the guessed parameters described above (Average 1.150, ≤.33: 0, ≤.5: 6, >2: 42, >3: 7). Reducing the bubble size to 128 µm gave only a slightly better distribution (Average 1.00, ≤.33: 2, ≤.5: 21, >2:20, >3:5).

The third attempt was similar to the above, but the pressure drop was calculated treating the flow rate as independent. This required an iterative solution of the equations, which increased the run time by a factor of about 8, but was still feasible for manual optimisation. It did not prove possible to obtain a solution for this, as the residual was extremely sensitive to yield stress and bubble size. It was necessary to converge the results to a physically unrealistic precision to get a minimum residual even for one parameter, and it would have taken an impossibly long time to optimise all four parameters. The behaviour was, in fact, similar to that shown by the NAG routine.

The final approach was to redefine the residual in terms of the ratio of experimental to theoretical pressure drop, and compute tables of this for a range of values of d, k and τ_y (holding n constant at 0.4). No absolute minimum of the residual could be found, but it was possible to find a minimum for particular values of k. This minimum always had d round about 60 µm, with τ_y falling from about 3.3 N/m^2 at k=2.0 to 0.1 N/m^2 at k=2.7. When these values were substituted back into the data, however, it was found that the average of the pressure ratios was considerably below unity, and many points lay outside the factor of 2 range. Increasing the bubble size to around 90 µm gave an average ratio around 1, and distributions similar to those in the table above, for values of k in the range 2.1 to 2.7. Generally, the distributions were narrower at the lower k end of the range.

The reasons for the failure of the optimisation processes are four-fold:

Firstly, the experimental data are subject to errors in all measured quantities, and none of the variables may be treated as truly independent.

Secondly, the model is non-linear in two important respects: it contains a fractional power which is one of the parameters, and it contains a discontinuity of slope at the (unknown) yield stress.

Thirdly, the particular combination of flow parameters, material characteristics and geometry used lead to the predictions being relatively insensitive to yield stress and (to a lesser extent) bubble size. While this is a virtue from the point of view of predicting behaviour, it is quite the opposite as regards optimisation.

Finally, the criteria of accuracy of fit are somewhat arbitrary and difficult to reconcile with each other. The optimum may be defined in terms of least-squares residual on flow rate, pressure drop or pressure drop ratio, or in terms of the arithmetic or geometric mean of pressure drop ratio or the shape of its distribution.

It is, however, possible to draw a number of qualitative conclusions from the optimisation attempt:

Results are very sensitive to the value of n, and in all cases the optimum was equal to or very close to 0.4;

Results are not very sensitive to the value of τ_y, which is normally around 1 N/m^2;

Results are not very sensitive to the value of d, which is normally around 90 to 100 µm. This is a considerable improvement over the original C-N model, where results were extremely sensitive to the value of the slip layer thickness δ;

The optimum value of k usually seems to be around 2 to 3;

The values of k and n are similar to those found by Calvert and Nezhati[6]. Since they were using a different foaming compound, a generator working on a different principle and a more restricted range of flow variables, it seems that these values could be of relatively wide application.

7 EFFECTS OF PRESSURE

The gas component of a foam is a compressible fluid, and the foam properties thus vary with pressure. The speed of sound in a foam can be very low (Calvert[7]), so that compressible flow effects may be important above a few metres per second. In the present work, all the velocities are below this range, and compressible flow effects have not been considered.

The volume flow rate and expansion ratio will both vary with pressure if the change in absolute pressure is significant. This is the case in the current work, and its effect must be considered.

The specific heat of foams is dominated by that of the liquid, even for high-expansion foams. We may thus assume that foam bubbles expand isothermally with changes in pressure. If we neglect the excess internal pressure of the bubbles which arises from surface tension, it is straightforward to calculate the effect of pressure on foam expansion ratio, mean bulk density, bulk flow rate and velocity.

Equation (3) relates the slip layer thickness δ to the bubble size and the expansion ratio. The variation of δ with pressure is therefore readily found.

The yield stress τ_y would be expected on dimensional grounds to depend on surface tension and either bubble size or lamella thickness. Surface tension is independent of pressure, so it should be possible to relate τ_y to pressure. However, as we have seen above, τ_y is not easy to evaluate accurately, and adjusting it for pressure may not be justifiable.

The parameters k and n must ultimately depend on the foam geometry. However, since there is no simple physical explanation for their magnitudes, it is by no means obvious how they might depend on pressure. The results above suggest no strong dependence, particularly in the case of n.

Given the dependence of the parameters on pressure, it should be possible to integrate along the length of the pipe to give the variation of all variables with position. However, the available experimental data is not adequate for such an exercise to be justifiable at present. In the data reported here, flow rate and expansion ratio were measured at pipe outlet (ambient) pressure. For the purposes of calculation, these were adjusted to the average of the inlet and outlet pressure. The bubble size, used to calculate δ, was adjusted from inlet pressure to the same average pressure, and all calculations were then carried out with these values.

A useful future development would be to make pressure measurements at intermediate points along the pipe, to enable an estimate of the sensitivity of k and n to pressure to be estimated.

8 CONCLUSIONS

A mixture of gas and liquid flowing as a foam has a very different behaviour from when it flows as two distinct phases. In particular, pressure drops are considerably higher and conventional two-phase calculation methods will dangerously under-estimate them.

The Calvert-Nezhati model, modified to calculate slip layer thickness in terms of bubble size and expansion ratio, is capable of predicting pressure drops to within a factor of two for most of the cases considered. It is relatively insensitive to assumptions regarding yield stress and bubble size. Standard values of consistency and index yield acceptable results.

Optimising the model for particular data sets may be possible, but may not be necessary or justified in many cases.

Although foam is a compressible fluid, the use of properties adjusted to an average pressure seems to yield acceptable results in the cases considered.

9 REFERENCES

(1) CALVERT, J.R. and NEZHATI, K. Bubble Size Effects in Foams. <u>Int J Heat and Fluid Flow</u>, 1987, 8, 102-106.

(2) KRAYNIK, A.M. and HANSEN, M.G. Foam Rheology, a model of viscous phenomena. <u>J Rheology</u>, 1987, 31, 175-201.

(3) HELLER, J.P. and KUNTUMUKKULA, M.S. A Critical review of the foam rheology literature. Ind Eng Chem Res, 1987, 26, 318.

(4) KAY, J.M. and NEDDERMAN, R.M. Fluid Mechanics and Transfer Processes. Cambridge University Press, 1985, p515.

(5) WHALLEY, P.B. Boiling, Condensation and Gas-Liquid Flow. Clarendon Press, Oxford 1987.

(6) CALVERT, J.R. and NEZHATI, K.A. A rheological model for a liquid-gas foam. Int J Heat and Fluid Flow, 1987, 7, 164-168.

(7) CALVERT, J.R. Foam in motion, Cahpter in Foams. Chemistry, Physics and Structure, (ed Wilson), Springer 1989, 27-37.

Table 1: Sensitivity of results to τ_y and d

τ_y (N/m^2)	d (μm)	Average ratio	Points outside factor of 2		Points outside factor of 3	
			≤ 0.5	>2	≤ 0.33	>3
0.1	87	1.002	10	7	1	1
1.0	94	1.002	10	8	1	1
2.0	101	1.004	7	10	1	1
3.0	106	1.001	7	10	1	2
0.1	100	1.067	6	15	1	2
1.0	100	1.032	6	9	1	1
2.0	100	0.999	8	9	1	1
3.0	100	0.970	10	8	1	1

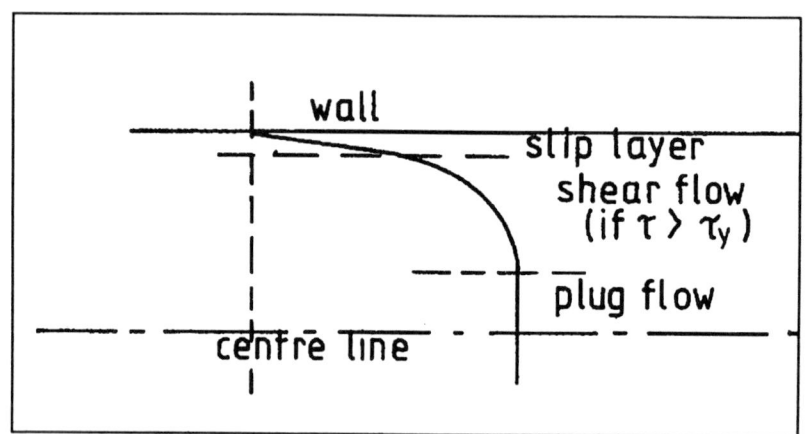

1. Velocity Profile for Foam Pipe Flow

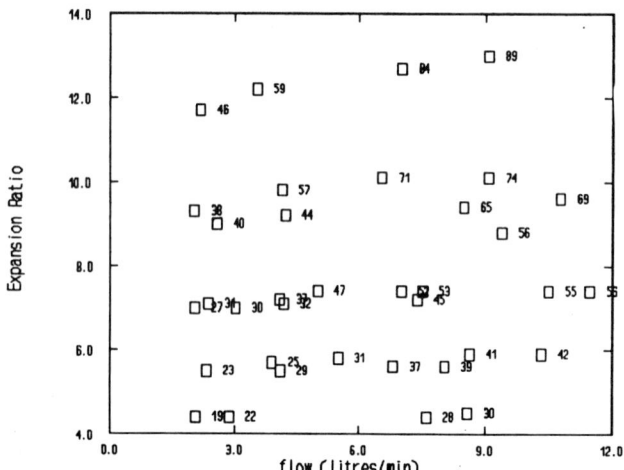

2. Pressure Drop (kPa) for Pipe 10 m long, 15.9 mm diameter

C413/020

Effect of confined geometry, using grids, on pool boiling curves

I RAJAB, BSc, MSc, PhD and R H S WINTERTON, BA, PhD,
Department of Mechanical Engineering, University of Birmingham

SYNOPSIS Transient pool boiling experiments in heating and cooling were conducted with refrigerant 113 on the same commercially pure aluminium boiling surface of 30 mm diameter. Two square grids of $0.89\lambda_d$ and $0.67\lambda_d$ were selected to see if enhancement of the heat transfer could be achieved. The results show that the confined geometry has no effect on the boiling curves, in contradiction of some results in the literature.

1 INTRODUCTION

The heat transfer rates possible in boiling are normally limited by the danger of exceeding the critical heat flux. Evidence, both theoretical and experimental, has in the past been presented to show that the critical heat flux in pool boiling can be increased by confining the surface, using solid barriers perpendicular to the surface with dimensions less than λ_d. λ_d is the fastest growing wavelength for disturbance of the liquid-vapour interface. More recent experimental work, by Egan and Westwater [1] and by the team at Birmingham [2], has not confirmed this. However the more recent work has used different surfaces (with nominally identical preparation) to study the effect of different dimensions of confining geometry; unfortunately boiling heat transfer is notoriously sensitive to surface preparation. The work to be described in this paper uses the same surface for all measurements, with different sizes of confining grids.

Transient experiments in heating and cooling were conducted with refrigerant 113 on the same commercially pure aluminium boiling surface of 30 mm diameter. Square grids of $0.89\lambda_d$ and $0.67\lambda_d$ were selected to see if enhancement of heat transfer could be achieved. Both polyester and stainless steel grids were used as an experimental check on whether there was a fin effect. In addition measurements were made with no grid present, i.e. a confined surface of $2.6\lambda_d$ diameter.

2 HYDRODYNAMIC THEORY

The hydrodynamic theory has dealt with the problem of instability in a liquid-vapour interface. Taylor [3] considered the problem with an acceleration directed perpendicular to the interface. Milne-Thompson [4] included the effect of surface tension and various other parameters. He ended with an expression for the most susceptible wavelength, λ_d, which corresponds to the fastest growth rate.

$$\lambda_d = 2\pi[3\sigma/g(\rho_\ell - \rho_v)]^{0.5} \quad (1)$$

Where σ is the surface tension, g is gravitational acceleration, and ρ_ℓ and ρ_v are the density of liquid and vapour respectively. This result was originally derived for a one dimensional wave. Two dimensional analysis increases the wavelength by a factor of $\sqrt{2}$. However the closest spacing of vapour columns on a 2D surface is given by equation (1); this has been confirmed experimentally.

In particular, Milne-Thompson [4] has found that surface tension tends to stabilize the interface, making disturbances of short wavelength unstable, and decreasing the growth rate for long wavelength disturbances. It has become conventional to refer to those instabilities in the absence of relative velocity effects as Taylor Instability. When relative velocity is important, the phenomenon is called Helmholtz Instability.

Zuber [5,6,7] proposed that transition boiling including the critical heat flux and the minimum heat flux in film boiling are characterized by the effect of hydrodynamic instabilities. Also, he postulated that the critical heat flux is limited by the combined effect of Taylor and Helmholtz instabilities, whereas the minimum heat flux in film boiling is limited by the effect of Taylor instability only.

3 LITERATURE REVIEW

Bernath [8] investigated the effect of the heater diameter on the critical heat flux in water. The heater was an electrically heated wire. The diameters used ranged from 0.0308 to 2.54 cm. He observed that the critical heat flux increases with the diameter.

Ishigai et al [9] boiled water in an open pool on the downward facing ends of copper rods of 2.5 and 5 cm diameter. The walls of the boiler were remote from the heater. The diameter of the heater was changed as well as the diameter of the boiler. The larger the rod diameter the lower the heat flux values for all the transition boiling region including the critical heat flux and the minimum film boiling point.

Costello and Frea [10] studied the effect of the boiling surface geometry on the critical heat flux. They used a maximum diameter of 1.27 cm for the boiling surface in a larger pool of liquid. They concluded that the smallest boiling surface gives the highest critical heat flux. Also, Costello et al [11] reported the critical heat flux data with water boiling on square stainless steel plates, face up, of 1.27, 2.54 and 5.08 cm sides. They observed that the critical heat flux increases as the heater size decreases. Moreover, they showed that the critical heat flux is greater when liquid inflow from the side is allowed, as compared to tests with restricting side walls.

Carne [12] measured the critical heat flux on wires, in organic liquids, of diameter ranging from 0.8 to 6.4 mm. He observed that the critical heat flux decreases as the diameter increases.

Kesselring et al [13] boiled Freon 113 on a flattened horizontal stainless steel tube. The tube was coated with epoxy resin except for a face-up bare strips of widths 0.635, 1.27 and 2.54 cm. The walls of the boiler were remote from the heater. Their results have shown no change near the critical heat flux with the strip width. But most of the transition boiling region and the minimum film boiling point have increased as the strip width decreased. Also, they concluded that a strip width in excess of $2\lambda_d$ would be needed to obtain results corresponding to those obtained using large plates.

Sauer and Ragsdell [14] studied film boiling of nitrogen on face-up, horizontal ribbons 10 cm long, having widths of 1.27, 2.54 and 5.08 cm. The heaters were made of inconel, electrically heated, and insulated from the bottom side with ceramic. The walls of the boiler were remote from the heater. They have shown that the minimum film boiling point is higher for the lower heater size.

A series of papers by Lienhard et al [15 to 21] presented the effect of the heater geometry on the critical heat flux. They reported the geometric shapes of hot wires having diameters from 0.00508 to 0.813 cm, flat ribbons having widths from 0.117 to 2.54 cm, square plates with edge lengths from 0.89 to 2.16 cm, spheres having diameters from 0.635 to 2.54 cm, and a circular disc of 6.35 cm in a centrifugal field from 1 to 7g.

Considering just the confined surface results, i.e. where the boiling surface is confined by perpendicular walls, they concluded that for a flat plate to be equivalent to an infinite plate, its diameter should be greater than $3\lambda_d$. However, for smaller diameters the critical heat flux increases as the diameter decreases. They argued that the hydrodynamic prediction of the peak heat flux on any finite body will be determined by the configuration of vapour jets above the body. They presented, with the help of the photographic study of Sun and Lienhard [22], the configuration of jets on different types of heaters. They related the areas of the jets and the number of jets to the most susceptible wavelength. Then they derived the corresponding equations to predict the values of the critical heat flux. For a confined cell of area A_H the critical heat flux is:

$$\frac{q_{max,finite}}{q_{max,Zuber}} = 1.14 \frac{N_j}{A_H/\lambda_d^2} \quad (2)$$

The number of vapour jets N_j issuing from the cell depends on the characteristic length in wavelength (L_c/λ_d; spacing of jets on an infinite surface is λ_d). Since N_j cannot be less than one there is enhancement of the critical heat flux once $L_c<\lambda_d$. $q_{max,Zuber}$ is the value predicted by Zuber [7].

Grigoryev et al [23] studied the boiling of nitrogen on upward facing, horizontal, copper surfaces of diameters from 0.8 to 3.6 cm. They have shown that the critical heat flux decreases as the heater diameter increases. This observation was for a diameter less than $2.5\lambda_d$ where for bigger diameters the critical heat flux becomes constant and equal to Zuber's prediction. It was not stated whether the boiler's walls were remote or close to confine the boiling surface.

Egan and Westwater [1] investigated the effect of the heater size on the whole boiling curve. They quenched in liquid nitrogen 9 circular copper plates, of 6.35 cm thickness, varying in diameter from 0.64 to 30.48 cm. The walls of the boiler restricted the liquid to the upper boiling surface of the plate. They concluded that the critical diameter of the heater is $2.5\lambda_d$ for nucleate boiling and the critical heat flux while for the film boiling and the minimum film boiling the critical diameter was $7\lambda_d$, for a confined heater to behave as an infinite plate. For transition boiling they considered the critical diameter to be in between the two previous values. On the other hand, for the diameters less than $2.5\lambda_d$, the critical heat flux was observed to increase with diameter.

Recently, Rajabi, Maracy and Winterton [2] studied the confined geometry effect of different circular surfaces on the boiling curves of methanol. They observed a very small change of the heat flux with diameter. Also they presented Liquid-Solid measurements which confirm their results.

4 APPARATUS AND DATA ANALYSIS

Fig. 1 shows the experimental apparatus. The experiments were performed, in heating and cooling, on a commercially pure aluminium surface of 30 mm diameter. The grids were inserted on the surface and covered with refrigerant 113. The grids were constructed from thin strips (0.2 mm thick for polyester and 0.4 mm for stainless steel). They extended 50 mm perpendicular to the boiling surface. They were not bonded to the surface.

The data were reduced by developing and applying the Kastelin method (Kudryavtsev [24]) at the upper level of the test section (3 mm below the boiling surface). A small correction is needed to give heat flux and temperature at the boiling surface rather than at the thermocouple position. Details of the experimental apparatus and data analysis technique have been given by Rajab [25,26].

Steady state tests were not conducted because of the length of time needed (around 9 hours each run) and the fear that surface conditions might change.

5 DISCUSSION

Figs. 2 and 3 show the boiling curves of refrigerant 113 with each of the two square polyester grids (10 and 7.5 mm sides) and without the grid (30 mm diameter) in heating and cooling respectively. The change in heat flux with the confined geometry is obviously insignificant. As reported elsewhere [27] the results are different in heating and cooling.

Similar results are observed with the stainless steel grids (Fig. 4). There is an increment of 4% in the critical heat flux after inserting the grids. This value is lower than expected because if the grid is welded to the surface (as opposed to just resting against it) it will work as a fin and an increment of 27% in the critical heat flux would be expected. The reproducibility of the results was good in general as long as the surface conditions did not change. Fig. 5 shows the insignificant change in the heat flux values after seven consecutive runs.

Lienhard et al [21] developed an equation (equation 3) for the critical heat flux proportional to the number of vapour jets on the boiling surface. In Fig. 6 the present experimental results for critical heat flux are compared with the Lienhard theory in a dimensionless plot of $q_{chf}/q_{chf,Zuber}$ versus L_c/λ_d. $q_{chf,Zuber}$ is Zuber's prediction. The predicted enhancement (for the smaller grid) is a factor of 2.5. There is no measured enhancement either in heating or in cooling.

On the other hand, Egan and Westwater [1], who quenched copper plates of different sizes in liquid nitrogen, reported that the critical heat flux increases slightly with the confined boiling surface dimension to an asymptotic value of 2.1×10^5 W/m^2 [Zuber's prediction] on a diameter of 2.8 cm (or 2.5 λ_d). The present work has scanned the wavelength of arcton 113 (2.6λ_d, 0.89λ_d, and 0.67λ_d) on the same boiling surface avoiding any systematic errors that might have arisen. It is very clear that in this range the dimension has no significant effect on any point of the boiling curve.

6 CONCLUSION

In the range 0.67 to 2.6λ_d the confining surface dimension has no effect on the boiling curves.

ACKNOWLEDGEMENT

We thank the Science and Engineering Research Council for a grant to support this work.

REFERENCES

(1) EGAN, J.P. and WESTWATER, J.W. Effect of horizontal plate diameter on boiling heat transfer from copper to nitrogen. *The Journal of Thermal Eng.*, 4, 1-12, 1985.

(2) RAJABI, A.A.A., MARACY, M. and WINTERTON, R.H.S. 2nd U.K. Nat. Heat Transfer Conf., Glasgow 1988, Vol. 1, 273-283, Transition pool boiling curves in confined geometry for various sizes of boiling surface. Published by Inst. Mech. Engineers.

(3) TAYLOR, G.I. The instability of liquid surfaces when accelerated in a direction perpendicular to their plane. 1.*Proc. Roy. Soc.* 201A, 192, 1950.

(4) MILNE-THOMPSON, L.M. *Theoretical hydrodynamics*. McMillan, p371, 1950.

(5) ZUBER, N. On the stability of boiling heat transfer. *Transactions of the ASME, J. of Heat Transfer*, pp.711-714, April 1958.

(6) ZUBER, N. and TRIBUS, M. Further remarks on the stability of boiling heat transfer. Rept. 58-5, Dept. of Eng., University of California at Los Angeles, January 1958A.

(7) ZUBER, N. On the stability of boiling heat transfer. *Transactions of the ASME*, pp.711-714, April 1958B.

(8) BERNATH, L. A theory of local-boiling burnout and its application to existing data. *Chem. Eng. Pro. Symp. Series*, Vol.56, pp.95-116, 1960.

(9) ISHIGAI, S., INOUE, K., KIWAKA, Z. and INAI, T. Boiling heat transfer from a flat surface facing downward. *Proc. 2nd Int. Heat Transfer Conf.*, Boulder and London, Vol.2, pp.224-229, 1961.

(10) COSTELLO, C.P. and FREA, W.J. The roles of capillary wicking and surface deposits in the attainment of high pool boiling burnout heat fluxes. *A.I.Ch.E. Journal*, Vol.10, pp.393-398, 1964.

(11) COSTELLO, C.P., BOCK, C.O. and NICHOLS, C.C. A study of induced convective effects on saturated pool boiling burnout. *Chem. Eng. Progr. Symp. Series*, Vol.61, pp.271-280, 1965.

(12) CARNE, M. Some effects of test section geometry in saturated pool boiling on the critical heat flux for some organic liquids. *Chem. Eng. Progr. Symp. Series*, Vol.61, pp.281-289, 1965.

(13) KESSELRING, R.C., ROSCHE, P.H. and BANKOFF, S.G. Transition and film boiling from horizontal strips. *A.I.Ch.E. Journal*, Vol.13, pp.669-675, 1967.

(14) SAUER, H.J. and RAGSDELL, K.M. Film pool boiling of nitrogen from flat surfaces. *Advances in Cryogenic Eng.*, Vol.16, pp.412-415, 1970.

(15) LIENHARD, J.H. and SCHROCK, V.E. The effect of pressure, geometry and the equation of state upon the peak and minimum boiling heat flux. *Trans. of the ASME J. of Heat Transfer*, pp.261-271, August 1963.

(16) LIENHARD, J.H. and WONG, P.T.Y. The dominant unstable wavelength and minimum heat flux during film boiling on a horizontal cylinder. *J. Heat Transfer*, Vol.86, pp.220-226, 1964.

(17) LIENHARD, J.H. and WATANABE, K. On correlating the peak and minimum boiling heat fluxes with pressure and heater configuration. *Trans. of the ASME J. of Heat Transfer*, pp.94-99, February 1966.

(18) LIENHARD, J.H. and KEELING, K.B. An induced convection effect upon the peak boiling heat flux. *Trans. of the ASME J. of Heat Transfer*, pp.1-5, February 1970A.

(19) LIENHARD, J.H. and SUN, K. Effect of gravity and size upon film boiling from horizontal cylinders. *Trans. of the ASME J. of Heat Transfer*, pp.292-297, May 1970B.

(20) LIENHARD, J.H. and DHIR, V.K. Hydrodynamic prediction of peak pool boiling heat fluxes from finite bodies. *ASME J. of Heat Transfer*, Vol.95, pp.152-158, 1973.

(21) LIENHARD, J.H., DHIR, V.K. and RIHERD, D.M. Peak pool boiling heat flux measurements on finite horizontal flat plates. *J. of Heat Transfer*, Vol.95, pp.477-482, 1973.

(22) SUN, K. and LIENHARD, J.H. The peak pool boiling heat flux on horizontal cylinders. *Int. J. Heat Mass Transfer*, Vol.13, pp.1425-1439, 1970.

(23) GREGORYEV, V.A., KLIMINKO, V.V., PAVLOV, YU, M., MAZOV, A.V. and POTHEKIN, S.A. An experimental study of crises and heat transfer in pool boiling of cryogenic liquids. *Heat Transfer Soviat Research*, Vol.10, pp.147-154, 1979.

(24) KUDRYAVTSEV, Y.V. Unsteady state heat transfer. American Elsevier Publishing Company, New York, 1966.

(25) RAJAB, I.K. Transition boiling in steady and transient states on the same boiling surface. Ph.D. Thesis, University of Birmingham 1989.

(26) RAJAB, I.K. and WINTERTON, R.H.S. The two transition boiling curves and solid-liquid contact on a horizontal surface. *Int. J. Heat and Fluid Flow* II 149-153, 1990.

(27) MARACY, M. and WINTERTON, R.H.S. Hysteresis and surface energy effects in transition pool boiling. *Int. J. Heat Mass Transfer*, 31, pp.1443-1449, 1988.

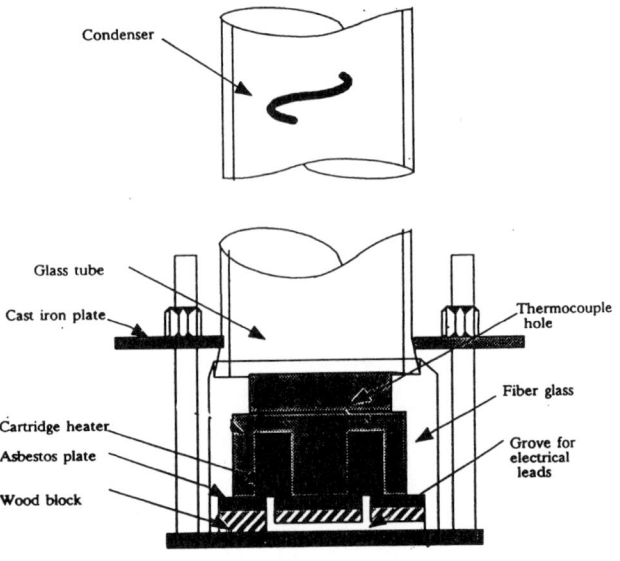

Fig 1 Pool boiling heat transfer appparatus.

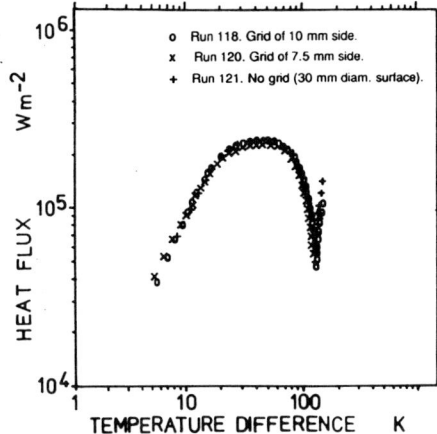

Fig 2

Boiling curves for refrigerant 113 in transient heating runs with various sizes of polyester grids.

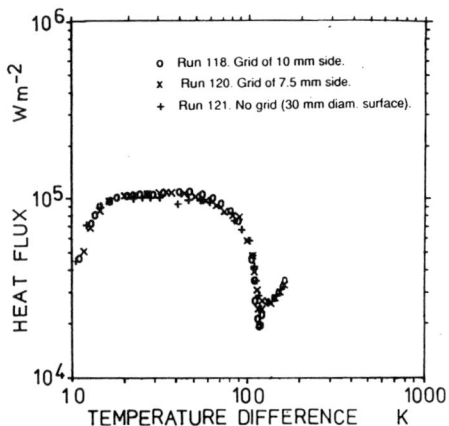

Fig 3

Boiling curves for refrigerant 113 in transient cooling runs for various sizes of polyester grids.

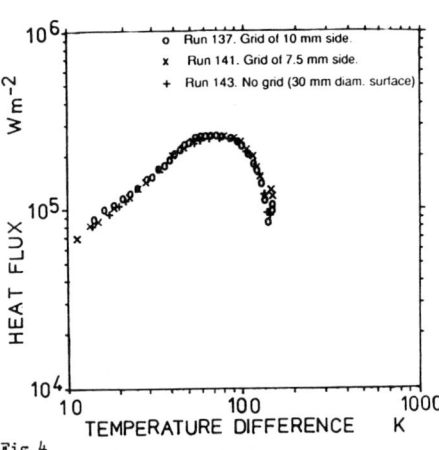

Fig 4

Boilings curves for R 113 in transient heating for various sizes of stainless steel grid.

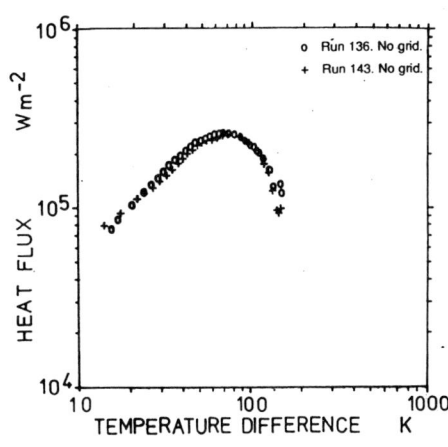

Fig 5

Reproducibility of runs under identical conditions (heating, before and after runs of Fig. 4).

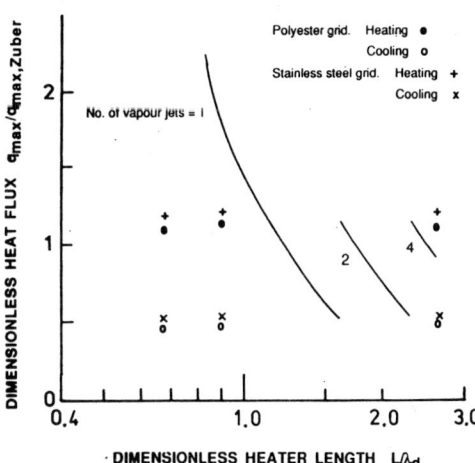

Fig 6

Comparison of present experimental results with theory (discontinuous solid curve). The theoretical prediction depends on the number of vapour jets.

C413/036

Electro-hydrodynamic enhancement of two-phase heat transfer

T KARAYIANNIS, PhD, AMIMechE
Institute of Environmental Engineering, South Bank Polytechnic, London
P H G ALLEN, PhD, CEng, MIMechE, FIEE
Thermo-Fluids Engineering Research Centre, Department of Mechanical Engineering and Aeronautics, The City University, London

ABSTRACT

In this paper the electrohydrodynamic enhancement mechanism is first outlined. A comprehensive review of past experimental work on EHD boiling and condensation is then presented. The paper includes recent experimental results for EHD boiling and condensation of refrigerants in single tube heat exchangers and a description of possible practical EHD electrode systems for applications in refrigeration and power production cycles.

NOTATION

c_A, c_B, c_C	:	empirical constants
c_p	:	specific heat
d	:	vapour film thickness
D	:	diameter
E	:	electric field strength
f	:	function defined in equation 14
F	:	body force
F_E	:	equivalent EHD force
g	:	acceleration due to gravity
h	:	surface height
ℓ	:	characteristic length
n	:	empirical exponent
Ne	:	defined in equation 17
Nu'	:	modified Nusselt Number (defined in equation 4)
q	:	electric charge density
$\dot{q}$	:	thermal flux density
Ra'	:	modified Rayleigh number (defined in equation 5)
Re	:	dimensionless group (defined in equation 11)
Re_0	:	dimensionless group (defined in equation 18)
Re_E'	:	dimensionless group (defined in equation 12)
t	:	charge relaxation time
T	:	temperature
ΔT	:	temperature difference, wall – saturated vapour
i_{lg}	:	enthalpy of evaporation
$\Delta i'$	:	thermal energy term (defined in equation 7)
V	:	dimensionless group (defined in equations 9)
α	:	heat transfer coefficient
α'	:	corrected (condensation) heat transfer coefficient
ε	:	fluid permittivity
η	:	dynamic viscosity
θ	:	contact angle
λ	:	thermal conductivity
λ^*	:	perturbation wavelength
ν	:	kinematic viscosity
π	:	dimensionless group (defined in equation 10)
ρ	:	density
σ	:	surface tension

Superscript

$-$	:	mean

Subscripts

c	:	critical
D	:	dielectrophoretic
E	:	due to electric field
g	:	saturated vapour

i	:	inner
l	:	liquid
ln	:	logarithmic mean
max	:	maximum
o	:	outer
0	:	of free space (permittivity), at zero electric field
w	:	solid surface

1. INTRODUCTION

The phase change that occurs during a boiling and a condensation process is generally a very effective mode of heat transfer. However, there is further need to develop methods of enhancing the heat transfer rates in both evaporators and condensers. A refrigeration/heat pump system usually operates with heat sink and heat source at predetermined or not controllable (e.g. atmospheric) temperatures. The working fluid operating temperature must be significantly lower than the source and higher than the sink in order for the evaporator and the condenser to be of reasonable, economic size. In a refrigeration/heat pump system these temperature differences, which constitute external thermal irreversibility, reduce the coefficient of performance (COP) of the refrigerator or heat pump. Similar arguments hold for the power producing cycles where the efficiency of the cycle is reduced as the greatest possible temperature difference is reduced by the external thermal irreversibility mentioned above. The beneficial result of heat transfer enhancement i.e. the increase in the boiling or condensing heat transfer coefficient, would be:
(a) the reduction in the size of evaporator and condenser for given ratings.
(b) the reduction in the temperature difference between evaporator and source and condenser and sink and thus greater COP or cycle efficiency. (Alternatively, the removal of greater rates of energy through a given size of evaporator/condenser while maintaining moderate temperature differences.)

Reducing the temperature differentials at evaporator and condenser becomes particularly important in the more recent development of power production using Ocean Thermal Energy Conversion (OTEC) plants, Geothermal Energy plants, Solar Pond generation plants and similar plants utilizing the Organic Rankine Cycle (ORC). In these plants the available temperature difference (between source and sink) is already small, e.g. in OTEC plants this is the difference between the temperature of surface (20-30° C) and deep (4-7° C) sea water. Plants operating between such small temperature differences must be larger than high temperature plants (nuclear or fossil fuel plants) in order to produce similar outputs. In addition, their performance is more sensitive to the temperature differential at the evaporator-source and the condenser-sink. A certain degree of superheat must be exceeded in the evaporator for nucleate boiling to commence. Such plants may, therefore, face additional "start-up" problems if this degree of superheat is not available. This places an additional limitation on the maximum possible operating fluid temperature difference and thus reduces the plant efficiency.

In the past, two-phase heat transfer enhancement was achieved using 'lo-fin' surfaces. More recently, manufacturers have introduced high heat flux surfaces. In boiling applications manufacturing techniques are employed to (a) deform the surface so as to create re-entrant type cavities or (b) to deposit a matrix of metallic particles on the surface. Either gives a porous layer which enhances boiling by providing sites (the pores) for bubble generation. Yilmaz et al [1] and Bergles [2] provide experimental results on the performance of these high heat flux surfaces in boiling. In condensation applications, in addition to 'lo-fin' tubes, tubes with sharp sawteeth type fins and trapezoidal-shaped extending fins are examples of enhanced tubes more recently introduced.

Heat transfer enhancement methods are categorized as "passive" or "active". The first category includes the high heat flux surfaces mentioned above. The second category includes techniques such as surface vibration, fluid vibration, vapor suction and electric fields, see Bergles [2] and [3]. The enhancing effect of a strong electric field on heat transfer rates (electrohydrodynamic -EHD- enhancement) has been known for over seventy years. Early work concentrated on the enhancement of single-phase convective heat transfer. In the last thirty years the greater potential of EHD in enhancing two-phase heat transfer rates has been realised by more industrial and academic researchers. This has led to the study of coupled electric field and pool boiling in the nucleate, transitional and film boiling regimes. The effect of the electric field on the peak nucleate and the minimum film boiling conditions has received particular attention. While work on EHD boiling proceeded steadily, relatively little was published on EHD condensation until the last decade, possibly due to the greater practical potential offered by EHD boiling. The present paper reviews the published work on EHD enhanced condensation and evaporation and presents the results of recent work on both.

2 EHD ENHANCEMENT MECHANISMS

The physical basis of much electrically enhanced heat transfer lies in the EHD force, F_E per unit volume, generated by an electric field, strength E, in a fluid of dielectric permeability ε, density ρ, at temperature T. This can be expressed [4]:

$$F_E = qE - \frac{1}{2} E^2 \nabla \varepsilon + \frac{1}{2} \nabla \left[E^2 \left(\frac{\partial \varepsilon}{\partial \rho} \right)_T \rho \right] \qquad (1)$$

where q is the electric charge density in the fluid. qE, its electrophoretic component, is the Coulomb force exerted by an electric field upon free charge in it. The other two terms are, respectively, the dielectrophoretic and electrostrictive forces on and within the fluid due to the nature of ε and its spatial distribution. In condensation and film boiling, the dielectrophoretic force acts to destabilise the layer of condensate or vapour, respectively, and thus reduce the thermal resistance at the heat transfer surface [5,6]. In the case of nucleate boiling, Rohsenow [7] has shown that the degree of superheat needed for bubble growth at a given bubble size depends directly on the surface tension of the liquid - vapour interface. Also, the contact angle θ is related to the surface tension by:

$$\cos\theta = \frac{\sigma_{wl} - \sigma_{gw}}{\sigma_{lg}} \quad (2)$$

where σ_{wl}, σ_{gw} and σ_{lg} are, respectively, the surface tensions at interfaces solid-liquid, vapour-solid and liquid-vapour [8].

Both θ and surface tension have been observed to vary with electric field strength [9,10] and this would explain the initiation of ebullition by an electric field at reduced superheat.

3 CONDENSATION EXPERIMENTS

Publications on the EHD enhancement of condensation began appearing in 1965 with the work of Velkoff and Miller [11] and Choi and Reynolds [12] using Freon 113. This literature is reviewed in Table 1.

Several themes can be discerned. First, condensation has been studied on surfaces that are flat or curved, horizontal or vertical. In all cases, the disposal of the condensate, once formed, can be a problem. On a vertical surface, as the condensate film thickens with increasing distance downwards, heat transfer coefficients reduce. In banks of horizontal tubes, the same effect results from condensate dripping from those above. Smirnov and Lukanov [24] report a >60% reduction in heat transfer coefficient between rows numbers one and five for Freon 11 condensing on a bundle of 'lo-fin' tubes.

As shown in Fig.1, even the best EHD assisted performance falls short of that given by using a single 'lo-fin' profile tube. EHD has little effect on 'lo-fin' tube condensation as it and the Grigorig effect act in opposition. However, in addition to destabilising the condensate film, EHD can strip condensate off the surface and overcome the effects described in the previous paragraph. Fig. 2 shows that EHD enhancement of Freon 114 condensing on a smooth, 506 mm long, 19.1 mm O.D. brass tube with vapour to tube wall mean temperature difference 25K is of much the same magnitude whether the tube is horizontal or vertical. Values of

$$\frac{\overline{\alpha}'_E}{\overline{\alpha}_0}$$

from the measurements of Cooper [25] are given as a function of voltage applied to a 38 mm I.D. cylindrical copper gauze electrode surrounding the tube. (Measured mean heat transfer coefficients α_E at electric field strength E have been corrected to primed values α'_E by taking into account the increase due solely to reduced ΔT [21].

Secondly, from the earliest work, relationships have been suggested for correlating the results of different experiments. This subject is dealt with in detail later.

4 BOILING EXPERIMENTS

By contrast with condensation, boiling appears to have benefited from EHD since 1916 when a threefold increase in water evaporation rate was claimed by Chubb [26]. However, the present era of such EHD usage began with the work of Bonjour, Weil et al [27] [28], published in 1960 and 1962, closely followed by Watson [29], Choi [30], Jalaluddin and Sinha [31] and Markels and Durfee [9, 32], all except [32] for pool boiling. The work used a wide variety of fluids and, with subsequent work, is reviewed in Table 2.

To define terms used in Table 2, Fig. 3 shows a typical boiling characteristic curve in terms of thermal flux per unit area, $\dot{q}$, versus superheat, ΔT. Starting from ΔT=0, the heat transfer process proceeds along curve AB, by natural or forced convection, to C, where ΔT is sufficient to initiate nucleate ebullition, thus giving greatly increased $\dot{q}$ (at D). $\dot{q}$ then continues to increase with ΔT until (at E) it reaches a value $\dot{q}_c$, known as peak nucleate boiling, or critical, heat flux density where the coalescence of the bubbles forming at the heat transfer surface into an unstable film impedes energy transfer and decreases $\dot{q}$ from E to F through a transition region of partial film boiling until, at F, a minimum film boiling condition is given by a stable, blanketing, vapour film. Increasing ΔT then gives stable film boiling, with increasing $\dot{q}$, beyond F. If from beyond D, ΔT is decreased, nucleate boiling continues along DB' towards A, giving "boiling hysteresis".

EHD eliminates the $\dot{q}$ decrease between E and G and, in addition, increases $\dot{q}_c$. These phenomena have received very much more attention than EHD's other action, viz the initiation of nucleate boiling at smaller ΔT and hence the virtual elimination of the hysteresis portion (ABCDB'A) of the characteristic e.g. by applying volts at ΔT values between b and c. This initiation process has been reported only by Jalaluddin and Sinha [31], Basu [35] and Allen and Cooper [39] who have also made a video record of it [47]. In this, observation is made of a tube, immersed in Freon 114. Hot water passes along the tube losing heat to the Freon. Thus, near the inlet the tube surface gives sufficient superheat for bubbles to evolve at a well-defined set of sites. However, towards the outlet ΔT is only sufficient to give convective heat transfer and thus a well-defined "front" is seen between boiling and non-boiling regions. On application of electrode voltages greater than about 8kV there is immediate - and more vigorous - ebullition everywhere on the tube surface. (Once initiated, it is not necessary to <u>maintain</u> the voltage to maintain ebullition.) In at least one case (Baboi et al [44]), the phenomenon was deliberately avoided by reducing the thermal loading to the experimental value required (i.e. along EDB' in Fig.3) "so that most of the vapour-formation centres were active". Also, as remarked by Markels and Durfee [9]: "It appears likely however that differences exist between voltage effects when a relatively large diameter heat transfer surface is used and when the boiling takes place on a thin wire". A fine wire fed by a constant electric current cannot instantaneously provide the additional thermal energy demanded by suddenly promoted ebullition. The energy could only come from a considerable decrease in temperature, and hence in ΔT, thus suppressing ebullition before it became established. By contrast, the more usual industrial case of a source heated by fluid flow maintains a much more nearly constant temperature, as can a high thermal capacity source (see Table 2 notes).

The most recent work in this area (based on data from [25] and [41]) is summarised in Fig.4 on a "tube rating" (kW per meter of tube versus ΔT) basis which, for a given tube diameter, governs heat exchanger design. As, without EHD, the mean temperature drop tube water to tube wall was about three times $T_w - T_g$, this "rating" is the characteristic that governs the rate of energy transfer.

5 EHD ENHANCEMENT CORRELATIONS

Correlation of heat transfer data in terms of non-dimensional quantities for specific geometries has long proved a valuable tool for heat exchanger design. In the case of EHD assisted condensation, correlation has been attempted from the earliest work. Having analysed the disruption of the condensate film in terms of a perturbation wavelength λ^*, Choi and Reynolds [12][13] used this as the characteristic length in Nusselt Number Nu' and modified Rayleigh number Ra' for the correlation.

$$Nu' = C_A (Ra')^{0.25} \qquad (3)$$

where C_A is an empirical constant and:

$$Nu' = \frac{\alpha \lambda^*}{\lambda} \qquad (4)$$

$$Ra' = \frac{F_\xi \rho (\lambda^*)^3 \Delta i'}{\lambda \eta \Delta T_{ln}} \qquad (5)$$

where λ and η are fluid thermal conductivity and viscosity, respectively, ΔT_{ln} is the logarithmic mean value of ΔT,

$$F_\xi = \frac{3}{4\sigma} \left[\left\{ 1 - \frac{\varepsilon_g}{\varepsilon_l} \right\}^4 \varepsilon_g^2 E_g^4 \right] \qquad (6)$$

and

$$\Delta i' = i_{lg} + C_{pg} \Delta T + \frac{3}{8} C_{pl} \Delta T_{ln} \qquad (7)$$

By contrast, Bologa and Didkovskiy [16] in their earlier work used the simpler relationship:

$$\frac{\overline{\alpha}_E}{\overline{\alpha}_0} = 0.28 \left(\frac{E}{E_c}\right)^{2.5} \qquad (8)$$

where E_c is the electric field strength at which the EHD effect becomes significant. However, in later work [17] they used the relatively complex correlation

$$\log \left[\frac{\overline{\alpha}_E}{\lambda} \left\{ \frac{\nu^2}{g} \right\}^{\frac{1}{3}} \right] \text{ versus } \log \left[\frac{V^{1.2} \pi^{0.2}}{Re^{1/3}} \right] \text{ where:}$$

$$V = \frac{\varepsilon_0 \varepsilon_g E_g^2 \ell}{\sigma} \qquad (9)$$

$$\pi = \frac{\ell^2}{t \nu} \qquad (10)$$

and $\quad Re = \frac{\dot{q} h}{i_{lg} \rho \nu} \qquad (11)$

the characteristic length ℓ being the annular gap while h is the surface height and t the charge relaxation time for the liquid. Subsequently [19], they correlated their own results together with those of Velkoff and Miller [11] and Smirnov and Lunev [18] in this way but with the index of π decreased to 0.175. Smirnov and Lunev [18] themselves proposed the correlation of

$$\frac{Nu_E}{C_C} - \frac{C_B Nu_O}{\left(\frac{D_i}{D_o}\right)^3} \quad \text{(Nusselt numbers using heat transfer surface height as characteristic)}$$

versus Re'_E where:

$$Re'_E = \frac{h (\varepsilon_1 - \varepsilon_g) E}{\nu \varepsilon_1} \sqrt{\frac{\varepsilon_0}{\rho} \left[1 - \left\{ \frac{E_c}{E} \frac{D_i}{D_o} \right\}^2 \right]} \qquad (12)$$

In spite of these later refinements, Cooper [25] found the original (Choi and Reynolds [12] [13]) correlation valid for his own results as well as for those of Didkovsky and Bologa [19]. With $C_A = 0.35$ (rather than 0.5 as originally proposed) the scatter is about ±30% for $F_\xi > 60000$ which is no worse than with the correlations suggested in [18] and [19].

In boiling, the effect of the electric field upon the maximum ($\dot{q}_c$) and minimum (point F in figure 5) heat fluxes and the correlation of this effect has received particular attention. In the case of $\dot{q}_c$ this is important since it could represent the maximum allowable heat transfer. Increasing the heat flux to a value greater then $\dot{q}_c$ gives rise to very high ΔT leading to burnout. Johnson [48] considered the influence of the electric field on hydrodynamic stability. He amended the hydrodynamic theory of boiling heat transfer to include the effect of a perpendicular electric field across the vapour-liquid interfaces that exist at the maximum and minimum heat flux and correlated the ratio $(\dot{q}_c)_E / (\dot{q}_c)_O$ as a function of fluid properties and electric field strength. The resulting equation for maximum heat flux condition was

$$\frac{(\dot{q}_c)_E}{(\dot{q}_c)_0} = f \left\{ \frac{[1 + \sqrt{1 + 3\frac{(\rho_1 - \rho_g)g\sigma}{f^2}}]}{\sqrt{3(\rho_1 - \rho_g)g\sigma}} \right\}^{\frac{1}{2}} \quad (13)$$

where

$$f = \varepsilon_0 \frac{\varepsilon_g}{\varepsilon_1} \frac{\left(\varepsilon_1/\varepsilon_0 - \varepsilon_g/\varepsilon_0\right)^2}{\varepsilon_1/\varepsilon_0 + \varepsilon_g/\varepsilon_0} E^2 \quad (14)$$

A similar relationship was obtained for the minimum heat flux. These relationships were compared both in Johnson [48] and Winer [33] with the limited experimental results of Winer (for a radial field and a cylindrical heat transfer surface) and found to be in good agreement.

Berghmans [49] considered a flat heater located in a conducting liquid and performed a stability analysis including the effect of a uniform DC field. This analysis resulted in

$$\dot{q}_c = \rho_g^{\frac{1}{2}} i_{1g} \frac{\pi}{2} \frac{1}{8} \left(\frac{1}{3}\right)^{\frac{1}{4}} (\sigma\rho_1 g)^{\frac{1}{4}}$$

$$\left\{\frac{G^2}{3^{1/2}B} + \left(\frac{G^4}{3B^2} + 1\right)^{\frac{1}{2}}\right\}^{\frac{1}{2}} \quad (15)$$

where $G^2 = \frac{\varepsilon_g E^2 d}{\sigma}$

is the ratio of the electric forces to the surface tension forces

and $B^2 = \frac{\rho_1 g d^2}{\sigma}$

B^2 being the Bond number which is the ratio of inertia to surface tension forces. d is the vapour film thickness which can be obtained from experiments. This equation compared well with the data of Markels and Durfee [9]. A correction was made to allow for the cylindrical surface used in [9]. The choice of d which varies in the experiment was considered to be a possible reason for any deviations of the theory from the experiment.

Zhorzholiani and Shekriladze [5] have successfully correlated their own results and those of Markels and Durfee [9] on a log plot of

$\frac{\dot{q}}{\dot{q}_c}$ versus $\frac{F_D + g\rho}{g\rho}$

F_D being the volume force density (given by summing the second and third terms of equation 1). More recently, Cooper [50] has proposed a model for EHD nucleate boiling that leads to the relationship:

$$\frac{\alpha_E}{\alpha_0} Ne^{-\frac{n}{2}} = 0.3 Re_0^{-0.16} \quad (16)$$

where n is found experimentally.

$$Ne = 1 + \left\{\frac{1.5\varepsilon_1(\varepsilon_g - \varepsilon_1)}{(\varepsilon_g + 2\varepsilon_1) g (\rho_1 - \rho_g)}\right\} \nabla E^2 \quad (17)$$

and

$$Re_0 = \frac{\dot{q}}{i_{1g} \eta_1} \sqrt{\frac{\sigma}{g(\rho_1 - \rho_g)}} \quad (18)$$

Equation 16 correlates the results of [28], [29], [30] and [44] to within ±10% for the majority of 37 observations graphed with only one outside ±30%.

6 ELECTRODE SYSTEMS

Little of the work reviewed here claims to include proposals for the practical implementation of EHD enhancement in industrial plant although most commend its potential for material savings etc. However, Yabe et al [20, 23] describe continuing work on the development of electrode systems for vertically aligned shell/tube condensers with particular applications in heat pump systems. The electode took the form of varying pitch helices. These use EHD to remove the condensate film at points along the tube surfaces and to spray it on to the high voltage electrodes down which it trickles. Fig.5 shows one such arrangement, optimised with respect to extraction and removal angles (defined in the figure) and dimensions.

This arrangement has been incorporated into a 50kW, 102 tube, EHD condenser for a prototype heat pump and has shown heat transfer coefficient improvements compared with no EHD of between 4 and 5 times for Freon 113 and of about 3 times for fluorohexane [51]. (This work has also confirmed the validity of applying EHD results obtained in single-tube rigs to multi-tube heat exchangers). To enhance boiling, the Japanese have also employed EHD pumping [46] or fine wire electrode systems [40] but neither has been applied in the prototype heat pump mentioned above.

By contrast, the electrode geometry developed by Allen and Cooper [52] specifically for shell/tube heat exchangers and shown in Fig. 6 is mechanically simpler and more compact. It has been shown to be effective for both condensation and boiling. It comprises a combination of rod and perforated plane electrodes giving a fair approximation to the ideal, viz. a concentric cylinder around each tube as shown in Fig. 7 [53].

7 DISCUSSION

Obstacles to the developments of improved heat transfer systems such as practical EHD heat exchangers have been reviewed by Butterworth [54]. He points out that plant construction is organised so "that the people who can see the advantages of improved heat transfer systems, and indeed benefit from them, may not be the same people charged with designing those systems. The latter have no incentive to make changes and will therefore design the system the same way as they did the last time". He concludes that "there are, however, possible ways to improve the shell and tube, and we should use as much ingenuity to making further developments in the shell and tube as we have to developing alternative types".

One must, of course, consider the relative merits of EHD and "passive" enhancement techniques, remembering that the latter are effective only in the nucleate boiling region (up to E on Fig. 5). By comparison, EHD can enhance every part of the boiling curve. Kajikawa et al [55] gave results for tubes comparable in size to those used in the present study coated with a "Highflux" - type porous copper coating and with fine metal fibre used in a Freon 114 boiler.

Apart from a graph showing a step increase of about three times in tube row heat transfer coefficients as the warm water temperature increased to 30 °C and above (no freon saturation temperature is quoted) which is described as "peculiar", there is no data on hysteresis. Maximum $\bar{\alpha}$ values of about 11kW/m²K can be construed from their results. Results for other freons are given by Czikk et al [56] (Freon 11, maximum $\bar{\alpha}$ 42.5kW/m²K) for coated tubes and by Arai et al [57] (Freon 12, maximum $\bar{\alpha}$ 24kW/m²K) for "Thermoexcel-E" with no mention of hysteresis. Hysteresis with a porous metallic coating has, however, been thoroughly investigated by Bergles and Chyu [58] for the case of Freon 113. They found that about 8K superheat was needed for ebullition, which then increased $\bar{\alpha}$ by about 20 times, a degree of boiling curve hysteries that they describe as "dramatic". They conclude that although greatly enhancing heat transfer, a porous metallic matrix surface is to be expected to give hysteresis and that this effect had not been presented previously in the literature.

Regarding enhanced condensation, Arai et al [57] report improvements using Freon 12 and "Thermoexcel" tube amounting to between 25% and 70%. This, however, could well be no better than the improved performance of tube banks given by EHD condensate stripping.

The excellent boiling performance of EHD combined with a 'lo-fin' surface is probably connected with Cornwell's observation [59] that ".... in the higher quality (boiling) regions this may become churn flow where bubbles coalesce to form elongated vapour regions". The electric field combines with buoyancy and the tube profile shape to produce the elongation.

Lastly, the many other possibilities inherent in EHD enhancement should not be ignored. For example, the work of Rutkowski [45] suggests that the evaporation of recalcitrant cryogenic liquids may well be assisted by EHD.

8 CONCLUSIONS

The authors hope that they have justified the following conclusions:

1. EHD can be applied effectively to the enhancement and control of condensation and boiling of very many fluids.

2. In condensers, the EHD effect upon smooth tubes cannot compete with the degree of enhancement offered by a 'lo-fin' surface. Nevertheless, it is expected to improve the performance of banks of 'lo-fin' tubes.

3. In boilers, the EHD effect can increase $\dot{q}_c$ and eliminate both boiling hysteresis and decreases in the boiling heat transfer coefficient due to transition and film boiling.

4. Electrode systems have been developed for the application of EHD in shell/tube heat exchangers. These have been proved, the Japanese one for vertical tube condensers, the UK one for vertical and horizontal condensers and for horizontal tube pool boilers.

9 ACKNOWLEDGMENTS

The results given in Figs 1, 2 and 4 are from work supported, successively, by the Science and Engineering Research Council and the British Technology Group.

10 REFERENCES

1. YILMAZ, S., HWALEK, J.J. and WESTWATER, J.W., Pool boiling heat transfer performance for commercial enhanced tube surfaces. ASME, 1980 Paper no.80-HT-41.

2. BERGLES, A.E., The challenge of enhanced heat transfer with phase change, Trans VII Congresso Nazionale Sulla Transmissione Del Calore, Florence, 1989, 1-12.

3. BERGLES, A.E., Techniques to Augment Heat Transfer, Handbook of Heat Transfer Applications, Rohsenow, W.M., Hartnett, J.P. and Ganic, E.N. eds, 1985. McGraw Hill, 2nd edition.

4. ALLEN, P.H.G., Heat transfer enhancement by dielectrophoresis, Proc. 2nd UK National Conference on Heat Transfer, Glasgow, 1988, 1, 861-870.

5. ZHORZHOLIANI, A.G., and SHEKRILADZE, I.G., Study of the effect of an electrostatic field on heat transfer with boiling dielectric fluids. Heat Transfer-Soviet Research, 1972, 4, (4), 81-98.

6. JONES, T.B. and SCHAEFFER, R.C. Electrohydro-dynamically coupled minimum film boiling in dielectric liquids. A.I.A.A.J., 1976, 14, (12), 1759 - 1765.

7. ROHSENOW, W.M. Nucleation with boiling heat transfer. Ind. Eng. Chem, 1966, 58, (1) 40-47.

8. ROHSENOW, W.M. A method of correlating heat-transfer data for surface boiling of liquids. Trans Amer Soc. mech. Engrs, 1952, 74, 969-976.

9. MARKELS, M. and DURFEE, R.L. The effect of applied voltage on boiling heat transfer. A.I. Ch. E.J., 1964, 10, 106-109.

10. OLINGER, J.L. and COLVER, C.P. A study of the effect of a uniform electric field on nucleate and film boiling. Chem. Eng. Prog. Symp. Ser., 1971, 67, (113), 19-29.

11. VELKOFF, H.R. and MILLER, J.H. Condensation of vapour on a vertical plate with a transverse electrostatic field. J. Heat Transfer, Trans Amer. Soc. mech. Engrs, Series C, 1965, 87, 197-201.

12. CHOI, H.Y. and REYNOLDS, J.M. Study of electrostatic effects on condensing heat transfer. Air Force Technical Report AFFDL-TR-65-51, Air Force Flight Dynamics Laboratory, Wright-Patterson Air Force Base, Ohio, 1965.

13. CHOI, H.Y. Electrohydrodynamic condensation heat tranfer. J. Heat Transfer, Trans Am. Soc. mech. Engrs, Series C, 1968, 90, 98-102.

14. HOLMES, R.E. and CHAPMAN, A.J. Condensation of Freon-114 in the presence of a strong nonuniform alternating electric field. J. Heat Transfer, Trans Amer. Soc. mech. Engrs, Series C, 1970, 92, 616-620.

15. SETH, A.K. and LEE, L. The effect of an electric field in the presence of noncondensable gas on film condensation heat transfer. J. Heat Transfer, Trans Amer. Soc. Mech. Engr, 1974, 96, 257-258.

16. BOLOGA, M.K. and DIDKOVSKIY, A.B. Enhancement of heat transfer in film condensation of vapors of dielectric liquids by superposition of electric fields. Heat Transfer - Soviet Research, 1977, 9, (1), 147-151.

17. DIDKOVSKII, A.B. and BOLOGA, M.K. Intensification of heat exchange upon condensation of a vapour in an electric field. High Temperature, 1978, 16, 490-496.

18. SMIRNOV, G.F. and LUNEV, V.G. Heat transfer during condensation of vapour of dielectric liquids in electric fields. App. Elec. Phen. (USSR), 1978, 2, 37-42.

19. DIDKOVSKY, A.B. and BOLOGA, M.K. Vapour film condensation heat transfer and hydrodynamics under the influence of an electric field. Int. J. Heat Mass Transfer, 1981, 24, 811-819.

20. YABE, A., KIKUCHI, K., TAKETANI, T., MORI, Y. and HIJIKATA, K. Augmentation of condensation heat transfer by applying non-uniform electric fields. Proc. 7th Int. Heat Transfer Conf., 1982, 5, 189-194.

21. COOPER, P. and ALLEN, P.H.G. The potential of electrically enhanced condensers. Proc. 2nd Int. Symp. on the Large Scale Application of Heat Pumps, 1984, 295-309.

22. TROMMELMANS, J. and BERGHMANS, J. Influence of electric fields on condensation heat tranfer of non-conducting fluids on horizontal tubes. Proc. 8th Int. Heat Transfer Conf., 1986, 6, 2969-2974.

23. YABE, A., TAKETANI, T., KIKUCHI, K., MORI, Y. and MAKI, H. Augmentation of condensation heat transfer by applying electro-hydro-dynamical pseudo-dropwise condensation. Ibid, 2957-2962.

24. SMIRNOV, G.F. and LUKANOV, I.I. Study of heat transfer from Freon-11 condensing on a bundle of finned tubes. Heat Tranfer-Soviet Research, 1972, 4, (3), 51-56.

25. COOPER, P. Electrically enhanced heat transfer in the shell/tube heat exchanger. Ph.D. Thesis, University of London, 1986.

26. CHUBB, L.W. Improvements relating to methods and apparatus for heating liquids. UK Patent No.100, 796, 1916.

27. BOCHIROL, L., BONJOUR, E. and WEIL, L. Etude de l'action de champs électriques sur les transferts de chaleur dans les liquides bouillants. C.R. Heb. des Seances de l'Acad. des Sciences (Paris), 1960, 250, 76-78.

28. BONJOUR, E., VERDIER, J. and WEIL, L. Electroconvection effects on heat tranfer. Chem. Eng. Progr., 1962, 58, (7), 63-66.

29. WATSON, P.K. Influence of an electric field upon the heat transfer from a hot wire to an insulating liquid. Nature, 1961, 189, (4764), 563-564.

30. CHOI, H.Y. Electrohydrodynamic boiling heat transfer. Ph.D. Thesis, Dept of Mech. Eng., M.I.T., 1962.

31. JALALUDDIN, A.K. and SINHA, D.B. Effect of an electric field on the superheat of liquids. Nuovo Cimento, Supplement to vol.26, Series X, 1962, 234-237.

32. MARKELS, M. and DURFEE, R.L. Studies of boiling heat transfer with electric fields Part I. Effect of applied A.C. voltage on boiling heat transfer to water in forced convection. A.I. Ch.E.J., 1965, 11, 716-719.

33. WINER, M. An experimental study of the influence of a nonuniform electric field on heat transfer in a dielectric fluid. TRW Systems Report, Aug. 1967, Report No. EM17-14.

34. LOVENGUTH, R.F. and HANESIAN, D. Boiling heat transfer in the presence of nonuniform, direct current electric fields. Ind. Eng. Chem. Fundam., 1971, 10, (4), 570-576.

35. BASU, D.K. Effect of electric field on boiling hysteresis in carbon tetrachloride. Int. J. Heat and Mass Transfer, 1973, 16, 1322-1324.

36. JONES, T.B. and SCHAEFFER, R.C. Electrohydrodynamically coupled minimum film boiling in dielectric liquids. A.I.A.A.J., 1976, 14, (12), 1759-1765.

37. JONES, T.B. and HALLOCK, K.R. Surface wave model of electrohydrodynamically coupled minimum film boiling. J. Electrostatics, 1978, 5, 273-284.

38. ZHELTUKIN, V.A., SOLOMATNIKOV, YU. K., MIKHAYLOV, D.M. and USMANOV, A.G. Effect of electric-field frequency on heat tranfer. Heat Transfer-Soviet Research, 1978, 10, (6), 10-13.

39. ALLEN, P.H.G. and COOPER, P. The potential of electrically enhanced evaporators. Proc. 3rd Int. Symp. on the Large Scale Application of Heat Pumps, 1987, 221-229.

40. KAWAHIRA, H., KUBO, Y. and YOKOYAMA, T. The effect of an electric field on boiling heat transfer of refrigerant-11. Proc. IEEE/IAS Annual Meeting, Atlanta, October 1987, 1446-1454.

41. KARAYIANNIS, T.G., COLLINS, M.W. and ALLEN, P.H.G. Electrohydrodynamic enhancement of nucleate boiling in heat exchangers. Chem. Eng. Comms, 1989, 81, 15-24.

42. BLACHOWICZ, R.A., BROOKS, B.W. and TAN, K.B. Boiling heat transfer in an electric field. Chem. Eng. Sci., 1980, 35, 761-762.

43. ASCH, V. Electrokinetic phenomena in boiling "Freon-113". J. App. Phys., 1966, 37, (7), 2654-2658.

44. BABOI, N.F., BOLOGA, M.K. and KLYUKANOV, A.A. Some features of ebullition in an electric field. Applied Electrical Phenomena (USSR), 1968, 20, 126-141.

45. RUTKOWSKI, J. The influence of electric field on heat transfer in boiling cryogenic liquid. Cryogenics, 1977, 17, 242-243.

46. YABE, A. and MAKI, H. Augmentation of convective and boiling heat transfer by applying an electro-hydrodynamical liquid jet. Int. J. Heat Mass Tranfer, 1988, 31, (2), 407-417.

47. Electrohydrodynamic (EHD) effects on condensing and evaporating Freons. Video held by ASME Heat Transfer Films Library, Mechanical and Aerospace Engineering. The University of Tennessee, Knoxville, Te. 37996-2210.

48. JOHNSON, R.L., Effect of an electric field on boiling heat transfer. AIAA Journal, 1968, 6, no. 8, 1456 - 1460.

49. BERGHMANS, J., Electrostatic fields and the maximum heat flux. Int. J. Heat Mass Transfer, 1978, 10, 791-797.

50. COOPER, P. EHD enhancement of nucleate boiling. J. Heat Transfer, Trans Am. Soc. Mech. Engrs, 1990, 112, 458-464

51. The development of Super Heat Pump Energy Accumulation System, Progress Report, New Energy and Industrial Technology Development Organisation, Tokyo, November 1988, 108-124.

52. ALLEN, P.H.G. and COOPER, P. Improvements in or relating to heat exchangers, UK Patent No. 8522680, 1985.

53. DARLEY, V. Private Communication.

54. BUTTERWORTH, D. Changes in heat exchangers for process applications: incentives and barriers. Heat Transfer Engineering, 1987, 8, (4), 19-22.

55. KAJIKAWA, T., AGAWA, T., TAKAZAWA, H., AMANO, M., NISHIYAMA, K. and HOMMA, T. Studies on OTEC power system characteristics and enhanced heat transfer performance. Proc. 6th OTEC Conf., Washington (USA), 1979, Paper 11-5.

56. CZIKK, A.M. GOTTZMANN, C.F., RAGI, E.G., WITHERS, J.G. and HABDAS, E.P. Performance of advanced heat transfer tubes in refrigerant-flooded liquid coolers. ASHRAE Trans, 1970, 76, (1), 96-119.

57. ARAI, N., FUKUSHIMA, T., ARAI, A., NAKAJIMA, T., FUJIE, K. and NAKAYAMA, Y. Heat transfer tubes enhancing boiling and condensation in heat exchangers of a refrigerating machine. ASHRAE Trans, 1977, 83, (2), 58-70.

58. BERGLES, A.E. and CHYU, M.C. Characteristics of nucleate pool boiling from porous metallic coatings. J. Heat Transfer, 1982, 104, 279-285.

59. CORNWELL, K. The influence of bubbly flow on boiling from a tube in a bundle. Proc. Eurotherm No.8, Advances in Pool Boiling Heat Transfer, Paderborn, FRG, May 11 - 12, 1989, 177 - 184.

60. DAMIANIDIS, C., COLLINS, M.W., KARAYIANNIS, T.G., and ALLEN, P.H.G. EHD effects in condensation of dielectric fluids. Proc. 2nd Int. Symp on COndensers and condensation, Bath, 1990; 505-518

61. ALLEN, P.H.G. and KARAYIANNIS, T.G., Electrohydrodynamic enhancement of boiling and condensation of dielectric fluids. Eurotherm No. 8, Advances in Pool Boiling Heat Transfer, Paderborn, Germany, 1989, 98-110.

TABLE 1

Review of Published Experimental Work on EHD augmented condensation heat transfer

Author(s)	Heat Transfer Fluid(s)	System	$\left(\dfrac{\bar{\alpha}_E}{\bar{\alpha}_0}\right)_{max}$	Notes
Choi and Reynolds [12]	Freon 113	Vertical stainless steel tube, 1346 mm long, 23.8 mm I.D., cooled on outside and coaxial with inner electrodes 6.3, 12.7 and 19.1 mm O.D.	2	(1)(2)(3)
Velkoff and Miller [11]	Freon 113	Vertical cooled cooper plate, 152 mm high, 229 mm wide, facing:- (1) 102 μm dia. horizontal wire (2) Mesh screens (3) Aluminium plate	~ 1.4 ~ 3 ~ 1.7	(3)(4)(5)(6)
Choi [13]	---as Choi and Reynolds [12]---			
Holmes and Chapman [14]	Freon R114	Flat silver-plated copper cooled and electrode plates (each 152 mm - horizontal - by 38 mm - sloping at different angles to give variable wedge shaped enclosure	Up to times 10 (overall)	(6) (7)
Seth and Lee [15]	Freon R113	Horizontal copper water tube with coaxial outer electrode	1.6	(5)(8)
Bologa & Didkovskiy [16]	Freon R113, hexane, diethyl ether	Vertical copper plate, 220 mm by 120 mm, spaced 7 mm from parallel (plain or horizontally slotted) electrode.	10 (for Freon & hexane) 20 (for ethene)	(2)
Didkovski & Bologa [17]	as [16]	As [16] together with vertical annulus, cooled tube 300 mm high, 21 mm O.D., electrode tube 35 mm I.D.	As [16]	(2)(5)(6)
Smirnov & Lunev [18]	Freon R113 diethyl ether	Vertical annulus, cooled outer jacket, 215 mm high (H) 30 mm I.D. (D_i), around central electrode (O.D. = D_o) with smooth, screw threaded or perforated surface, gaps 2.5 to 4 mm	3.6	(2)(3)(6)(7)
Didkovsky & Bologa [19]	---As [16]---			(2)(3)(7)(9)

Author(s)	Heat Transfer Fluid(s)	System	$\left(\dfrac{\bar{\alpha}_E}{\bar{\alpha}_0}\right)_{max}$	Notes
Yabe, Kikuchi, Taketani, Mori & Hijikata [20]	(a) Water	Needle electrode above liquid surface		(10)
	(b) Freon	Cooled brass plate, 600 mm high 100 mm wide with shaped copper wire electrode systems	2.24	(3)(11)
Cooper & Allen [21]	Freons R12 and R114	Single cooled horizontal tubes 514 mm long with smooth (19.1 mm O.D.) and 'lo-fin' (19.0 O.D.) surfaces, Electrode systems 1) Concentric copper gauze (38 mm I.D.) (also vertical with R114) 2) Parallel brass plates with auxiliary 6.25 mm O.D. rods (to simulate [1])	2.9	(3)(5)(7)(12)
Trommelmans & Berghmans [22]	Freons R11, R113 & R114	Horizontal smooth copper cooled tube 1100 mm long, 18 mm O.D., concentric with copper wire spiral electrode, 38 mm I.D.	1.1	(3)(14)
Yabe, Taketani Kikuchi, Mori and Maki [23]	a) Silicone oil	Thin liquid film on and between electrically conducting glass sheets.	–	(3)(5)
	b) Freon R113	Vertical smooth brass cooled tube 540 mm high, 18 mm O.D. with copper helical (condensate removal) wire, 0.5 mm O.D. and curved (stress applying) plate electrodes, 21.2 mm I.D.	4.5	(12)

NOTES:
(1) Analysis in terms of perturbation wavelength
(2) Results correlated
(3) D.C. electric field
(4) Corona discharge gave electrophoresis
(5) condensate film perturbation observed
(6) Condensate spraying observed
(7) A.C. electric field
(8) Inhibiting effect of non-condensable gas partly neutralised by EHD.
(9) Condensate film thickness measured
(10) Gas-liquid interface stability studied theoretically and experimentally
(11) Improved heat transfer due to condensate removal by electrodes measured
(12) α_E values corrected for reduced ΔT.
(13) EHD effect with 'lo-fin' tube <10%.
(14) Effect of EHD on shape of pendant condensate droplets discussed

TABLE 2
Review of published experimental work on EHD augmented pool boiling heat transfer

Author(s)	Heat transfer fluid(s)	Heated surface diameter, μm	$\left(\dfrac{\dot{q}_{c,E}}{\dot{q}_{c,0}}\right)_{max}$	$\left(\dfrac{\alpha_E}{\alpha_0}\right)_{max}$	Notes
Coaxial cylindrical systems:					
Bochirol, Bonjour and Weil [27]	Benzene, hexane, toluene, trichlorethylene, ethyl ether, liquid nitrogen	200	–	2.7 (for trichlorethylene and ethyl ether)	(3)(6)
	Pure water, methyl ethyl ketone, acetone, methyl alcohol	100	–	6 (for methyl alcohol)	
Watson [29]	n-hexane	100	–	2.6	(2)(6)
Bonjour, Verdier and Weil [28]	---------- as Bochirol, Bonjour and Weil [27] ----------				
Choi [30]	Freon 113	254 or 508	>2	–	(1)(4)(6)
Markels and Durfee [9]	Isopropanol, distilled water	9525	6 >1.4	–	(2)(5)(7)
Winer [33]	Freon 114	5570	–	–	(1)(2)(6)
Lovenguth and Hanesian [34]	Freons 21, 113, carbon tetrachloride, chloroform	510	2.9	–	(1)(2)(6)
Basu [35]	Carbon tetrachloride	40	–	–	(2)(3)(6)(8)
Jones and Schaeffer [36]	Freon 113	127 or 406	–	–	(1)(4)(6)
Jones and Hallock [37]	Freon 113 Freon 113/methanol	812, 1150 or 1830	–	–	(4)(5)(6)
Zheltukhin, Solomatnikov, Mikhaylov and Usmanov [38]	Acetone, benzene, n-diethyl ether	5000 to 6500	–	1.82 (for acetone)	(5)(6)
Allen and Cooper [39]	Freon 114	18600 ('lo-fin' O.D.)	–	Up to 60	(2)(7)(8)
Kawahira, Kubo and Yokoyama [40]	Freon 11	22400	–	4	(2)(3)(6)(8)(10)(11)
Karayiannis, Collins and Allen [41]	Freon 114	19100	–	4	(2)(7)(8)

Author(s)	Heat transfer fluid(s)		$\left(\dfrac{\dot{q}_{c,E}}{\dot{q}_{c,0}}\right)_{max}$	$\left(\dfrac{\bar{\alpha}_E}{\bar{\alpha}_0}\right)_{max}$	Notes
Parallel plate systems:		Electrode gap, μm			
Olinger and Colver [10]	Deionized water	6350 or 25400	–	–	(2)(3)(6)(9)(10)
Blachowicz, Brooks and Tan [42]	Benzene	5000 to 20000	–	2	(3)(7)(9)
Kawahira, Kubo and Yokoyama [40]	Freon 11	3000	–	4.4	(2)(3)(6)(10)(11)
Other systems:		System(s)			
Jalaluddin and Sinha [31]	Methanol, isopropanol, methyl ethyl ketone, benzene	Heated sphere and plane electrode	–	–	(2)(3)(6)(8)(9)
Asch [43]	Freon 113	Heated wire (356 μm dia.) between 25400 μm dia. discs, spaced 25400 μm or (asymmetrically) 12700 μm dia. spheres with 60300 μm gap	6	–	(2)(3)(6)(11)
Baboi, Bologa and Klyukanov [44]	Benzene, toluene, liquid argon	Heated wires, 50, 70, 100 and 500 μm dia., parallel to 6000 μm dia. cylindrical electrode	–	–	(1)(2)(3)(6)(11)
Zhorzholiani and Shekriladze [5]	Acetone, benzene, n-pentane	Heated horizontal tubes, 6000 and 8000 μm dia., parallel to plane electrode Vertical semi-cylindrical heated channel facing plane electrode	3 (for acetone)	–	(1)(2)(3)(6)(12)
Rutkowski [45]	Liquid nitrogen	Heated horizontal wire, 100 μm dia., parallel to 3000 μm cylindrical electrode	–	–	(3)(6)
Yabe and Maki [46]	Freon 113/ ethanol ("Furonsuburu AE")	Heated plane beneath toroidal EHD "pump" electrode	2.1	–	(2)(7)(8)(13)

NOTES:
(1) Dielectrophoresis analysed
(2) D.C. electric field
(3) A.C. electric field
(4) Effect of EHD on minimum film boiling
(5) A.C. electric field frequency varied
(6) Electrically heated source
(7) Fluid heated source
(8) Ebullition initiated by EHD
(9) High thermal capacity source
(10) Fine wire mesh or array high voltage electrode
(11) Observations of bubble dynamics
(12) Correlation suggested
(13) Enhancement by liquid jet (but bubble dynamics also affected).

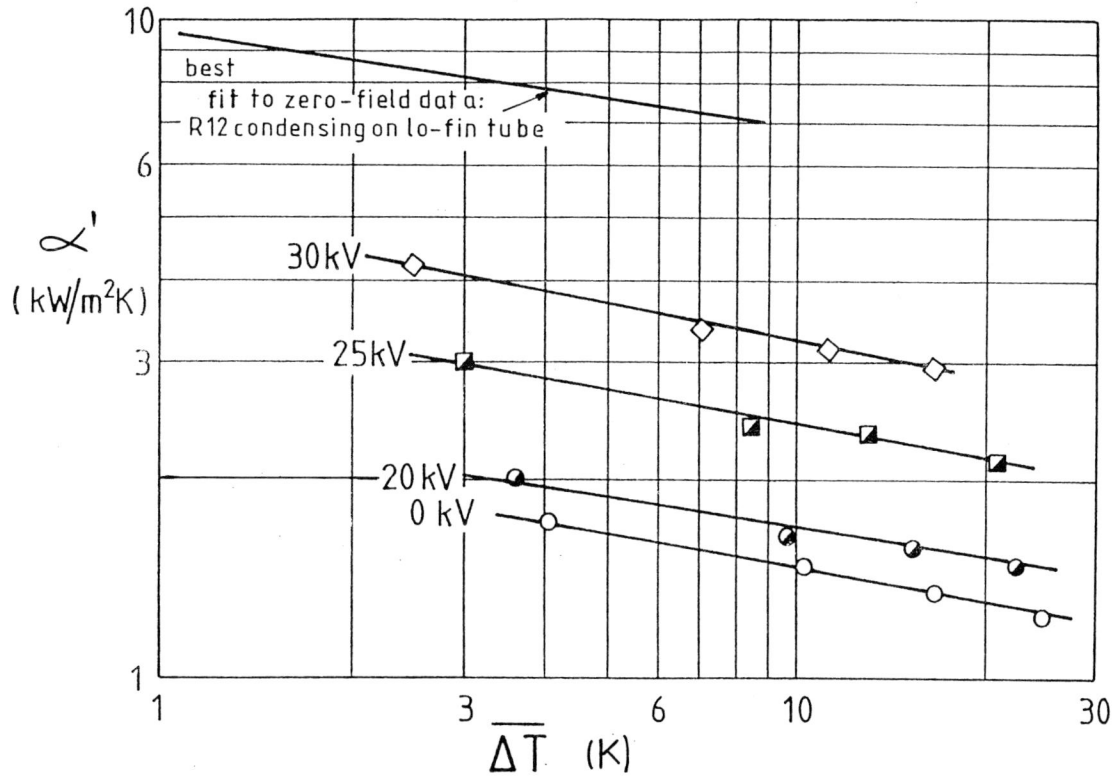

Fig. 1. EHD enhancement of Freon 12 condensation with electrode voltages 20, 25, and 30kV at outer surface of smooth horizontal tube compared with EHD-free smooth and 'lo-fin' surface tubes, ($T_g=30°C$), [25].

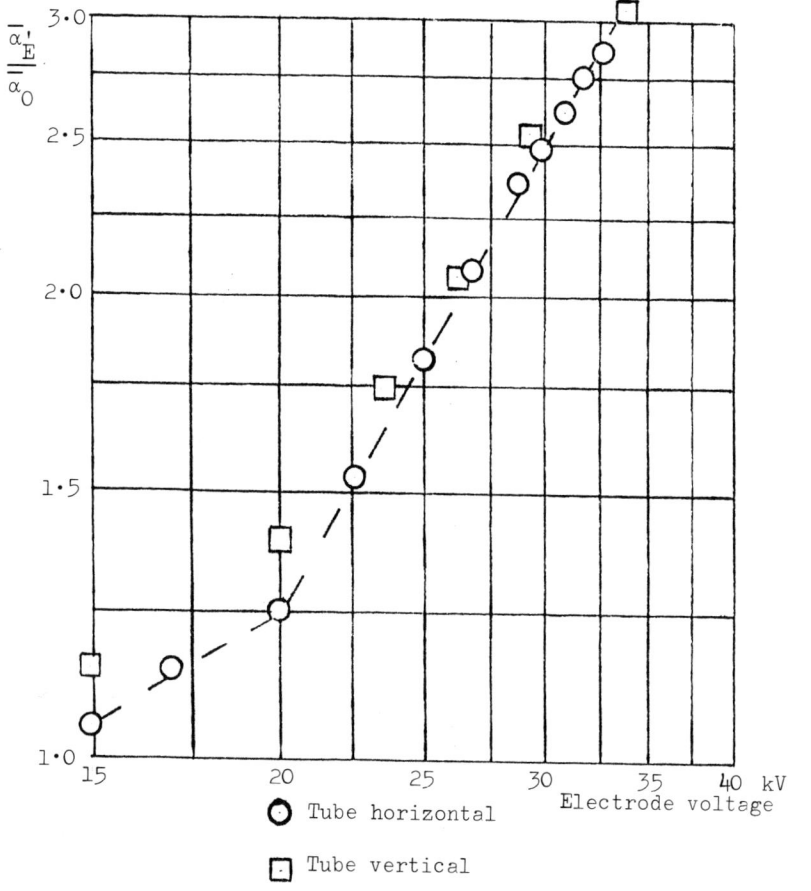

Fig. 2. EHD enhancement of Freon 114 condensation at outer surfaces of vertical and horizontal tubes ($\Delta T=25K$, $T_g=90°C$), [25].

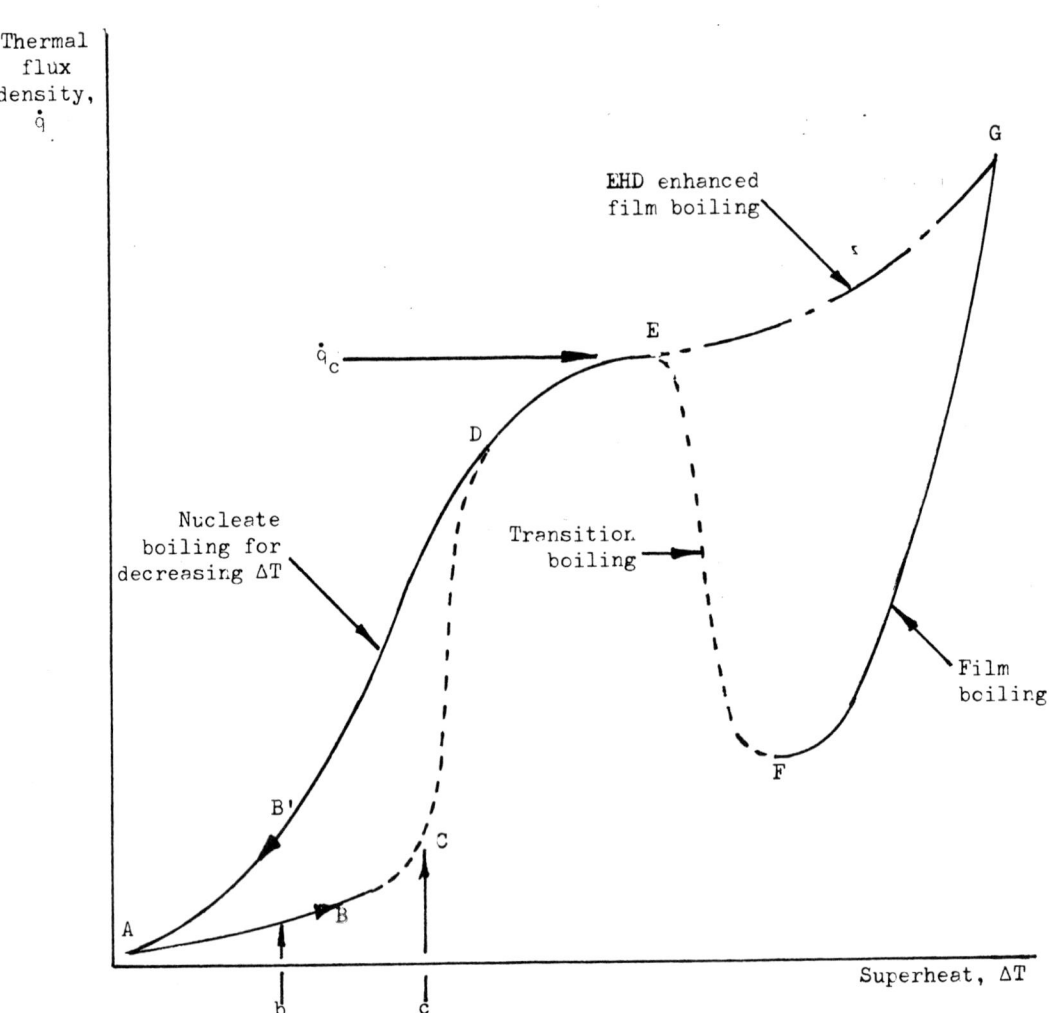

Fig. 3. Heat flux density versus superheat boiling curve.

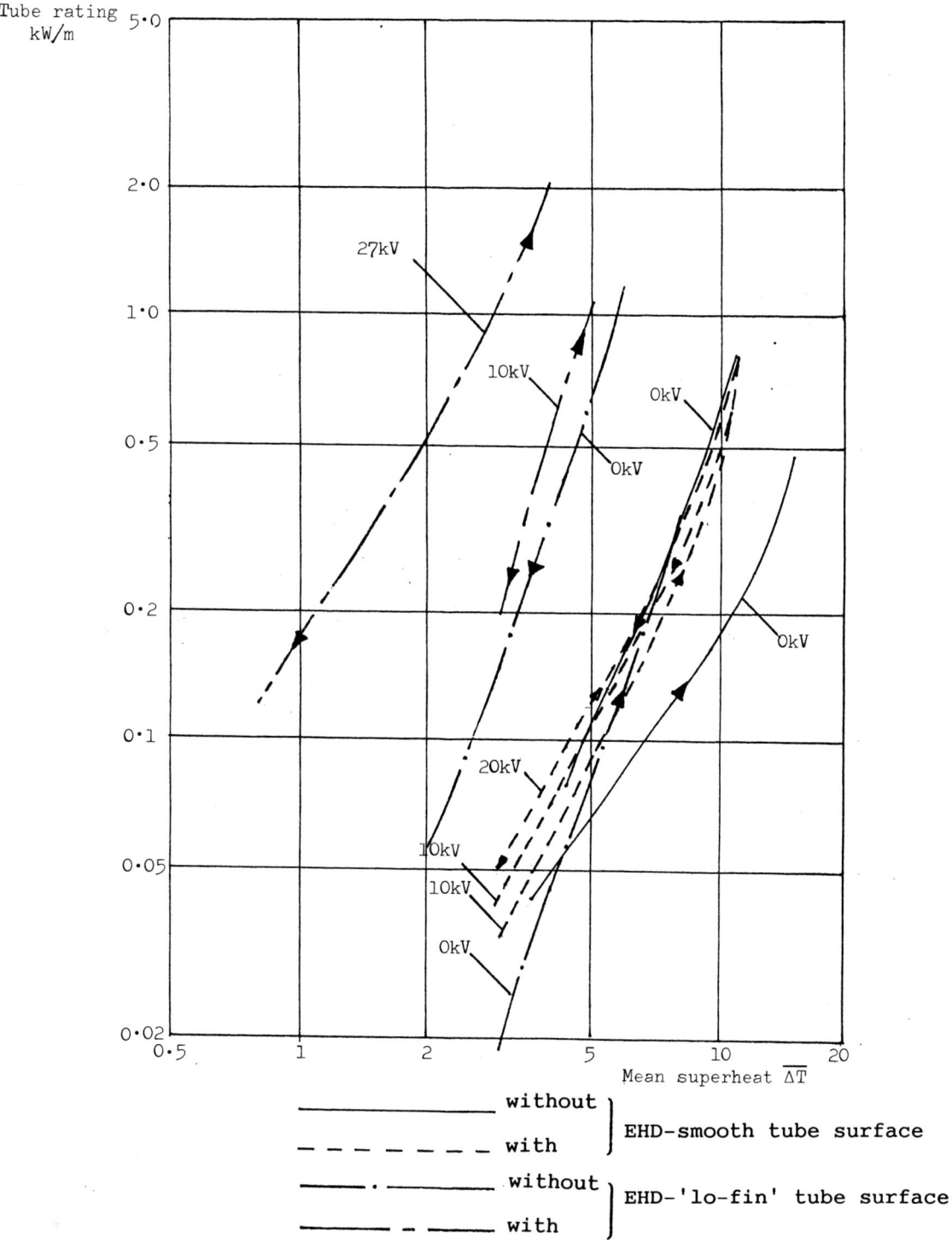

Electrode voltages as indicated, T_g=21.5°C, tube diameter≈19mm.

Fig. 4. Results for boiling Freon 114 on basis of tube rating.

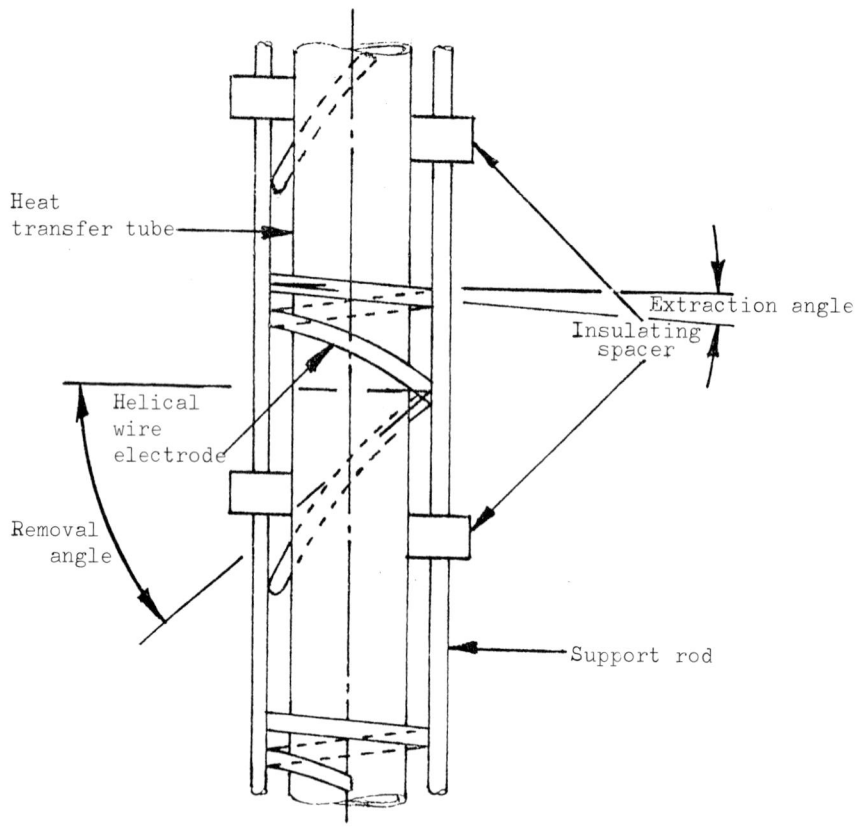

Fig. 5. Electrode system for EHD enhancement of vertical condenser tubes, [23, 51].

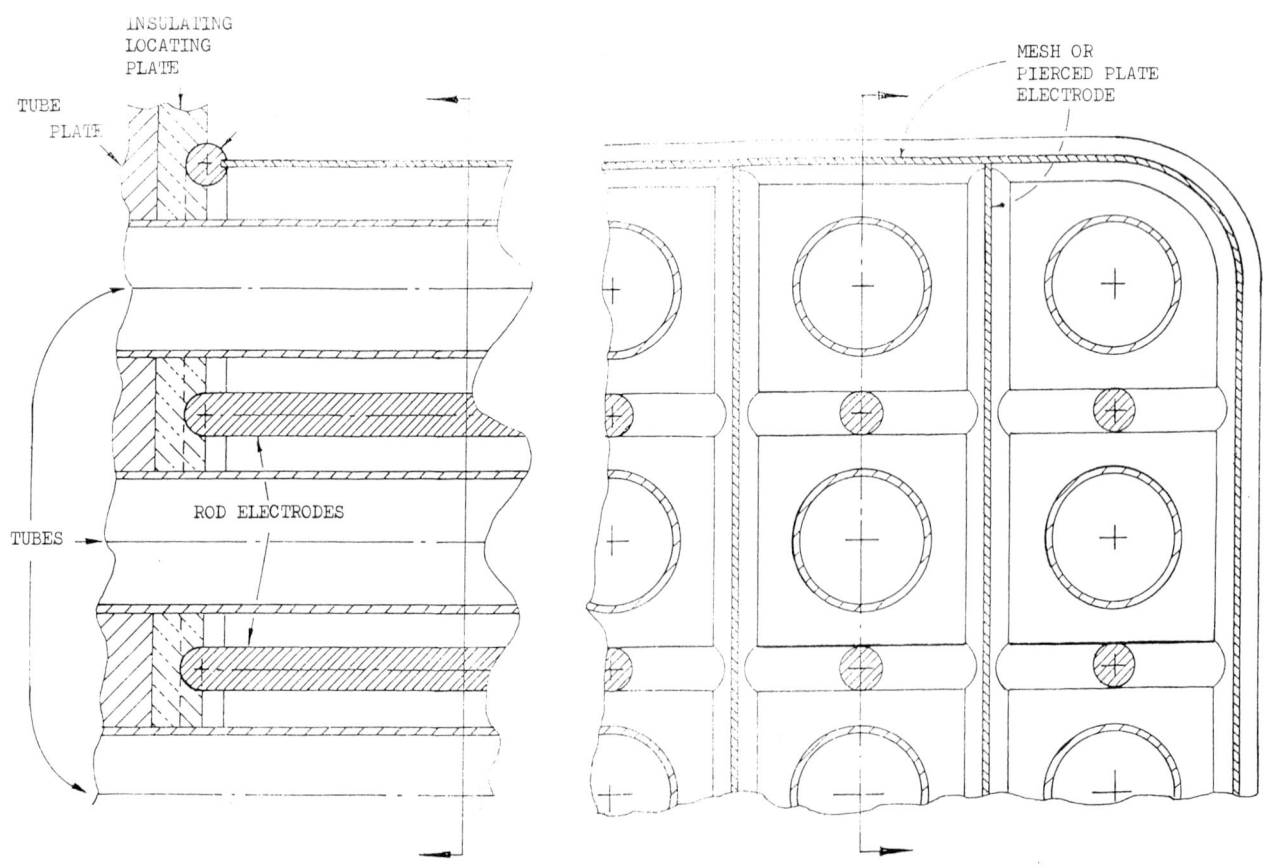

Fig. 6. Electrode system for EHD enhancement of vertical and horizontal shell/tube condensers and boilers, [52].

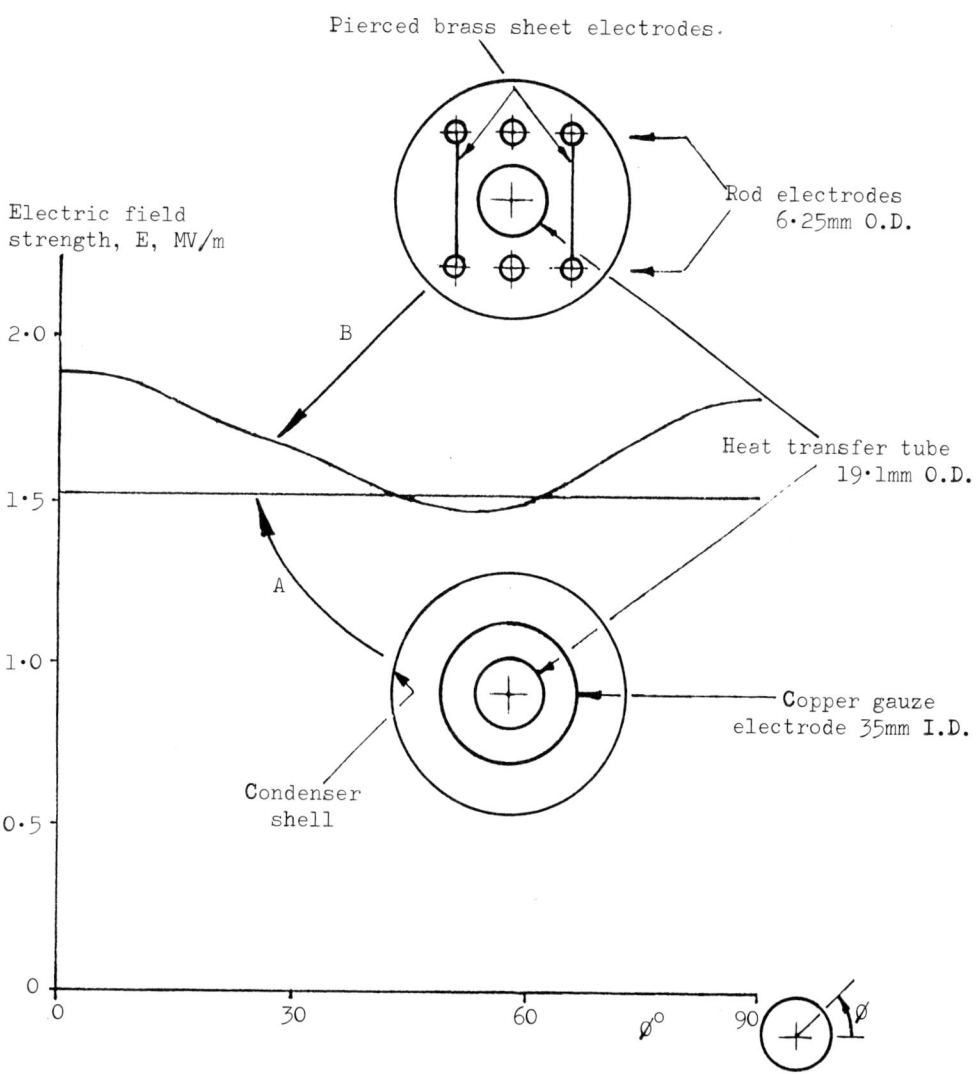

Fig. 7. Distribution of electric field strength, E, normal to heat transfer tube surface with electrode voltage 10kV for coaxial cylinder (A) and for optimised rod and plane (B) electrode systems. (Supersedes similar graphs published previously, [21, 60, 61].)

C413/038

Numerical analysis of transient two-dimensional bubbly flow with size and number density distribution

Y MATSUMOTO, PhD, MJSME, J MATSUI, MSc, MJSME and H OHASHI, PhD, MJSME, MASME
Department of Mechanical Engineering, University of Tokyo, Japan

SYNOPSIS A dispersed two phase flow consisting of a suspension of discrete gas bubbles, which have distributed size and number density, is investigated numerically. The numerical results reveal that the transverse motions of the bubbles have significant influence on the flow characteristics.

NOTATION

C_D : drag coefficient
K : number of bubble groups
R : Reynolds number
g : gravitational acceleration
h : height of free surface
m : unit outward tangential vector
n : unit outward normal vector
p : pressure
r : bubble radius
t : time
u, v : velocity components
w : slip velocity between liquid and a bubble
x, y : coordinates
α : void fraction
β : coefficient of virtual mass
μ : viscosity
ρ : density

Subscripts

d : bubble
g : gas phase, non-condensable gas in bubbles
k : k-th group of bubbles
l : liquid phase
m : two phase condition
s : surface
v : vapour

1 INTRODUCTION

Bubbly flow is observed in many situations, such as nuclear power plants, chemical plants and bio-reactors. Bubbly mixtures, which have initially uniform size and number density have been analyzed theoretically and numerically by a considerable number of researchers. Wijngaarden[1,2] proposed a set of equations for wave propagation in a bubbly flow. Cook and Harlow[3,4] developed a computer code and calculated the behavior of vortex shedding from a rectangular cylinder in a bubbly flow. Nigmatulin[5] reviewed the modeling of bubbly flow and hydrodynamical effects in wave propagation.

However, real bubbly flows have distributed size and number density of the bubbles. Matsumoto et al.[6] calculated numerically such a bubbly flow in a cascade of blades. In this paper, the bubbly flow in a rectangular tank with a free surface two dimensional domain is investigated numerically. The governing equations for the flow are formulated, the assumptions made being clearly stated, and the boundary conditions discussed. A numerical method, which can be used for both compressible and incompressible flows, is outlined. The numerical results reveal that the relative motions of the bubbles to the surrounding liquid have significant influences on the characteristics of the bubbly flow.

2 BASIC EQUATIONS

2.1 Assumptions

The following assumptions are employed to formulate the governing equations.
(1) The two phase flow is treated as a continuum fluid consisting of a suspension of bubbles which are small compared to the characteristic length of the flow such as the depth or breadth of the tank.
(2) The mass and momentum of the gas phase are ignored as they are negligibly small compared to those of the liquid phase. In the present calculations, the ratio of the mass of gas phase to that of liquid phase is of the order of 10^{-5}.
(3) The liquid is incompressible. The compressibility of two phase flow is caused by the gas phase.
(4) The two phases are in a state of thermal equilibrium and upon fluid compression or expansion, all changes are isothermal.
(5) The liquid mass variation due to phase change is ignored as it is negligibly small compared to the mass of liquid phase. However, the phase change has a contribution to the change of void fraction because the vapour pressure inside the bubbles is assumed to be constant.
(6) Bubble size is in equilibrium with the surrounding pressure.
(7) Coalescence and fragmentation of the bubbles are neglected.
(8) The gases inside the bubble, consisting of air and water vapour, obey the perfect gas law.
(9) There is no diffusion of air through the bubble wall, so that the amount of air inside the bubble is assumed to be constant.

2.2 Governing Equations

Under the preceding assumptions, the following equations are formulated.

(1) The conservation equation for mass is

$$\frac{\partial}{\partial t}\{(1-\alpha)\rho_l\} + \frac{\partial}{\partial x}\{(1-\alpha)\rho_l u_l\} + \frac{\partial}{\partial y}\{(1-\alpha)\rho_l v_l\} = 0 \quad (1)$$

where t is time, ρ_l is density of liquid, u_l and v_l are the liquid velocities. The void fraction, α, is defined as,

$$\alpha = \sum_{k=1}^{K} \alpha_k = \sum_{k=1}^{K} \frac{4}{3}\pi r_{dk}^3 n_k, \quad (2)$$

where r_{dk} is bubble radius, n is number density of bubbles, and k is number of bubble groups.

(2) The conservation equations for momentum are

$$\frac{\partial}{\partial t}\{(1-\alpha)\rho_l u_l\} + \frac{\partial}{\partial x}\{(1-\alpha)\rho_l u_l^2\} + \frac{\partial}{\partial y}\{(1-\alpha)\rho_l u_l v_l\}$$

$$= -\frac{\partial p}{\partial x} + \frac{\partial}{\partial x}[\mu(\frac{4}{3}\frac{\partial u_l}{\partial x} - \frac{2}{3}\frac{\partial v_l}{\partial y})] + \frac{\partial}{\partial y}[\mu(\frac{\partial u_l}{\partial y} + \frac{\partial v_l}{\partial x})], \quad (3)$$

$$\frac{\partial}{\partial t}\{(1-\alpha)\rho_l v_l\} + \frac{\partial}{\partial x}\{(1-\alpha)\rho_l u_l v_l\} + \frac{\partial}{\partial y}\{(1-\alpha)\rho_l v_l^2\}$$

$$= -\frac{\partial p}{\partial y} + \frac{\partial}{\partial x}[\mu(\frac{\partial u_l}{\partial y} + \frac{\partial v_l}{\partial x})] + \frac{\partial}{\partial y}[\mu(\frac{4}{3}\frac{\partial v_l}{\partial y} - \frac{2}{3}\frac{\partial u_l}{\partial x})]$$

$$-(1-\alpha)\rho_l g, \quad (4)$$

where p is the averaged pressure in the two phase flow. The viscosity of the two phase flow, μ, is taken as $\mu = (1+\alpha)\mu_l$.

(3) Assuming local equilibrium for the bubble the relationship between the average two-phase pressure and the bubble radius is derived as

$$p_v + p_{gk} - \frac{2\sigma}{r_{dk}} - p + \frac{\rho_l w^2}{4} = 0, \quad (5)$$

where p_v is the vapour pressure, p_{gk} is the non-condensable gas pressure in the bubble of k-th group, σ is the surface tension and ρ_l is the liquid density. The non-condensable gas pressure p_{gk} is taken as $p_{gk} = [p_{gk}r_{dk}^3]_s / r_{dk}^3$. The slip velocity, w, is defined as

$$w = \sqrt{(u_{gk} - u_l)^2 + (v_{gk} - v_l)^2}$$

(4) The equations for relative motion between bubble and liquid is

$$\frac{D_g}{Dt}(\beta\rho_l V_k u_{gk}) - \frac{D_l}{Dt}(\beta\rho_l V_k u_l) = -V_k \frac{\partial p}{\partial x}$$

$$-C_D \pi r_{dk}^2 \rho_l \frac{w(u_{gk} - u_l)}{2} - 3.23\mu_l r_{dk}(v_{gk} - v_l)\sqrt{R_{Gk}}\frac{\omega}{|\omega|}, \quad (6)$$

$$\frac{D_g}{Dt}(\beta\rho_l V_k v_{gk}) - \frac{D_l}{Dt}(\beta\rho_l V_k v_l) = -V_k \frac{\partial p}{\partial y}$$

$$-C_D \pi r_{dk}^2 \rho_l \frac{w(v_{gk} - v_l)}{2} - 3.23\mu_l r_{dk}(u_{gk} - u_l)\sqrt{R_{Gk}}\frac{\omega}{|\omega|}, \quad (7)$$

where the substantial derivatives, the coefficients of virtual mass, β, the drag coefficient of bubble, C_D, Reynolds numbers, R_d, R_{Gk}, and vorticity, ω are defined as

$$\frac{D_g}{Dt} = \frac{\partial}{\partial t} + u_{gk}\frac{\partial}{\partial x} + v_{gk}\frac{\partial}{\partial y}, \quad \frac{D_l}{Dt} = \frac{\partial}{\partial t} + u_l\frac{\partial}{\partial x} + v_l\frac{\partial}{\partial y},$$

$$\beta = \frac{1}{2}, \quad V_k = \frac{3}{4}\pi r_{dk}^3, \quad C_D = \frac{24}{R_d} + \frac{6}{1+\sqrt{R_d}} + 0.4,$$

$$R_d = \frac{2\rho_l r_{dk} w}{\mu_l}, \quad R_{Gk} = \frac{4\rho_l r_{dk}^2 |\omega|}{\mu_l}, \quad \omega = \frac{\partial v_l}{\partial x} - \frac{\partial u_l}{\partial y}.$$

The last terms of the equations express the Saffman's force(7).

(5) The conservation equation for bubble number is

$$\frac{\partial n_k}{\partial t} + \frac{\partial}{\partial x}(n_k u_{gk}) + \frac{\partial}{\partial y}(n_k v_{gk}) = 0. \quad (8)$$

(6) The height of the liquid surface is expressed as follows,

$$\frac{\partial h}{\partial t} + u_{ms}\frac{\partial h}{\partial x} = v_{ls}, \quad (9)$$

where u_{ms} is volumetric mean velocity which is defined as $u_{ms} = (1-\alpha)u_{ls} + \sum_{k=1}^{K} \alpha_k u_{gs,k}$ and v_{ls} is the liquid velocity at the surface.

2.3 Boundary Conditions

At the free surface, assuming that the compressibility does not have a significant role, then the boundary condition for the liquid surface takes the form(8),

$$2n_x m_x \frac{\partial u}{\partial x} + (n_x m_y + n_y m_x)(\frac{\partial u}{\partial y} + \frac{\partial v}{\partial x})$$

$$+ 2n_y m_y \frac{\partial v}{\partial y} = 0 \quad (10)$$

where n and m are the unit outward normal and tangential vectors respectively. They are expressed as

$$n_x = \frac{\partial h}{\partial x}[1 + (\frac{\partial h}{\partial x})^2]^{-\frac{1}{2}}, \quad n_y = [1 + (\frac{\partial h}{\partial x})^2]^{-\frac{1}{2}},$$

$$m_x = n_y, \quad m_y = -n_x.$$

The bubbles disappear outward from the liquid surface due to the buoyancy. At the surface cell, the gas velocities are set the same as those of the cell under the surface cell. At the wall, the liquid velocity is assumed to be zero.

2.4 Numerical Procedure

A staggered mesh is used to differentiate the governing equations and a semi-implicit method, which is similar to the HSMAC method, is applied for numerical integration of the above mentioned equations according to the following procedures.

(1) The new pressure is estimated to be equal to the old pressure.

(2) The new velocities of bubbles are calculated using the new pressure by Eqs.(6),(7) and then the bubble number density is calculated by Eq.(8).

(3) The bubble radius is calculated by Eq.(5) using the new pressure and then the new void fraction is calculated by Eq.(2) using the new radius and the new number density.

(4) The new void fraction is also calculated by Eqs.(1), (3) and (4) using the new pressure.

(5) The pressure is corrected by an iterative relaxation method to make the two void fractions obtained by procedures (3) and (4) coincide.

(6) Using this new pressure, the new void fraction, velocities and bubble radius are determined.
(7) By these new velocities, the new height of liquid surface is calculated by Eq.(9) to satisfy the boundary conditions shown by Eq.(10).

3 NUMERICAL RESULTS AND DISCUSSION

The bubbly flow in a two dimensional rectangular tank with a free surface is simulated. The computational domain and the grid points are shown in Fig 1. The pressure at the surface is 100 kPa, which is almost atmospheric pressure. The initial height of the liquid surface is 0.45 m and the width of the tank is 0.50 m. The bubbles are injected from the half part of the bottom by keeping the void fraction constant at the injection cells. They are classified into three groups whose radii are 1.0mm, 0.5mm and 0.2mm at the surface and each group occupy 1/3 of the void fraction respectively.

Because of the gravitational acceleration, the bubbles have an upward slippage relative to the surrounding liquid. The buoyancy works on the bubbly liquid due to the mean density difference caused by the void fraction distribution. By this natural convection, the bubbly liquid portion starts to move upward and a vortex is formed near the interface between the bubbly region and the liquid region. The vortex develops in a rather complicated manner.

At time 0 the injection of bubbles is started. Fig. 2 shows the bubbly flow in a tank with free surface at 0.5 second after the start. The liquid velocity vector, the void fraction contour, and the bubble number density contours and the velocity vectors of the bubbles whose radii are 1.0 mm and 0.2mm at 100 kPa are shown. The liquid velocity is not developed and a few short vectors are observed near the center of the bottom, where the interface between the bubbly region and the pure liquid region exists. The flow expands the distribution of the void fraction to the right side. The slippage velocities of the bubbles to the liquid due to the gravitational acceleration depend on the sizes. The large bubble has large rising velocity and the small bubble has small one, so that the void fraction has a wide distribution to the

(a) Liquid velocity vector

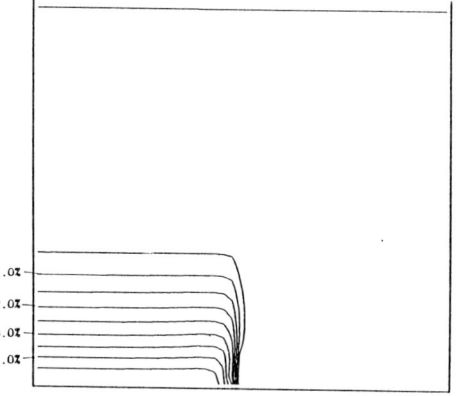

(b) Void fraction contour

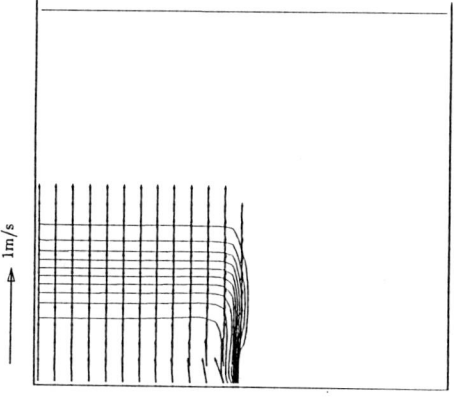

(c) Bubble velocity vector and number density contour for $r_d = 1.0$mm

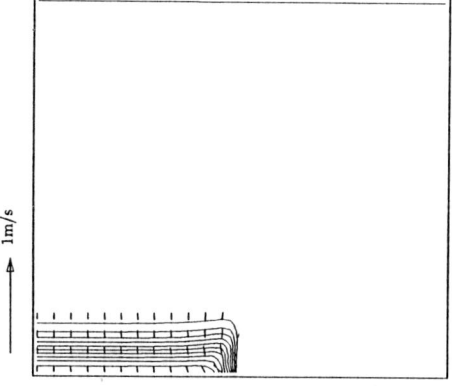

(d) Bubble velocity vector and number density contour for $r_d = 0.2$mm

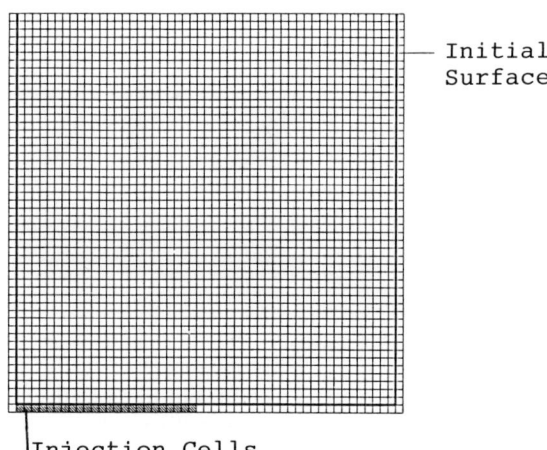

Figure 1. Computational domain in a rectangular tank

Figure 2. Bubbly flow in a tank with free surface. The time t= 0.5 second after the bubble injection is started.

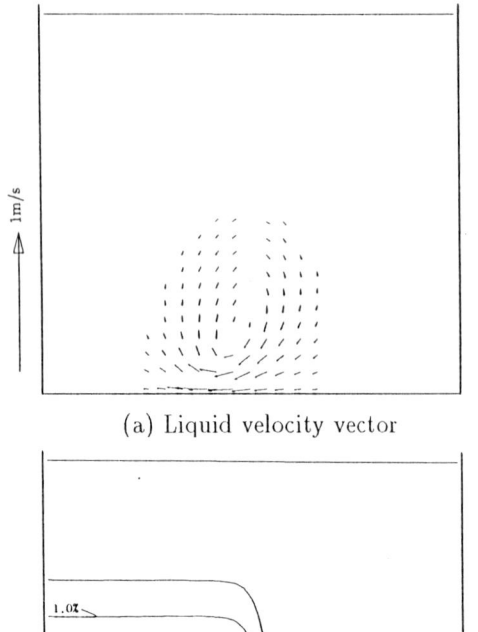

(a) Liquid velocity vector

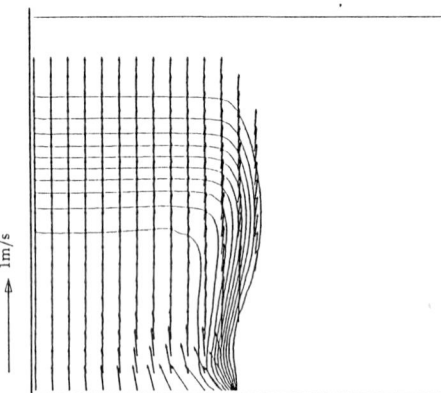

(b) Void fraction contour

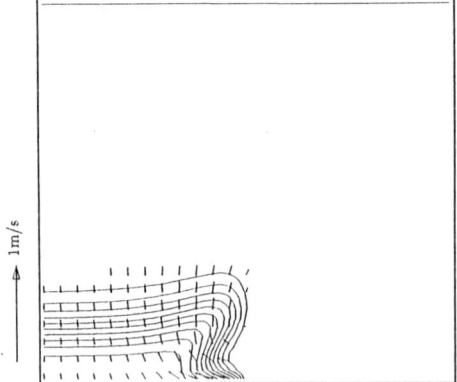

(c) Bubble velocity vector and number density contour for $r_d = 1.0$mm

(d) Bubble velocity vector and number density contour for $r_d = 0.2$mm

Figure 3. Bubbly flow in a tank with free surface. The time t= 1.0 second after the bubble injection is started.

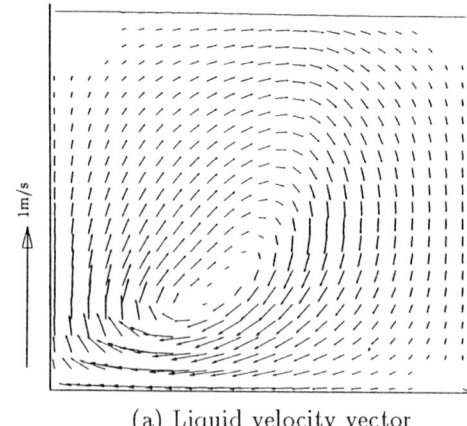

(a) Liquid velocity vector

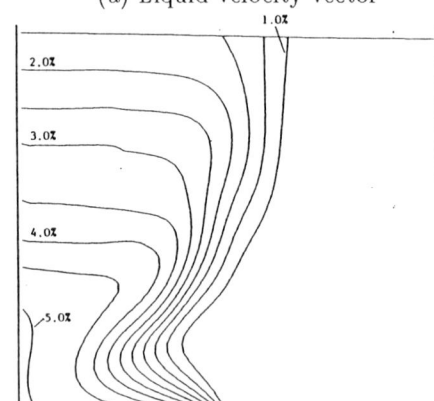

(b) Void fraction contour

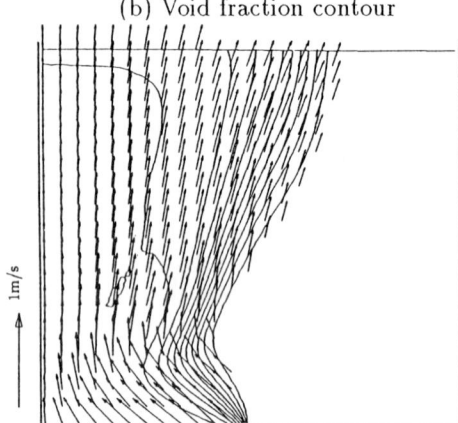

(c) Bubble velocity vector and number density contour for $r_d = 1.0$mm

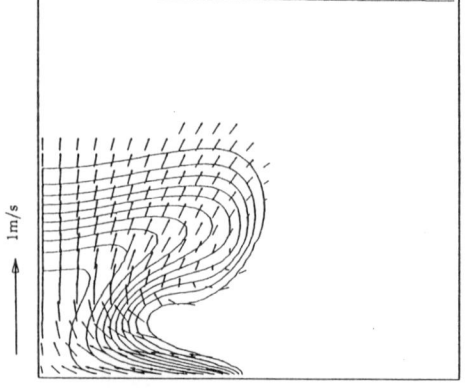

(d) Bubble velocity vector and number density contour for $r_d = 0.2$mm

Figure 4. Bubbly flow in a tank with free surface. The time t= 2.0 second after the bubble injection is started.

vertical direction. The bubbles are put stationarily in the injection cells beneath the bottom so that the rising velocity of the bubble increases with time to the terminal one. The size of the bubble is larger in the shallow region compared to that in the deeper region due to the surrounding pressure, so that the terminal rising velocity of the bubble is larger in the shallow region. The vertical number density distribution becomes wider as the bubbles rise in this case where the bubbles have no interaction with each other. In this simulation, the number density of the bubble is calculated using donor cell differencing scheme, which may cause considerable numerical diffusion. This numerical diffusion also has some contribution to the wide distribution of the number density.

After 1 second, the convective flow is observed clearly as shown in Fig 3. Due to this vortex, the void fraction is constricted near the bottom and expands in the middle. The vertical distribution of the void fraction develops due to the rising velocity difference between the bubbles. The large bubbles rise straight upward compared with the small ones and the contours of the number density are almost horizontal in the left side, because the liquid flow is not yet developed. On the contrary, the small bubbles still remain near the bottom due to the slow rising velocity and their movements are affected by the convective flow of the liquid. The number density distribution is much deformed.

The convective flow is well developed in the whole field after 2 seconds as shown in Fig 4. The center of the vortex moves upward according to the rise of the bubbles. The large bubbles rise up to the surface and disappear from the liquid. The small bubbles do not reach the surface yet and they follow the liquid flow. The shape of the liquid surface starts to incline from left to right.

Figure 5 shows the bubbly flow after 4 seconds from the start. The liquid surface is deformed very much by the convective flow caused by the bubble injection. The liquid flow is accelerated near the left wall by the steep gradient of the void fraction distribution. The fast part turns horizontal at the upper left corner and this fast flow makes the free surface low above the vortex center. The down flow is also well developed near the right wall. The accumulation of the large bubbles is observed at the left wall, because the Saffman's force becomes significant due to the large slip velocity and the liquid velocity gradient near the wall. The large bubbles rise diagonally to the surface and all the injected large bubbles are removed from the liquid through the surface. It is observed that the injected small bubbles are pressed against the left wall and the small bubbles are carried downward by the convective low.

The vortex center is shifted downward to the right side and the well developed down flow is observed near the right wall after 6 seconds as shown in Fig 6. By this downward movement of the vortex the transverse and downward liquid velocity near the surface decreases, so that the bubbles easily escape through the surface. This may accelerate the vortex again. Considerably high void fraction is observed in the vortex center, which is caught by the transient convective flow and remains in the liquid. The flow is not in steady state. The center of the vortex shifts right and left according the complicated

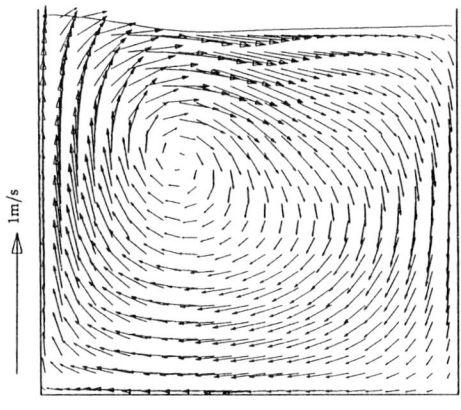

(a) Liquid velocity vector

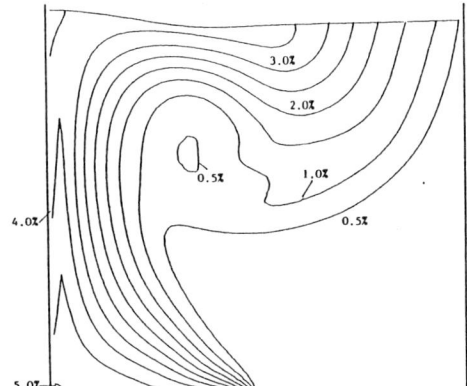

(b) Void fraction contour

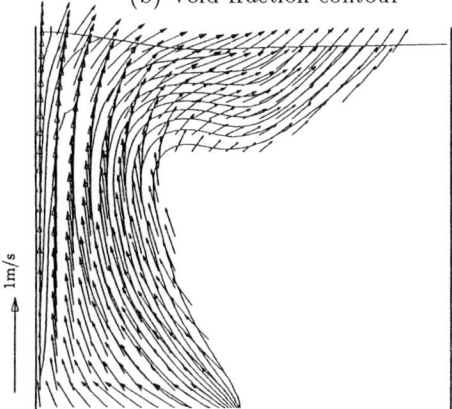

(c) Bubble velocity vector and number density contour for $r_d = 1.0$mm

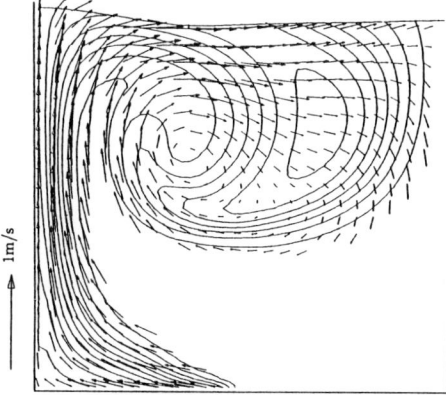

(d) Bubble velocity vector and number density contour for $r_d = 0.2$mm

Figure 5. Bubbly flow in a tank with free surface. The time t= 4.0 second after the bubble injection is started.

movement of the bubbles.

4 CONCLUDING REMARKS

The flow characteristics of the dispersed two phase flow consisting of a suspension of discrete gas bubbles, which have distributed size and number density, in a rectangular tank with free surface was investigated numerically. The governing equations for the two phase bubbly flow are formulated accounting for the slippage between two phases. A numerical method for this set of equations has been developed. The bubbly flow field in a tank with free surface are calculated successfully by this numerical method. The numerical results reveal that the transverse motions of the bubbles have significant influences on the flow characteristics. These characteristics are affected by the sizes of the bubbles and their size distribution.

REFERENCES

(1) Wijngaarden,L.van, "On the equations of motion for mixture of liquid and gas bubbles," J. Fluid Mech. 33, p.465, 1968.

(2) Wijngaarden,L.van, "One-Dimensional Flow of Liquids Containing Small Gas Bubbles,"
Ann. Rev. Fluid Mech. 4, p.369, 1972.

(3) Cook,T.L. and Harlow,F.H., "VORT: A computer code for bubbly two-phase flow," Los Alamos National Laboratory report LA-10021- MS, 1984.

(4) Cook,T.L. and Harlow,F.H., "Vortices in Bubbly Two-Phase Flow," Int. J. Multiphase flow, 12-1, p.35, 1986.

(5) Nigmatulin,R.I., "Mathematical modelling of bubbly liquid motion and hydrodynamical effects in wave propagation phenomenon," Mechanics and Physics of Bubbles in Liquids, Nijhoff, p.267, 1982.

(6) Matsumoto, Y., Nishikawa, H. and Ohashi, H., Performance of Cascade of Blades in Two Phase Flow, JSME Int. J., Ser.II, 13-4, p.652, 1988.

(7) Saffman, P.G., "The lift on a small sphere in a slow shear low," J. Fluid Mech. 22-2, p.385, 1965.

(8) Hirt C.W. and Shannon, J.P., Free-Surface Stress Conditions or Incompressible-Flow Calculations,
underlineJ. Comp. Physics 2, p.403, 1968.

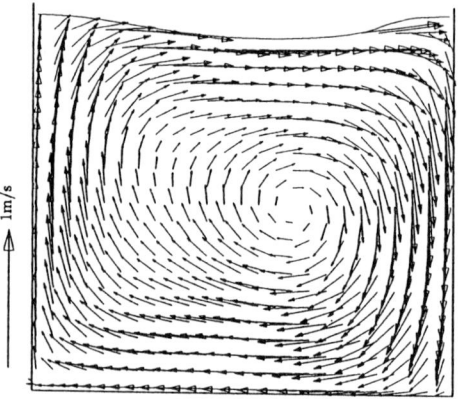

(a) Liquid velocity vector

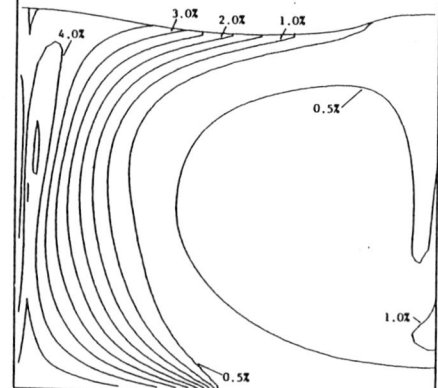

(b) Void fraction contour

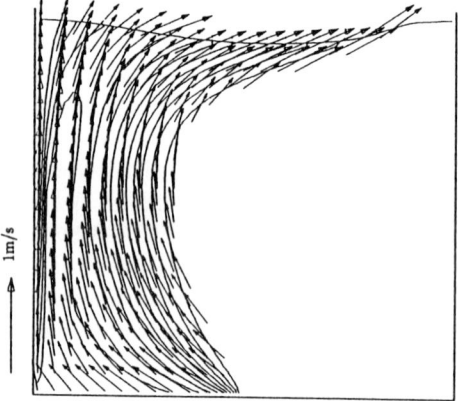

(c) Bubble velocity vector and number density contour for $r_d = 1.0$mm

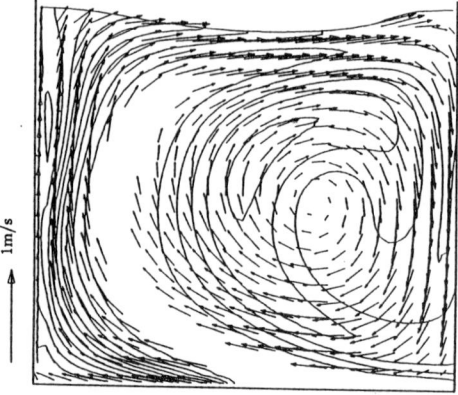

(d) Bubble velocity vector and number density contour for $r_d = 0.2$mm

Figure 6. Bubbly flow in a tank with free surface. The time t= 6.0 second after the bubble injection is started.

C413/006

A transient technique for determining average heat transfer coefficients in finned surfaces

N LAWRENCE, BSc, MSc, PhD, CEng, FIMechE, FIMarE,
Australian Maritime College, Tasmania, Australia
T COWELL, BSc, PhD
Department of Mechanical and Manufacturing Engineering, Brighton Polytechnic

SYNOPSIS A technique is described which enables the average surface heat transfer coefficient of finned surfaces to be determined from experimental cooling curves. A finite difference method is used to predict a theoretical cooling curve for the finned assembly for a range of heat transfer coefficient values. The experimental value is taken as that theoretical value which gives the best fit with the experimental curve. The results for a range of fin geometries are presented in the form of a correlating equation for Nusselt number as a function of Reynolds number and the non-dimensional geometric parameters of the fin. Values of friction factor are also presented.

NOTATION

d	hydraulic diameter
f	friction factor
K_c, K_e	entrance and exit pressure loss coefficients
L	duct length
Nu	Nusselt number
Pr	Prandtl number
Re	Reynolds number
v	air velocity in finned passages
Δp_{ce}	pressure drop from contraction /expansion
Δp_f	pressure drop from finned passages
ρ	air density

1 INTRODUCTION

The methods and results reported here form part of a larger investigation into heat flow in air-cooled diesel engine cylinder heads (1). This comprehensive study involved the measurement of heat transfer coefficients inside the cylinder, the numerical modelling and experimental measurement of temperature distributions within the cylinder head and the experimental measurement of external heat transfer coefficients for a wide range of possible fin configurations. The results of the metal temperature modelling and measurement are reported further in Lawrence and Edmonds (2). The paper presented here is concerned with the measurement of heat transfer coefficients between the cooling air and the fins and in particular with the transient method of measurement used.

The reason for the study was that, although some information is available on the performance of cylinder head cooling fins, it is not of a form that can be used for design purposes. In particular there is no information on the characteristics of an array of rectangular section passages as is typical of modern air-cooled diesel engines. The existing literature is briefly reviewed in Section 2.

This paper reports the results of an exercise to obtain the desired information. A decision was taken to use a transient method of measurement and the range of test samples, experimental equipment and test procedure are described in Section 3. The extraction of the required values of heat transfer coefficient involved the development of a numerical model of the fin for comparison with the experimental cooling curve. The required result was assumed to be the value of heat transfer coefficient in the numerical model that gave the best fit with the experimental curve. The data analysis procedure is described in Section 4 and the experimental results are presented in Section 5.

2. LITERATURE REVIEW

In theory cooling fins on cylinder heads can have a variety of different sections. Rectangular, trapezoidal, triangular and parabolic have been proposed. Mackerle (3) has stated that a parabolic fin section gives the best cooling/weight ratio and that a triangular section fin is second best. However, he acknowledges that neither is practical for reasons of manufacturing cost and in reality fins are essentially rectangular in section with a slight taper from base to tip to facilitate removal of dies from the mould.

A series of research studies on heat transfer from air cooled engines was commissioned by the National Advisory Committee for Aeronautics between 1920 and 1940. This work has been summarised by Ellerbrock and Biermann (4) who were unable to develop a correlating equation for Nusselt number as a function of Reynolds number and the geometric parameters. They developed a dimensional equation of limited value.

Since the passages are essentially rectangular in shape it is possible to use relationships derived for this simple geometry to describe the heat transfer in the fin array. At the most basic level, the relationships for flow in circular ducts can be used by taking a hydraulic diameter as the duct characteristic dimension. The best known of these is the Dittus-Boelter equation (5)

$$Nu = 0.023\, Re^{0.8} Pr^n \qquad (1)$$

where Nu is Nusselt number, Re is Reynolds number and Pr is Prandtl number. $n = 0.3$ for cooling and 0.4 for heating.

A number of refinements to this equation have been proposed (6,7,8).

An alternative method is to use Reynolds analogy between heat and momentum transfer (9). At its most basic this assumes that Prandtl number equals unity and Nusselt number is then given by:-

$$Nu = f\, Re/2 \qquad (2)$$

where f is friction factor.

A development of this is presented by Petukhov (10) and for gases it has the form:-

$$Nu = f\, RePr/2x \qquad (3)$$

where $\quad x = 1.07 + 12.7(Pr^{2/3}-1)\,(f/2)^{1/2}$

Results for rectangular ducts in turbulent flow are limited in number. Berzhinskii et al (11) analysed fully developed flow in a rectangular duct with an isothermal perimeter. They concluded that:-

$$Nu = 0.0186(1.42 - 1.37a + 0.95a^2)\, Re^{0.8} Pr^{0.43} \qquad (4)$$

where a is the value of rectangular passage width to height ratio.

Lowdermilk (12) recognised that entry length effects are relevant and proposed the following for rectangular ducts:-

$$Nu = 0.034\, Re^{0.8} Pr^{0.4} (d/L)^{0.1} \qquad (5)$$

where d is hydraulic diameter and L is duct length.

Weidenmuller (13) proposes the following similar equation specifically for air cooled diesel engine fin design:-

$$Nu = 0.024[1+(d/L)^{2/3}]\, Re^{0.786} Pr^{0.45} \qquad (6)$$

3. EXPERIMENT

Several possible techniques were considered for the determination of heat transfer coefficients. Steady state direct measurement was rejected because of the very high heat fluxes that were required - as high as 500 kW/m² in some cases, and because steam for heating was not available. Mass transfer analogue methods can have the attraction of simplicity, but do not lend themselves to automatic data collection by computer without very sophisticated profile measurement equipment. The method finally chosen was a transient method that relied upon the monitoring of fin metal temperature variation as the pre-heated sample was cooled by sudden exposure to a cooling air stream. The method has the advantages that the experimental equipment is relatively simple and that instrumentation and data collection are straightforward. Its disadvantage is that special data analysis methods had to be developed. These are described in Section 4.

3.1 Fin array samples

The dimensions of a typical sample are shown in Figure 1. In all cases the fins were designed to project into the wind tunnel duct section of 45x44 mm. The geometry models the critical valve bridge region in typical engines (1). Details of the ten variants, all with even fin spacing are given in Table I.

Sample	No. Fins	Fin Thickness - mm	Flow passage hydraulic diameter - mm
1	5	3	8.70
2	6	3	6.84
3	7	3	5.41
4	3	3.36	14.22
5	4	3.36	10.73
6	5	3.36	8.21
7	4	3.8	10.19
8	6	3.8	5.67
9	4	4.2	9.68
10	4	4.8	8.92

Table I Dimensions of Fin Samples

The samples were made from copper, but were not machined from solid metal. The fins were cut from sheet of appropriate thickness and soft soldered into accurately machined grooves in the base plates. Some checks were made to ensure a good metallurgical bond. Microscopic inspection of the interfaces on sectioned samples found consistent bonds with an average thickness of 0.025 mm. Ultrasonic and eddy current techniques were both able to detect the interface, but in neither case were they able to detect flaws. It is believed that these very thin metallurgical bonds led to no significant disturbance of thermal continuity.

The bases of the blocks were each instrumented with five thermocouples at the locations shown in Figure 2. The thermocouples were made from 0.2 mm diameter PVC coated Chromel and Alumel wires welded together to form a fine bead. The wires were located with Araldite in grooves machined down either side of the threaded plugs as shown in the enlarged detail of Figure 2. The plug threads were coated with Thermpath to maximise thermal contact. A simple test of the thermal response of the thermocouples was made by plunging the preheated junction into iced water. Time constants of less than 0.1 seconds were measured.

3.2 Experimental apparatus

The essential components of the experimental equipment are shown in diagrammatic form in Figure 3. The fan was a two stage Keith Blackman driven by a 3.75 kW DC motor fitted with speed control. The air flow rate was measured using a viscous flowmeter. A 1m length of duct of cross-section 45x44 mm was used to ensure that the air flow was fully developed before entry to the test section containing the finned block. An over-the-centre clamp was fitted to the test section to facilitate fast clamping of the sample block and to eliminate leakage. The test section was well insulated to minimise heat loss. Water manometers were used to measure pressure drop across the viscous flow-meter for evaluation of air flow rate and for measurement of the sample flow resistance. Thermocouples before and after the test sample and the five thermocouples in the base of the sample under test were taken to a 12 channel thermocouple selection unit. This was constructed from a printed circuit board supplied by CIL Electronics that is usually part of their PCI 1002 thermocouple converter. It contained a platinum resistance thermometer for determination of cold junction temperature. The data was collected and stored on disc using a BBC microcomputer with a 12 bit A/D converter. A temperature controlled oven was located adjacent to the rig for heating of the samples to test temperature.

3.3. Experimental procedure

The finned block was placed cold into the test section and the air flow adjusted to give the required flow rate and the air side pressure drop measured. The block was then placed in the oven and its temperature raised to approximately 120°C. Once the block reached this temperature, the computer was activated for data collection and the block was then removed from the oven and inserted in the test section as rapidly as possible. The insulating cover was slid over the test section and the temperature time history stored on disc as the finned block cooled down to the temperature of the air stream.

Cooling curves for all samples were obtained for ranges of air cooling velocities between 10 and 60 m/s.

4. DATA ANALYSIS

The geometry under study, with its long thin fins attached to a relatively solid base meant that the lumped capacity method of analysis of the cooling curves was unlikely to be adequate. Estimation of typical Biot numbers confirmed that this was the case and that a more sophisticated method of analysis was required. The solution chosen was to develop a two-dimensional finite difference model of the fin and to compare the cooling curves generated by this model with those given by experiment. The value of heat transfer coefficient in the model was adjusted until the best fit with the experimental data was found. This value of heat transfer coefficient was taken to be the experimental result.

4.1 The finite difference model

The finite difference grid used to model the fins is shown in Figure 4. Symmetry of the samples means that only half of a single fin and its associated base need to be modelled. The nineteen node grid was chosen to minimise computing time without sacrificing solution accuracy. Finer grids were tested and found to give results not significantly different from the grid chosen. The other important influence on computing time was choice of time step. Use of values of time step less than 0.1 seconds gave no further increase in accuracy. A time step of 1 second gave nodal temperatures that differed by only 0.03 K from the 0.1 second values. It was considered that savings in computational time were worth this slight loss of accuracy and a time step of 1 second was chosen. The need for computational efficiency arose because for each run the nodal equations had to be solved for each time step and typical tests lasted 180 seconds. Furthermore for each experimental test the computer model had to be run a minimum of 17 times to find the best curve fit, and this for each of the 5 thermocouples. The time saving factor of 10:1 was thus significant.

4.2 Curve fitting technique

The curve fitting technique required a value of heat transfer coefficient to be chosen as input data for the finite difference model. A measure of the agreement of the model with the experimental result was obtained by summing the differences in the temperatures between the two curves over all time steps. A second value of heat transfer coefficient was chosen so that these two initial values straddled the required solution and a second measure of the agreement with experiment obtained. The choice of the next and subsequent values of heat transfer coefficient for insertion in the model was made using the Regula Falsi method (14). In this, the algebraic error sums calculated in the previous two cases are used to estimate the next value of heat transfer coefficient. This method was chosen because of its rapid convergence. The procedure was repeated until the acceptable level of agreement was achieved. In most cases it was found possible to match the curves with sufficient accuracy for the difference to be barely discernible when plotted on the same graph. (see (1) for details.)

4.3 Validity of two-dimensional models

A typical experimental result with the cooling curves for the five thermocouples is displayed in Figure 5. It shows that there are some differences amongst the five cooling curves and that there is a significant increase in air temperature on passing through the sample. Both these phenomena demonstrate that the analysis problem is really three dimensional in nature. As one would expect, the upstream thermocouples demonstrate more rapid cooling because of the higher value of local temperature difference. (That this effect is genuine and not due to experimental inaccuracy of the thermocouples was demonstrated by reversing the sample and seeing the order of the cooling curves reversed as expected). The model analysis took partial account of these effects by using the experimental value of air outlet temperature at each time step to interpolate the appropriate value of air temperature to be used in the model analysis for each of the five thermocouple locations.

The temperature gradient in the base of the samples in the air flow direction demonstrates that there must be some heat flow in this direction of which account is not taken in the two-dimensional model. However the temperature differences along the block are at no time greater than 8 K and the heat transfer coefficients from the five cooling curves generally agreed with each other to within 5%. Any longitudinal conduction effects within the block will tend to increase heat transfer coefficient values obtained at the upstream end of the block and decrease those obtained at the other end. This phenomenon is demonstrated in the results displayed on Figure 5. The analysis here uses the average of the values from all five curves. The relatively small differences amongst the values from the five thermocouples and the self compensating effects of the extreme values demonstrate that this inherent approximation has little significance for the accuracy of the reported results.

5. EXPERIMENTAL RESULTS

The experimental results for the 10 samples are presented in Figures 6 and 7. Figure 6 displays curves of Nusselt number plotted against Reynolds number separated into two groups for clarity of presentation. Curves of friction factor against Reynolds number are plotted in Figure 7 in a similar fashion.

5.1 Nusselt number curves

The curves in Figure 6 are all straight lines on the double logarithmic plot and all have similar slopes. The Reynolds number ranges are different for the different samples because a similar volume air flow rate range was used in each test and hydraulic diameter values varied by a factor of three.

The curves do not all lie in the same location as the simple relationships of the Dittus-Boelter type would suggest. However visual inspection of the curves suggest that the larger the hydraulic diameter of the basic flow passage through the sample, the higher the curves tend to lie. This is easily explained in terms of entry length effects since the samples all had the same test length of 100 mm. This is confirmed by correlation of the experimental data which yielded the following equation for Nusselt number:-

$$Nu = 0.18 Re^{0.70} Pr^{0.33} \left[\frac{d}{L}\right]^{0.42} \qquad (7)$$

This equation represents the experimental data well with a coefficient of determination $R^2 = 99\%$.

A correlation exercise was also carried out using the parameter $(1+(d/L)^{2/3})$ given by Weidenmuller (13) and although agreement was again quite good, equation (7) above proved to be more accurate.

It is possible that the variation in location of the Nusselt number curves comes to some extent from a passage aspect ratio effect rather than one of entry length. However, these effects are difficult to separate out from this data since the samples all had the same flow length and the same passage height. This results in the two parameters being quite closely related to each other. Correlation against passage aspect ratio yields a less accurate result than equation (7) and correlation against both parameters yields a result no more accurate. In reality the entry length effect, particularly with these relatively short passages is likely to be more significant and is assumed to be the cause of the experimental behaviour.

5.2 Friction factor curves

Curves of friction factor plotted against Reynolds number are presented in Figure 7. Friction factor values were derived from the experimental results using the expression:

$$f = \frac{\Delta p_f}{2\rho v^2} \cdot \frac{d}{L} \qquad (8)$$

where Δp_f is the pressure drop due to the finned surface after correction for entry and exit effects, ρ is the air density and v the air velocity.

The methods described in Kays and London (15) were used to correct the measured pressure loss due to contraction and expansion of the air as it enters and leaves the fin passages. The effect was to reduce the measured pressure drop by an amount:

$$\Delta p_{ce} = (K_c + K_e) \rho v^2/2 \qquad (9)$$

where K_c and K_e are the contraction and expansion pressure loss coefficients, curves for which are also given in (15).

It can be seen that the friction factor curves are not in general straight lines with a single slope on the double logarithmic plot and that the experimental scatter is somewhat greater than for the Nusselt number curves. The location of the curves again shows a tendency to depend upon hydraulic diameter, but the relationship is not as clear as for the Nusselt number curves.

It must be pointed out that a degree of uncertainty about these curves results from the fact that the contraction/expansion component of the total pressure loss could be as much as 50%. The contraction/expansion coefficients themselves cannot be evaluated with great accuracy and hence the friction factor values presented here cannot be claimed as more accurate than ± 20%. This relatively large inaccuracy arises because of the short flow length of the fins that are typical of diesel engine cooling passages. It was considered inappropriate to try and produce correlating equations for these data because of their scatter, because of the non-linear shape of the curves and because of the inherent inaccuracy. Nonetheless the curves represent experimental data for the geometries tested which can be reconstructed for geometries of similar flow length.

6. CONCLUSION

The heat transfer and pressure drop results presented here are suitable for use as a component of the air-cooled diesel engine design process. The heat transfer results have demonstrated that there is a strong entry length effect with these relatively short flow passages. It has also been demonstrated that the configuration leads to a degree of uncertainty about friction factor data because of the large component of pressure drop that arises from entry contraction and exit expansion losses. It has been demonstrated how transient methods can be applied to the determination of heat transfer characteristics of fin surfaces by matching experimental results with those from a transient finite difference model of the configuration.

REFERENCES

(1) LAWRENCE, N. The prediction of temperature distribution in air-cooled diesel engine cylinder heads.
PhD Thesis, CNAA, Brighton Polytechnic, 1988.

(2) LAWRENCE, N. and EDMONDS, M.J. A numerical and experimental study of the valve bridge temperature in an air-cooled diesel engine cylinder head.
Ninth International Heat Transfer Conference, Jerusalem, August, 1990.

(3) MACKERLE, J. Air cooled automotive engines.
Charles Griffin and Co. Ltd., 2nd Edition, 1972.

(4) ELLERBROCK, H.H. and BIERMANN, A.E. Surface heat transfer coefficients of finned cylinders.
NACA Report No.676, 1939.

(5) DITTUS, F.W. and BOELTER, L.M.K. University of California Berkeley, Publications in Engineering, 2, 443, 1930.

(6) COLBURN, A.P. A method of correlating forced convection heat transfer data and a comparison with fluid friction.
Trans. AIChE, vol.29, 1933, pp174-210.

(7) SIEDER, E.N. and TATE, G.E. Heat transfer and pressure drop of liquids in tubes.
Ind.Eng.Chem. 28, 1429, 1936.

(8) MIKHEEV, M.A. Fundamentals of heat transfer.
Gosenergoizdat, 1956.

(9) REYNOLDS, O. On the extent and action of the heating surface for steam boilers.
Proc. Manchester Lit. Phil. Soc. 1874.

(10) PETUKHOV, B.S. Heat transfer and friction in turbulent pipe flow with variable physical properties.
See Hartnett, J.P. and Irvine, T.F. (eds.) Advances in heat transfer, Academic, New York, 1970, pp 504-564.

(11) BERZHINSKII, R.A. and KONOPATOVA, E.A. Heat transfer in rectangular ducts with an isothermal perimeter.
Teploenergetika, No.24(1) 75-77, 1977, pp 61-63.

(12) LOWDERMILK, W.H., WEILAND, W.F. and LIVINGOOD, J.N.B. Measurement of heat transfer and friction coefficients for flow of air in non circular ducts at high surface temperatures.
NACA. RM. E53J07, 1954.

(13) WEIDENMULLER, M. The state of development of the air cooled diesel.
MTZ 34 No.4, 1973, pp 115-121.

(14) CHAPRA, S.C. and CANALE, R.P. Numerical methods for engineers with personal computer applications.
McGraw Hill, 1985, pp 109-153.

(15) KAYS, W.M. and LONDON, A.L. Compact Heat Exchangers.
McGraw Hill, 1984.

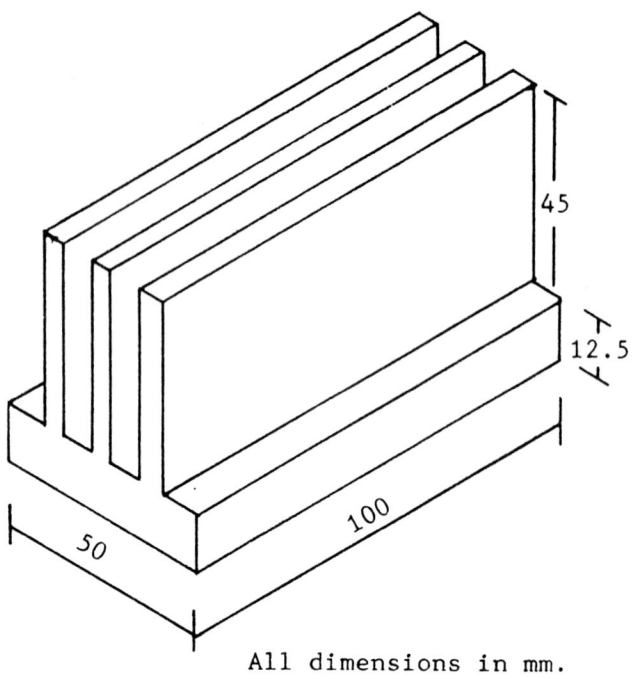

Figure 1 Dimensions of Typical Sample.

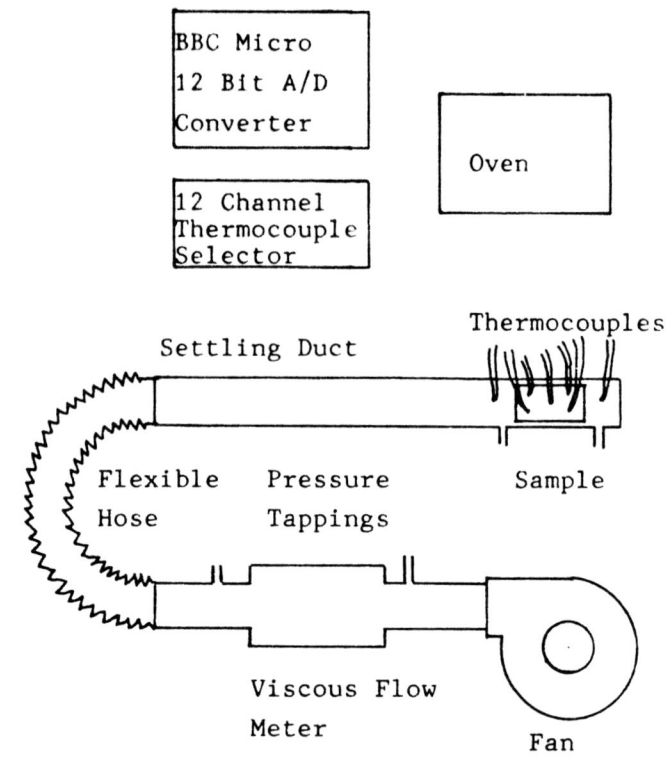

Figure 3 Experimental Apparatus.

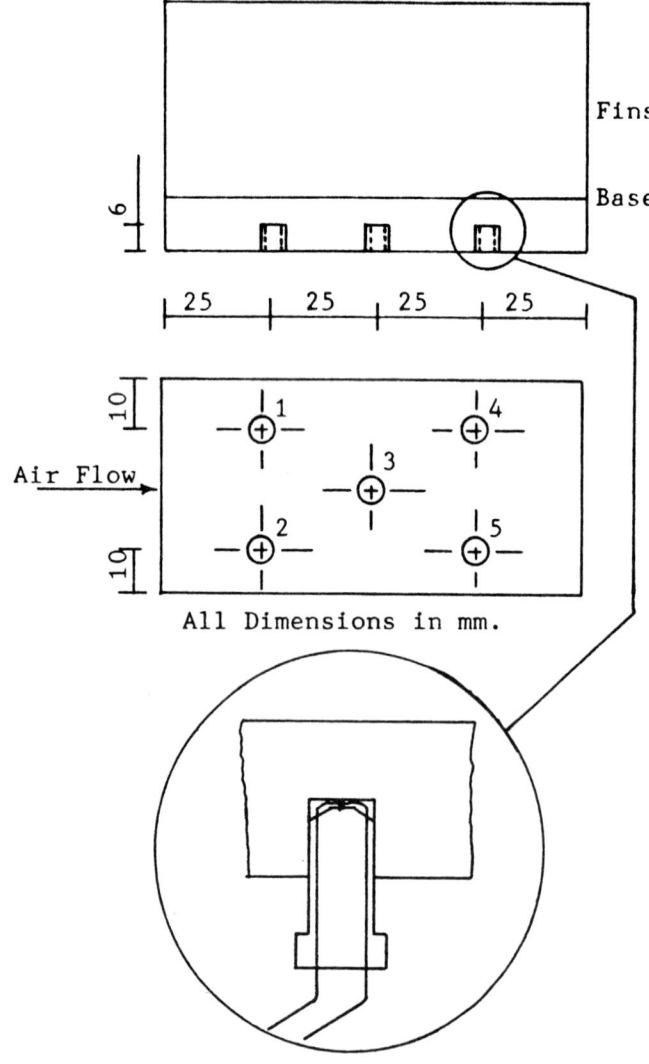

Figure 2 Thermocouples in Fin Samples.

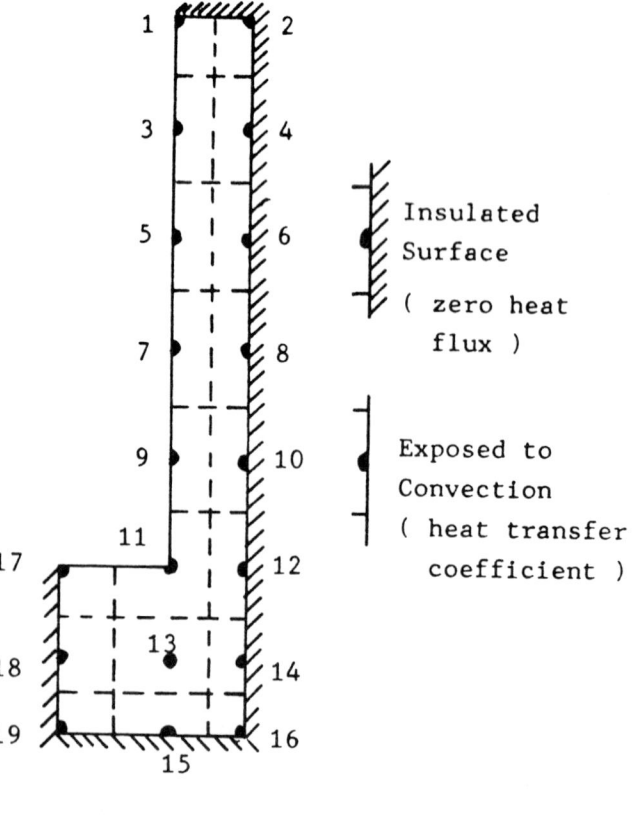

Figure 4 Finite Difference Model of Fin Element.

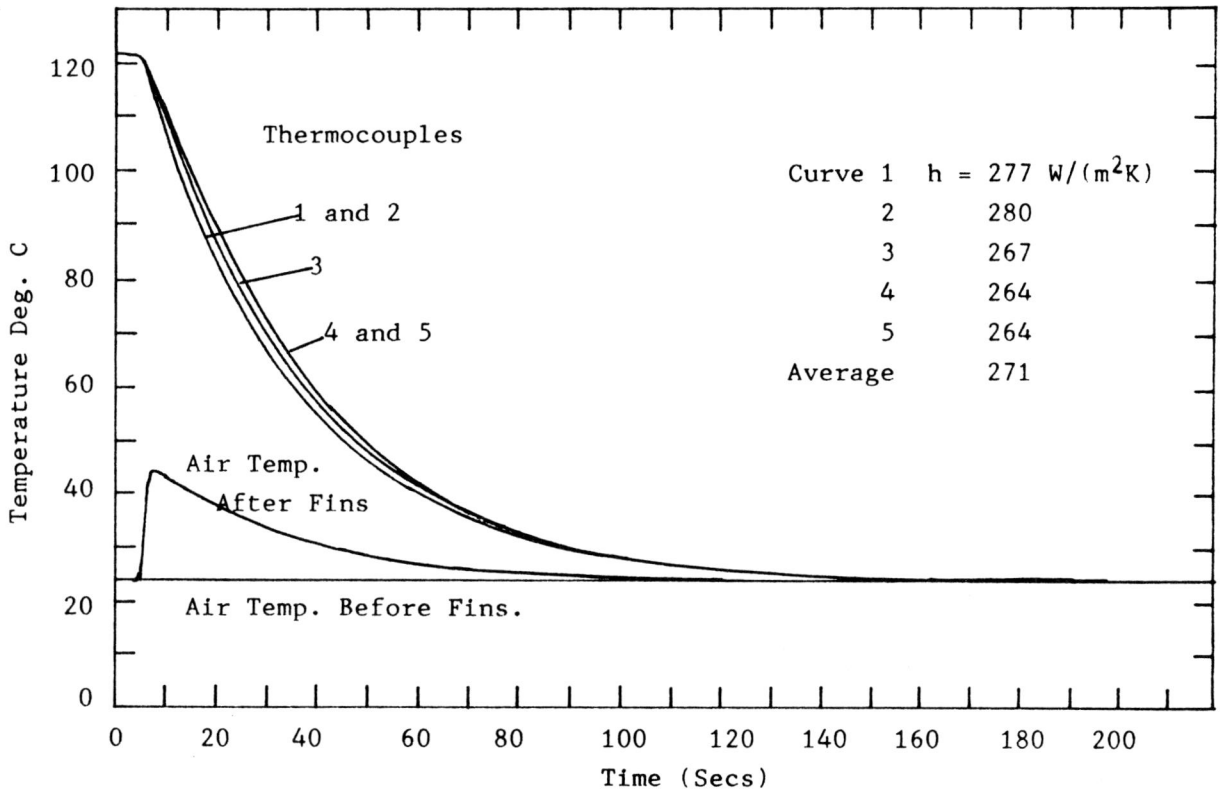

Figure 5 Typical Experimental Results Showing Cooling Curves For 5 Thermocouples And Variation of Air Outlet Temperature.

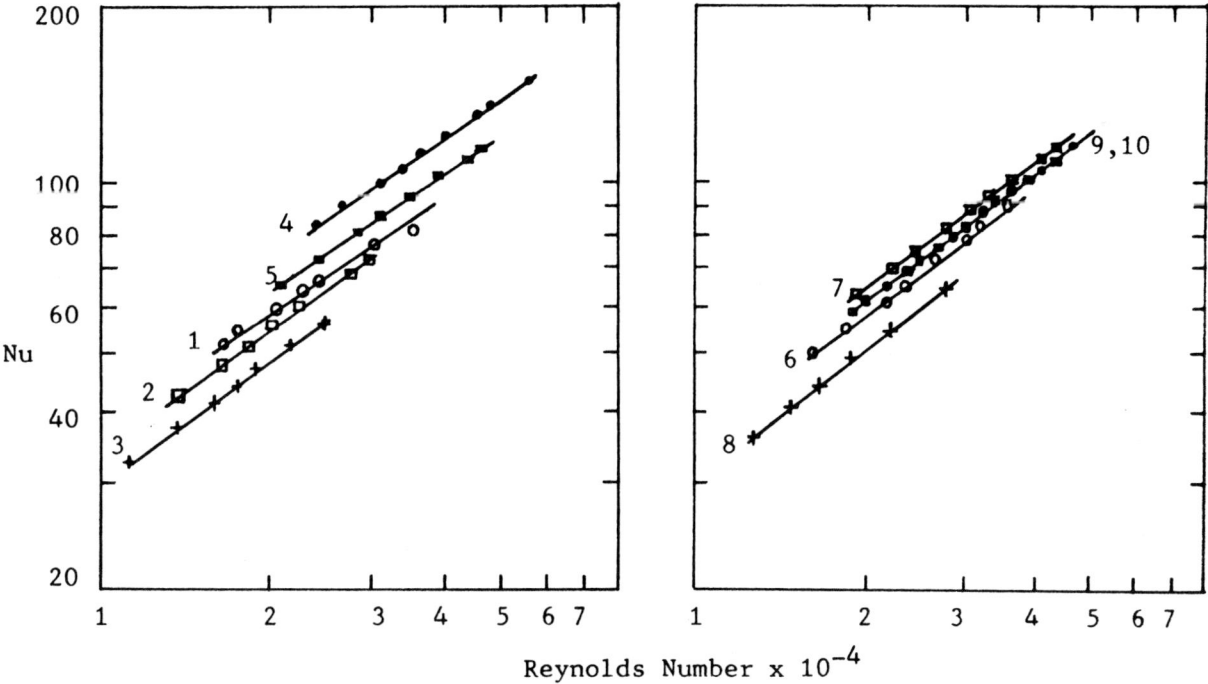

Figure 6 Variation of Nusselt Number With Reynolds Number

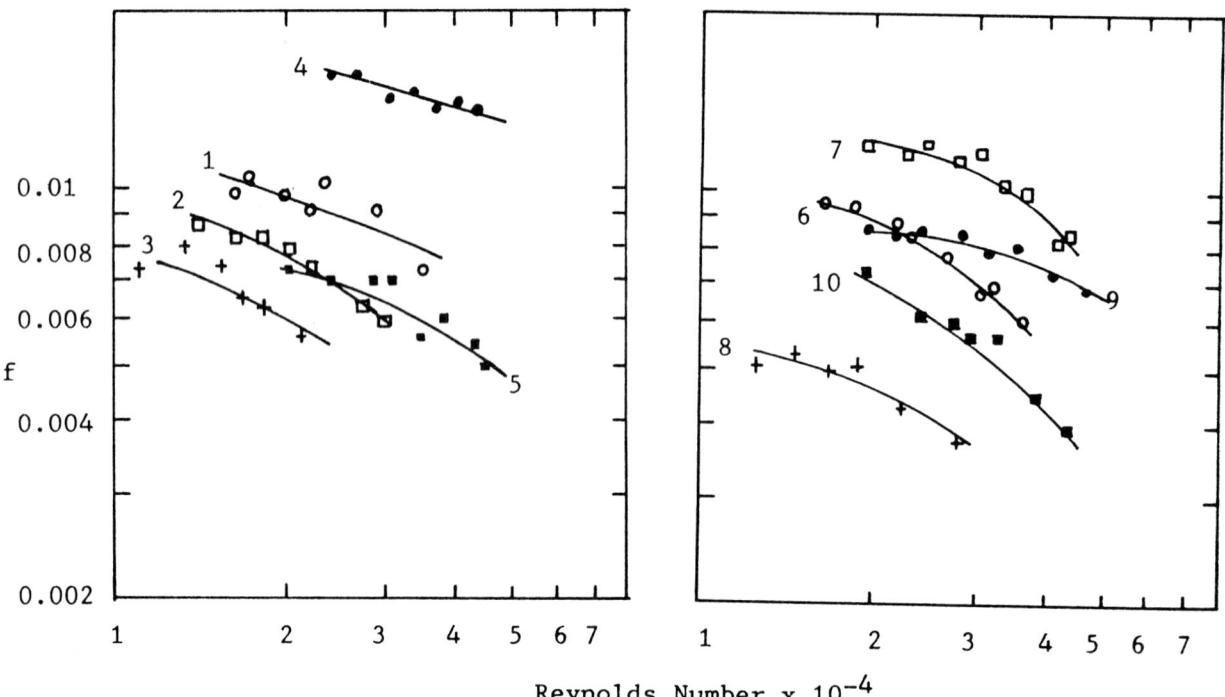

Figure 7 Variation of Friction Factor With Reynolds Number

C413/062

Flow resistance in circular tubes rotating about a parallel axis

A R JOHNSON, BEng, PhD, CEng, MIMechE
Department of Mechanical and Process Engineering, University of Sheffield

SYNOPSIS Theoretical and numerical predictions for laminar and turbulent isothermal flow through a circular tube rotating about a parallel axis indicate a Coriolis generated axisymmetric cross-stream swirling flow superimposed upon a developing axial profile, the effects of flow resistance being negligible. This swirling flow builds up rapidly after the entrance and then decays away, the flow then being identical to fully developed flow through a stationary tube. Experimental results for smoothed inlets confirm the predictions for both laminar and turbulent flow. However, for unsmoothed inlets very significant increases in flow resistance occur in the entry region of the tube, these increases persisting for long lengths along the tube.

NOTATION

$\underline{a}_o$	acceleration of tube axis
C_f	friction factor
d	tube diameter
e	eccentricity ratio
$\underline{e}_\theta, \underline{e}_r, \underline{k}$	unit vectors for cylindrical co-ordinates
H	tube eccentricity
$\underline{i}, \underline{j}, \underline{k}$	unit vectors for cartesian co-ordinates
J	rotational Reynolds number
p	pressure
p_h	cross-stream hydrostatic pressure
$\underline{r}$	position vector of fluid particle relative to cylindrical co-ordinates
r, θ, z	cylindrical co-ordinates, fixed to tube
Re	Reynolds number
u, v, w	radial, tangential and axial fluid velocities relative to tube
v_p	peak swirl velocity at an axial location
v_{max}	maximum swirl velocity
$\underline{v}$	fluid velocity vector relative to tube
w_m	mean axial fluid velocity
z	axial length along tube
μ	fluid viscosity
ν	fluid kinematic viscosity
$\underline{\xi}$	vorticity vector
ρ	fluid density
$\underline{\omega}$	angular velocity vector
$\underline{\Omega}$	angular velocity

1. INTRODUCTION

Coolant passages in rotating components are often found in modern highly rated power generating machinery such as large turbo-generators and gas turbines. In large turbo-generators the efficient generation of electricity requires that the electrical windings are highly loaded both electrically and magnetically. To ensure a long, reliable service life the resultant heat generated must be effectively removed to keep the insulation below its long term degradation temperature. In large turbo-generator rotors this is achieved by passing coolant, typically hydrogen or water, through passages within the rotor or through hollow windings (1)(2). In gas turbines the efficiency is improved by operating at as high a maximum cycle temperature as the constructional materials will permit. The high temperature turbine blades in particular are exposed to this severe environment combined with high rotational and aerodynamic loadings. High temperature creep and other material problems thus impose a limit on the maximum cycle temperature for a satisfactory service life. One way of reducing material temperature is through internal cooling passages within the rotating turbine blades.

The rotating coolant circuits will consist of the coolant passage within the blade or winding and also the necessary ducting, plenum chambers, joints, branches, etc to deliver and distribute the coolant within the rotating component. For a stationary coolant circuit it is possible, if by no means easy, to perform a network type analysis to determine the pressure losses and the mass flow rates throughout the coolant circuit and hence evaluate the cooling performance. When the circuit is rotating the fluid flowing through the circuit is constrained to rotate with the component. It is, therefore, necessary to allow for the effects of this rotation on the flow, pressure losses and heat transfer in such components. This information is often not available in the literature, although some work has been performed on certain elements, see for example the review by Morris (3). It has for example been shown for a radial passage that rotation can in some circumstances improve cooling performance and in others impair it (4). It is important to stress that the uncertainty in predicting the cooling performance in rotation machinery can be the cause of premature machine failure. For example a critical area may not have sufficient coolant flow due to rotation adversely affecting that branch of the flow circuit, thus reducing coolant flow and so producing a localised hot spot. These hot spots must be avoided if the integrity of the overall machine is to be preserved, for example a 10^0C temperature rise in the insulation of a turbo-generator winding can halve its service life (5).

This paper outlines the main findings from an investigation (6) which aimed to assist in filling certain gaps in the literature. The configuration investigated was that of a circular tube rotating about a parallel axis as shown in Figure 1a. Such passages are frequently found in the direct cooling of large turbo-generator rotor windings and the experimental flow circuit shown in Figure 1b approximates to one of these. The scope of the investigation outlined here was to consider isothermal flow in both the fully developed and entrance regions of the circular tube analytically, numerically and experimentally. The investigation also complements heat transfer investigations which had previously been conducted on this geometry (7), and more recently combined heat transfer and pressure drop measurements on a larger scale apparatus (8)(9).

2. THEORETICAL OBSERVATIONS

In order to illustrate the physical effects of rotation the fluid motion inside the tube is referred to a polar co-ordinate frame osz which is fixed to the rotating tube as shown in Figure 1a. This reference frame is non-inertial and so corrections terms for any translational acceleration of the origin together with centripetal and Coriolis components of the acceleration must be included in the momentum conservation equations. To permit the interpretation of the numerical and experimental investigations presented later the effect of these correction terms on the momentum equations will be considered for laminar constant property flow. The resulting general observations will also be applicable to turbulent flow.

The flow system illustrated in Figure 1a is geometrically described by the tube diameter, d, the tube length, z, and the eccentricity of the centre line of the tube from the axis of rotation, H. The tube rotates about the axis of rotation with a constant angular velocity, Ω, and the flow through the tube is taken to be steady and laminar with invariant properties. The fluid motion is referred to the fixed cartesian co-ordinate system XYZ shown in Figure 1a so that the momentum conservation equation can be expressed vectorially by (see (3)(6))

$$\frac{D\underline{v}}{Dt} + 2(\underline{\omega}\times\underline{v}) + (\underline{\omega}\times\underline{\omega}\times\underline{r}) + \underline{a}_o = -\frac{1}{\rho}\nabla p + \nu\nabla^2\underline{v} \quad (1)$$

where $\underline{r}$ and $\underline{v}$ refer to the position vector and velocity of a particle in the flow measured relative to the rotating reference frame osz, $\underline{\omega}$ is the angular velocity of the reference frame (i.e. that of the tube) and $\underline{a}_o$ is the acceleration of the co-ordinate origin o (i.e. that of the tube centre line). Also ρ is the density of the fluid, ν its kinematic viscosity, the pressure is p and the operator D/Dt is the usual total derivative.

When equation 1 is applied to the system under consideration the following three component equations are obtained where u, v and w are the velocity components in the radial, tangential, and axial directions respectively relative to the rotating reference frame osz.

$$\frac{Du}{Dt} - 2\Omega v - \Omega^2(r+H\cos\theta) = -\frac{1}{\rho}\frac{\partial p}{\partial r} + \nu\nabla^2 u$$

$$\frac{Dv}{Dt} + 2\Omega u + \Omega^2 H\sin\theta = -\frac{1}{\rho r}\frac{\partial p}{\partial \theta} + \nu\nabla^2 v \quad (2)$$

$$\frac{Dw}{Dt} = -\frac{1}{\rho}\frac{\partial p}{\partial z} + \nu\nabla^2 w$$

The non-dimensional parameters which control the flow and pressure fields and consequently the pressure loss characteristics may be determined by substituting the following transformations into equation set 2

$$U = \frac{ud}{\nu}, \quad V = \frac{vd}{\nu}, \quad W = \frac{wd}{\nu}$$

$$R = \frac{r}{d}, \quad Z = \frac{z}{d}, \quad P = \frac{p}{\rho w_m^2} \quad (3)$$

where w_m is the mean axial velocity through the tube. Equation set 2 then becomes

$$\frac{DU}{Dt} - 2J.V - J^2(R+e\cos\theta) = -Re^2\frac{\partial P}{\partial R} + \nabla^2 U$$

$$\frac{DV}{Dt} + 2J.U - J^2 e\sin\theta = -Re^2\frac{1}{R}\frac{\partial P}{\partial R} + \nabla^2 V$$

$$\quad (4)$$

$$\frac{DW}{Dt} = -Re^2\frac{\partial P}{\partial Z} + \nabla^2 W$$

where

$$Re = \frac{w_m d}{\nu} \quad \text{(Reynolds number)}$$

$$J = \frac{\Omega d^2}{\nu} \quad \text{(rotational Reynolds number)} \quad (5)$$

$$e = \frac{H}{d} \quad \text{(eccentricity ratio)}$$

Equation set 5 suggests that the flow field will be dependent upon the usual pipe flow Reynolds number, the eccentricity of the tube relative to the rotational axis, and the so-called rotational Reynolds number which is a measure of the relative strength of Coriolis forces to viscous forces. Additionally in order to integrate the overall pressure loss along a specified length, z, of the tube it is expected that the usual length/diameter ratio will also be required.

The conservation of momentum equation 1 may also be used to illustrate another important feature for the present rotating geometry. By taking the curl of the momentum conservation equation it is possible to derive the so-called vorticity equation which represents the rotations of the flow about the co-ordinate axes. When this is done for equation 1 it is interesting to note (see (3) (6) for details) that the centripetal acceleration terms in the equations vanish identically as a source term for the generation of vorticity relative to the tube. The centripetal terms are made manifest in a purely hydrostatic manner similar to the earth's gravitational field. This cross-tube hydrostatic pressure distribution is given by

$$p_h = \rho\Omega^2 \left(\frac{r^2}{2} + Hr\cos\theta \right) \quad (6)$$

This is not so for the Coriolis acceleration component. If the vorticity vector, $\underline{\xi}$, is defined as $\underline{\xi}$ = curl $\underline{v}$ then by taking the curl of equation 1 we get after some algebraic manipulation invoking the continuity equation that

$$\frac{D\underline{\xi}}{Dt} = (\underline{\xi}.\nabla)\underline{v} + \nu\nabla^2\underline{\xi} + 2(\underline{\omega}.\nabla)\underline{v} \quad (7)$$

With the exception of the term $2(\underline{\omega}.\nabla)\underline{v}$ which has its origin in the Coriolis acceleration component equation 7 is the usual vorticity equation for an inertial reference frame. Consequently provided $2(\underline{\omega}.\nabla)\underline{v}$ is not zero then Coriolis acceleration can generate vorticity relative to the tube. This implies that cross stream secondary flows can be created due to Coriolis accelerations. For the present geometry the Coriolis-induced vorticity source may be expanded by noting that $\underline{\omega} = \Omega\underline{k}$ and $\underline{v}=u\underline{e}_r+v\underline{e}_\theta+\omega\underline{k}$ where $\underline{e}_r,\underline{e}_\theta$, $\underline{k}$ are unit vectors in the radial, tangential and axial directions respectively. Thus

$$2(\underline{\omega}.\nabla)\underline{v} = 2\Omega \left(\frac{\partial u}{\partial z}\underline{e}_r + \frac{\partial v}{\partial z}\underline{e}_\theta + \frac{\partial w}{\partial z}\underline{k} \right) \quad (8)$$

This result implies that Coriolis-induced vorticity generation will vanish when when there are no axial gradients of velocity, that is when the flow is fully developed. In turn this suggests that, at distances well away from the entrance effects, the pressure loss characteristics described via the Blasius friction factor, C_f, will tend to be unaffected by rotation. Conversely in the immediate entrance region Coriolis-induced increases in frictional pressure loss may be expected. The above results, strictly speaking, apply only to isothermal laminar flow, however, a similar overall flow pattern is likely to exist with isothermal turbulent flow. It should perhaps also be emphasised that with heat transfer to the flow a buoyancy induced cross stream secondary flow exists when the flow is fully developed (3)(7)(8)(9) so increasing flow resistance.

It is against this foregoing physical discussion that the numerical and experimental investigations undertaken are described.

3. NUMERICAL INVESTIGATION

Analytic solutions of the governing continuity and momentum equations are not possible due to their simultaneous, non-linear partial differential form. In recent years numerical finite difference formulations for the solution have been proposed based mainly on the pioneering work of Spalding (10). The numerical predictions outlined below make use of the results of developments made on this work and which is available as a package (11).

The basic procedure required to model the present situation is to specify suitable boundary and initial conditions, fluid properties and to input special terms to specify the additional terms in the momentum equation not included in the formulation in the code. In this case these are the terms due to the Coriolis and centripetal acceleration in equation set 2. The other main aspect of the modelling is the choice of suitable grids and solution control parameters in order to ensure an accurate and correctly converged solution.

3.1 Defining the model

The model used the largest 14.17 mm diameter tube used in the experimental investigation described later. Air was used as the working fluid in the experimental investigation so the properties used were that for air at standard atmospheric pressure and temperature. For turbulent flow the two equation k-ε turbulence model proposed by Launder and Spalding (12) is available as a built-in option in the code and was thus employed, no attempt being made in this initial modelling to consider any effects of rotation on the turbulence structure. The inlet flow was assumed to be a uniform axial velocity without any swirl relative to the tube. The boundary conditions at the tube wall were zero velocity, and the inbuilt wall functions were used.

The centripetal terms in the governing momentum equation have been shown not to contribute to the generation of cross-stream secondary flows but only to produce a cross-stream hydrostatic pressure gradient given by equation 6. For this reason the centripetal terms were not included as source terms in the momentum equations in the numerical model. The hydrostatic pressure distribution would need to be added to the predicted pressure field for the complete pressure field to be specified. The Coriolis terms in the governing momentum equation were included in the source terms in the momentum equation in the code.

The solution control parameters to ensure stable convergence to the solution in an acceptably short number of iterations were found largely by trial and error, as is normally the case. Full details of setting up the numerical model are given in (6).

3.2 Laminar flow

To gain confidence laminar flow through a stationary circular tube was initially studied, the results agreeing closely with available data (6). For the rotating case it is not immediately apparent whether the flow will be axisymmetric or not. A fully three dimensional polar grid was initially used consisting of ten equal radial cells, ten equal tangential cells, and twenty equal longitudinal cells. The converged solution predicted a developing axial velocity profile on which was superimposed an axisymmetric swirling flow which built up rapidly close to the entrance of the tube, before decaying away as the flow became more fully developed, thus tending to confirm the earlier predictions. This overall flow pattern was also confirmed using a finite element based approach (13).

Since the swirling flow was axisymmetric considerable computational savings could be made using axisymmetric grids. Further, finer cells could be used near to the walls and the entrance to improve the accuracy. Grid refinement procedures were conducted to ensure solutions were essentially grid independent, and enough iterations were performed to ensure full convergence.

The development of the axisymmetric tangential velocity is shown in Figure 2 for a typical case with a Reynolds number of 973 and a rotational Reynolds number of 800. The peak swirl velocity, at each axial location along the tube, increases very rapidly to a maximum value of approximately 10 per cent of the mean flow velocity after 13 diameters. The peak swirl velocity then decays to 1 per cent of the mean flow after 129 diameters and to 0.1 per cent after 235 diameters. It can therefore be concluded that when the flow is fully developed there will be no swirl velocity, thus confirming the earlier predictions.

The maximum swirling velocity increases with the rotational speed, but for a given Reynolds number occurs at the same length/diameter ratio. Figure 3 shows how all the numerical results for a range of Reynolds number (486, 973, 1460, 1946) and Rotational Reynolds number (400, 800, 1400) were correlated using these two dimensional groups suggested by equation set 5. From Figure 3b the maximum swirl velocity is given by

$$v_{max} = 0.122 \frac{w_m J}{Re} \qquad (9)$$

at an axial location given by

$$\frac{z}{d} = 0.013\, Re \qquad (10)$$

Further the swirl velocity has effectively fully decayed by an axial position of

$$\frac{z}{d} = 0.3\, Re \qquad (11)$$

It should be noted that equation 11 gives an indication of the hydrodynamic entry length of the tube and is approximately five times greater than that for a stationary tube (14).

The axial development is shown in Figure 4a and is almost identical to that for a stationary pipe flow although rotation appears to lead to a very slightly faster development. The effect of rotation on the friction factor is negligible as shown in Figure 4b for a Reynolds number of 973 and for rotational Reynolds numbers from 0 to 1400. This is probably due to the negligible change in the development of the axial velocity profile and the relatively low swirling velocities compared to the through flow velocity.

3.3 Turbulent flow

Flow through a stationary tube again showed good agreement with published data (6). The three dimensional grid for the initial rotational case also showed that the flow was axisymmetric thus enabling the flow of finer axisymmetric grids and grid refinement.

A similar overall flow pattern to that found for laminar flow was obtained as shown in Figure 5 for a Reynolds number of 29200 and a rotational Reynolds number of 1200. For this case the maximum swirl velocity reached approximately 0.15 per cent of the mean axial velocity before decaying away, thus confirming the previous predictions. It should be emphasised that although this may appear to be only a small percentage of the mean velocity it nevertheless has a reasonable numerical value. For example with a 14.17 mm diameter tube, for a Reynolds number of 29200 and a rotational Reynolds number of 1200 the maximum swirl velocity is approximately 45 mm/s. The effect on axial velocity development and friction factor development were negligible, again probably due to the relatively small swirling velocities compared to the through flow velocity.

The results for all Reynolds numbers (19460, 29160, 48660) and rotational Reynolds numbers (400, 800, 1200) could again be correlated using these dimensional groups, the results are shown in Figure 5. The maximum swirl velocity being given by

$$v_{max} = 0.113 \frac{w_m J}{Re^{1.11}} \qquad (12)$$

at an axial position given approximately by

$$\frac{z}{d} = \frac{0.85}{Re^{-0.25}} \qquad (13)$$

In this case due to limitations of computer core and storage available a refined grid could not be used to ensure fully developed flow, that is for the swirling flow to have decayed away. However, consideration of Figure 5 would indicate that this would be in the region of

$$\frac{z}{d} \simeq \frac{15}{Re^{-0.25}} \qquad (14)$$

It should be noted that this gives an entry length about 24 times greater than that for stationary tube flow (14).

4. EXPERIMENTAL INVESTIGATION

4.1 Experimental Apparatus

A rotor system to support the test section was available as a result of previous investigations into the effect of rotation on heat transfer in cooling channels in turbo-generator windings. This rotor system has been described in detail elsewhere (6) (7). The rotating flow circuit portion of the rig is shown schematically in Figure 1b. Air was used as the working fluid and enters the rotor via a sealing chamber at A. The air enters the circular test section CD via an inlet plenum chamber B which is shown in Figure 1c and described later. The air exhausts from the test section CD into the exit plenum chamber E and leaves the rotor via another sealing chamber at F. Two different radius rotor support arms were available (457 mm and 305 mm) to enable different eccentricities to be investigated.

Four circular drawn brass test sections nominally 600 mm long were used with bore diameters of 5.3, 6.86, 10.23 and 14.17 mm. The test section was well supported by spacers inside a rigid aluminium tube to prevent any distortion during running. Each test section had five equi-spaced pressure tappings along the axial direction. This permitted pressure drops for tubes having four length/diameter ratios to be measured from the same experiment, thus permitting an assessment of entry length sensitivity. Further, the final quarter span portion of the test section gave the 'best approach' to fully developed conditions. The different diameter test sections also enabled a wide range of length/diameter ratios to be investigated.

The pressure signals were transmitted from the rotor via a five channel rotary seal unit attached to the main rotor by a flexible coupling. Rotary sealing was achieved by using a system of permanent magnets and a magnetic fluid. Full details of this leakproof method of transmitting the rotary pressure signals to stationary measuring instruments have been given elsewhere (6) (15). Pressure measurements were made using micromanometers or 'U' tube mercury manometers as appropriate. The air flow was measured after leaving the sealing chamber F using rotameter type flow meters. The air temperature in both the plenum chambers was monitored by thermo-couples the signals being transmitted via silver/silver-graphite slip rings. The pressure signals and thermocouple outputs were logged on a computer data logger which was also used for subsequent analysis.

In order to investigate the sensitivity to upstream flow conditions of the subsequent pressure loss in a rotating channel the bell mouth inlet shown in Figure 1c could be fitted with various flow straightening devices. The flow enters the annulus and then enters the main plenum chamber via holes around the annulus, before entering the test section via the bell mouth entrance. To ensure air tight joints 'O' ring seals (not shown) were used at all joints. For the different diameter test sections used the inlet diameter of the bell mouth was varied to keep a constant area reduction. A flush brass ring (not shown) enabled a woven brass wire gauze to be positioned at the entrance section XX of the bell mouth, and similar screens could also be trapped between the inlet of the test section and the end of the bell mouth at section ZZ. A flow straightening honeycomb made out of very thin plastic straws could also be inserted between sections XX and YY. By placing screen gauzes of varying mesh geometries at the locations XX and ZZ and using the honeycomb flow straightener between X and YY, it was possible to conduct flow resistance measurements over a range of flows and rotational speeds with a variety of inlet configurations. The range of inlet geometries studied is given in Figure 1c.

When interpreting the results it should be noted that the length/diameter ratios quoted are referred to the location of the first pressure tapping. Thus there was a short axial length (1.6 diameters) of tube between the true physical end and the location of the first pressure taps. An additional set of the smaller diameter tubes were made with pressure tappings at the same length/diameter ratios as the largest diameter tubes to enable an effective comparison of the effects of tube length on the pressure drop to be made.

4.2 Experimental Results

Tests were initially undertaken at zero rotational speed and a range of Reynolds numbers to commission the instrumentation and data processing software and to furnish a reference data base with which subsequent experiments with rotation could be compared. These initial tests showed that the results agreed with standard correlations within the accuracy of the correlations and the experimental results (6). For convenience in interpreting the experimental results the standard pipe flow correlations for fully developed flow are shown on Figures 6 to 9, and 10b in order to provide a suitable reference.

It was argued earlier from the examination of the momentum equation that the influence of rotation on flow resistance is likely to diminish as axial velocity gradients become increasingly small that is when fully developed flow is approached. This is confirmed in Figure 6a, constructed from earlier work (16), which shows the 'best approach' to developed friction factors for a circular tube having a length/diameter ratio of 98.8 prior to the pressure loss measurement. In the notionally laminar and turbulent regimes rotation has little influence on the friction factor. However in the transitional range of Reynolds number (say 2000 < Re < 7000) there is a marked increase in the friction factor, with a more gradual change from laminar to turbulent flow with the normal transitional dip being absent.

The influence of rotation on the 'best approach' to fully developed flow for the largest 14.17 mm diameter tube is shown in Figure 6b, constructed from earlier work (17), for the four inlet configurations detailed in Figure 1c. It is clear for this case that the influence of the entry plane velocity field has not yet decayed sufficiently with each inlet configuration for the claim to be made that rotation has no effect on developed flow resistance. It should however be noted that the overall length/diameter ratio of the 14.17 mm test section was 42.3 whereas the equivalent figure for the smallest 5.3 mm diameter test section shown in Figure 6a was double this. This is the most likely reason that rotation is still demonstrating a significant influence particularly with the inlet designated A (i.e. empty bell mouth). However for Reynolds numbers greater than about 6000 the data gradually tended to approach the individual zero speed data with inlets C and D. These are fitted with screen gauzes at the immediate entry plane ZZ to the test section as well as the straightening honeycomb. Nevertheless, for notionally laminar flow (Re < 2000) rotation has a significant effect on the friction factor. The tendency for a relatively smooth transition from laminar-like to turbulent-like flow is again apparent.

It is with relatively short aspect ratio tubes that the combined effect of rotation and inlet configuration produces the greatest effect on friction factor. In the axial regions immediately downstream of the entry plane, the departure from a fully developed velocity profile is most marked and hence the axial gradients of velocity are relatively high. This is the condition which gives rise to Coriolis action as dictated by the conservation of momentum principle discussed earlier. It is in these pipe locations that Coriolis induced secondary flows enhance mixing with a consequential increase in the friction factor. This is shown in Figure 7 where data from earlier work (17) for two length/diameter ratios is presented for each inlet.

Inlet configuration A is likely to produce the most disturbed entry plane profile because no attempt is being made to smooth out the velocity field resulting from the realignment of the flow through the entry plenum. It can therefore be expected that this inlet will produce the most significant effect on friction factor when the test specimen is rotating. This is shown in Figure 7a, where the increases in friction factor compared to the corresponding zero speed condition is severe. For this length/diameter ratio, progressive modification of the inlet configuration by the inclusion of gauzes and straightening honeycomb tends to suppress the overall increase in friction factor when rotating. Beyond a Reynolds Number of about 18000 data for all inlets having some element of flow smoothing tended to gradually approach its own zero speed characteristic and became virtually coincident at the maximum Reynolds number tested. Figure 7b shows similar effects for a length/diameter ratio of 42.3, the increases being less significant since it contains a greater proportion of more fully developed flow where the increases due to rotation are less as shown in Figure 6b.

Consideration of the results given in Figures 6b and 7 suggest that the inlet configuration D is the most successful at removing the rotational and entry effects upstream of the entrance to the test section at plane ZZ. For this reason inlet configuration D was chosen to extend the previously reported work (17) for all subsequent experimental tests which are now detailed.

It is particularly informative to note from Figure 7 that as the experimental inlet becomes more smoothed, that is removes upstream effects more, the friction factor increases due to rotation become significantly less. Indeed, there is a clear tendency for them to approach the stationary speed cases, and this is particularly true for turbulent flow where the increases become negligible. Even in the laminar region, where differences still exist, the more smoothed inlets are approaching those for the non-rotating case. This would tend to support the earlier numerical predictions where no significant increase in friction factor with rotation was noted. Further, it should be emphasised that these numerical predictions were for a truly smoothed inlet, that is constant axial velocity with no inlet swirl, even the most smoothed experimental inlet configuration, that is D in Figure 7, will certainly not be this smooth.

The examination of the governing momentum equation 1 detailed earlier indicated that the centripetal acceleration terms were made manifest solely by the cross-stream hydrostatic pressure distribution given by equation 6. It was also argued that the length/diameter ratio would be needed to define the axial development along the tube. These effects were investigated by performing tests using both the smaller 305 mm and the larger 457 mm radius rotor arms to give different eccentricity ratios for each tube. Also the different diameter test sections were made with the pressure tappings at constant length/diameter ratios of 10.6, 21.2, 32.4 and 42.3 and with a constant calming length of 1.6 diameters from the inlet plane at ZZ to the first pressure tapping. Typical results for a rotational Reynolds number of 300 are shown in Figure 8 from which two important observations can be drawn. Firstly, by considering each diameter test section in turn at both the eccentricities it can be seen that the eccentricity has a negligible effect on the friction factor. It is this eccentricity ratio which describes the effect of the centripetal acceleration terms which are thus confirmed to have a negligible effect on the flow resistance. Secondly, the results for all the tubes fall within a tight band at both the length/diameter ratios shown thus indicating that the axial development is defined by the length/diameter ratio for the full range of experimental Reynolds numbers. It should also be noted that similar results were obtained for the two length/diameter ratios not shown, and for the full range of rotational Reynolds numbers investigated of 0 to 1000.

The hydrostatic cross-stream pressure distribution was further investigated using a test section with the pressure tappings arranged around the circumference of the tube. Three axial locations being investigated with length/diameter ratios of 7.1, 22.8 and 38.5 from the tube inlet. In order to ensure the maximum hydrostatic pressure change across the tube the largest 14.17 mm diameter tube was used together with the larger 457 mm diameter rotor arms. Further the maximum rotational speed of 1025 rev/min was used. The measured cross-stream hydrostatic pressure distribution confirmed that predicted by equation 6 at each of the three axial locations and for Reynolds numbers from 1000 to 41000. These results also confirm the validity of excluding the centripetal terms from the earlier numerical analysis.

The typical effect of rotation on the friction factor at various rotational Reynolds numbers is shown in Figure 9 for the largest 14.17 mm diameter test section with the larger 457 mm rotor arms. The most smoothed inlet configuration was also used and results are presented for length/diameter ratios of 10.6, 31.8 and for the best approach to 'fully developed flow' that is with a calming length of 33.4 diameters. At the lowest rotational Reynolds number of 200 rotation has a negligible effect on the friction factor. There is also negligible effect in the turbulent flow regime (Reynolds numbers in excess of 12000) for all rotational Reynolds numbers, thus confirming the earlier numerical predictions.

In the entrance region shown in Figure 9a rotation has a significant effect at rotational Reynolds numbers above 600 particularly for Reynolds numbers less than 5000. It is however likely that at these higher rotational speeds some residual rotational effects from the upstream flow still remain after the smoothing inlet and these will probably account for the increase in the friction factor. The numerical predictions indicated that the peak swirling velocity would occur at an axial location given by equation 10, that is at 13 diameters for a Reynolds number of 1000 increasing to 26 diameters for a Reynolds number of 2000. It is probable then that rotational effects in the laminar region will still be increasing after 10.6 diameters for which the results are given in Figure 8a. This is clearly shown in Figure 8b for a length/diameter ratio of 31.8 which show larger increases at all rotational Reynolds numbers. Again for turbulent flow increases are negligible.

The effect on the best approach to fully developed flow is shown in Figure 8c. It has been argued earlier that residual inlet effects and rotational effects become less important as the flow becomes more fully developed. This is clearly seen in Figure 8c where all the data for rotational Reynolds numbers greater than 400 fall within a tight band, which is however significantly higher than that for a stationary tube or the low rotational Reynolds number of 200 thus indicating that the flow has not yet become fully developed. Again the smoother fairing from laminar-like to turbulent-like flow is in evidence.

Examination of the governing momentum equation 1 indicated that the rotational Reynolds number and the Reynolds number could be used to describe the flow and hence be used to correlate the data. This approach is shown in figure 10a for the length/diameter ratio of 31.8. The abrupt change of slope indicates a change from laminar-like to turbulent-like flow. The data for this length/diameter ratio of 31.8 may be correlated within approximately ± 10 per cent using the following equations

Laminar-like flow

(valid for $1.12 \times 10^{-3} < \frac{J^{0.34}}{Re} < 1.2 \times 10^{-2}$)

$$C_f = 2.30 \, J^{0.21} \, Re^{-0.62} \qquad (15)$$

Turbulent-like flow

(valid for $1.2 \times 10^{-4} < \frac{J^{0.34}}{Re} < 1.12 \times 10^{-3}$)

$$C_f = 0.108 \, J^{0.058} \, Re^{-0.17} \qquad (16)$$

Due to the mathematical structure of these equations they do not apply to zero rotational speed. Indeed examination of Figure 9 reveals that the results for a rotational Reynolds number of 200 are effectively coincident with those at zero rotational speed. It is therefore suggested that for rotational Reynolds numbers less than 200 the stationary correlations for this length/diameter ratio of 31.8 shown in Figure 10b are used, these are

Laminar-like flow

(valid for $900 < Re < 7000$)

$$C_f = 7.37 \, Re^{-0.63} \qquad (17)$$

Turbulent-like flow

(valid for $8000 < Re < 41,000$)

$$C_f = 0.138 \, Re^{-0.16} \qquad (18)$$

The accuracy of equations 17 and 18 are about ±5 per cent and it is interesting to note that the discrepancy for a rotational Reynolds number of 200 between using equations 15 and 16 and the corresponding one of equations 17 and 18 is under 5 per cent. Using equations 15 and 16 the increases in friction factor for a rotational Reynolds number of 1000 compared with 200 are

laminar-like flow +40%
turbulent-like flow +10%

Similar correlating equations were also obtained for other length/diameter ratios, they indicated similar structure but had different correlating coefficients and powers (6). For a length/diameter ratio of 10.6 for example they predicted increases of 30 per cent and 4 per cent respectively.

These results confirm that for turbulent flow the effects of rotations are small for a smoothed inlet. For laminar flow increases are predicted which conflict with the numerical predictions. However, as discussed earlier the experimental results shown in Figure 7 show that as the inlet is made smoother (A to D) the rotational results tend to those for the stationary case. The most smoothed inlet is still unlikely to have removed all upstream effects and it is considered that this is the most likely reason for the discrepancy between the experimental and numerical predictions. Further, it should be stressed that these observations are for the most smoothed inlet (inlet configuration D), for an unsmoothed inlet (such as A in Figure 1c) large increases in friction factor are apparent particularly in the laminar region as shown in Figure 7, these being due to the disturbed flow from the upstream flow circuit which is of course also rotationally modified.

5. CONCLUDING REMARKS

It has been demonstrated analytically that for constant property flow Coriolis acceleration is an important source for the generation of cross-stream secondary flow in the entrance region, and that this would decay away completely once the flow became fully developed. The flow would then be identical to that for fully developed flow through a stationary tube. A finite difference numerical procedure confirmed that for a uniform non-swirling inlet velocity an axisymmetric swirling flow developed rapidly and then decayed completely when the flow became fully developed to become identical to fully developed flow through a stationary tube.

In the experimental investigations results for tubes with large length/diameter ratios confirmed that for fully developed flow there is a negligible increase in the friction factor for both laminar and turbulent flow. However, rotation does appear to produce a more gradual transition from laminar to turbulent flow with the well known dip in the friction factor being absent. In this transitional region increases in friction factor are therefore evident. For the entrance region significant increases in friction factor due to rotation were noted, these being particularly severe where no particular attempt was made to smooth the inlet to the test section. When the inlet was smoothed with flow straightening devices the increases in the friction factor were considerably reduced. Indeed for the most smoothed inlets the friction factor for turbulent flow were almost indistinguishable from those for the stationary tube, whilst those for the laminar flow were tending towards those for the stationary tube. Even with the most smoothed inlets some entry effects are likely to remain and these will be modified further due to the Coriolis accelerations. It is therefore probable that with a completely smooth inlet that the results for laminar flow would approach very closely those for a stationary tube. This would then agree with the numerical predictions for a uniform non-swirling inlet velocity where no significant increases in friction factor with rotation were predicted.

It should however be emphasised that the numerical predictions indicated that the tube length for the swirling flow to decay completely was many times greater than the hydrodynamic entry lengths for stationary tubes. This was confirmed experimentally where rotational increases in friction factor were evident for length/diameter ratios for which the stationary results indicated fully developed flow. These increases in development length were particularly severe for the unsmoothed inlets. Further it should be emphasised that the swirling flow will affect the following features in the flow circuit, for example a bend or plenum chamber, and may thus significantly increase the flow resistance of these later features.

The analytical analysis suggested that the centripetal acceleration terms, and hence the tube's eccentricity were made manifest solely by a cross tube hydrostatic pressure distribution. This was confirmed experimentally.

Finally two important observations can be made for the design of rotating cooling circuits. Firstly, with unsmoothed inlet conditions significant increases in friction factor occur. The cross-stream swirling motion will extend for long lengths downstream of the entrance and may increase the flow resistance of later components. Flow smoothing devices can reduce these increases, in particular to a negligible level for turbulent flow, however these may be prone to blockages, and also the additional swirling motion will be beneficial in improving the heat transfer. Secondly, the gaps in knowledge about the effects of rotation inevitably mean uncertainties exist in the prediction of the performance of rotating cooling circuits. In particular, analysis of cooling circuits using stationary correlations, or experiments on a stationary flow circuit, may give a poor indication of the performance when the circuit is rotating. This could lead to failure of the cooling circuit to ensure that no part of the component exceeds its safe working temperature.

ACKNOWLEDGEMENTS

The author would like to express his appreciation to Prof. W. D. Morris who was his academic mentor during this programme of work, and to the Central Electricity Research Laboratories of the Central Electricity Generating Board and the Science and Engineering Research Council for financial support.

REFERENCES

(1) VICKERS, V. J., Recent trends in turbogenerators, Proc. IEE 1974, 121, no. 11 R, 1273.

(2) MARLOW, B. A., The mechanical design of large turbogenerators, Proc. I. Mech. E., 1986, 200, 1-13.

(3) MORRIS, W. D., Heat transfer and fluid flow in rotating coolant channels, 1981, Res Studies Press, John Wiley and Sons.

(4) MORRIS, W. D. and AYHAN, T., Observations on the influence of rotation on heat transfer in the coolant channels of gas turbine rotor blades, Proc. I. Mech. E., 1979, 193, no. 21, 303.

(5) HEARD, J. G. and BENNETT, R. B., Some considerations in the design of ventilating systems for large turbine generator rotors. IEE power division - Professional Group PI (Rotating electrical machines), Digest, 1976, no. 1976/5, 3.

(6) JOHNSON, A. R., Flow resistance in circular tubes rotating about a parallel axis, PhD thesis, 1988, University of Hull.

(7) WOODS, J. L. and MORRIS, W. D., An investigation of laminar flow in the rotor windings of directly cooled electrical machines. J. Mech. Eng. Sci., (IMechE), 1974, 16, no. 6, 408.

(8) STEPHENSON, P. L., An experimental study of flow and heat transfer in generator rotor cooling ducts. Int. Conf. on Elec Machines - Design and Applications, IEE, London, 1982.

(9) STEPHENSON, P. L., An experimental study of the pressure drop in generator cooling ducts. Int. Conf. on Elec Machines and Drives, IEE, London, 1989.

(10) SPALDING, D. B., A novel finite difference formulation for differential equations involving both first and second derivatives, Int. J. Numer. Methods Eng., 1972, 4, 551.

(11) PHOENICS User Manual, Concentration Heat and Momentum Ltd., (CHAM) Bakery House, 40 High Street, Wimbledon, London.

(12) LAUNDER, B. E. and SPALDING, D. B., The numerical computations of turbulent flows. Computer Methods in App. Mechs. and Eng., 1974, 3, 269.

(13) D. T. GETHIN and A. R. JOHNSON, Numerical Analysis of the Developing Fluid Flow in a Circular Duct Rotating Steadily About a Parallel Axis. International Journal for Numerical Methods in Fluids, 1989, 9, 151.

(14) ROSEHOW, W. M. and HARTNETT, J. P., Handbook of heat transfer, 1973, McGraw-Hill Book Co., New York.

(15) MORRIS, W. D., A pressure transmission system for flow resistance measurements in a rotating tube. J. Phys. E: Sci. Instrum., 1981, 14, 208.

(16) A. R. JOHNSON and W. D. MORRIS, Pressure Loss Measurements in Circular Ducts Which Rotate About a Parallel Axis. Proceedings XIV ICHMT Symposium, Duvbovnik, 1982, Hemisphere Corporation.

(17) A. R. JOHNSON and W. D. Morris, An Experimental Investigation into the Effects of Rotation and Entry Conditions on the Flow Resistance in Circular Tubes Rotating About a Parallel Axis. International Journal of Heat and Fluid Flow, Butterworth Scientific, 1984, 5, (2), 121.

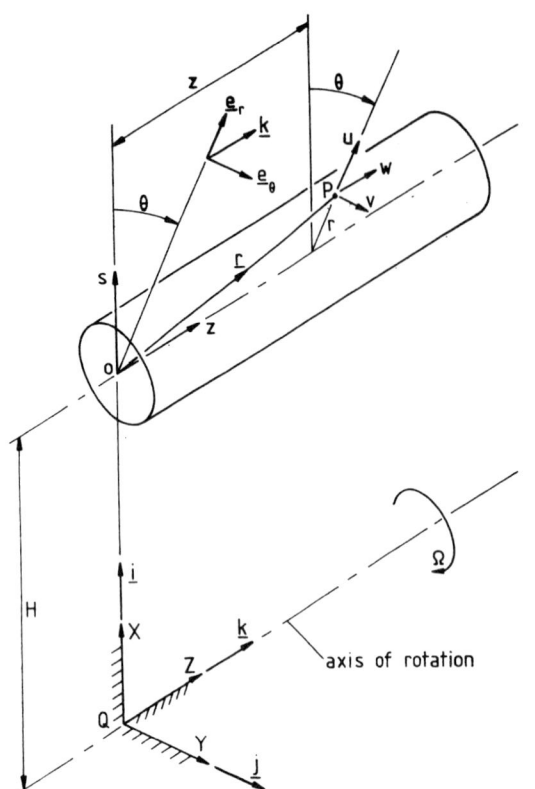

Fig 1a Definition of the flow geometry

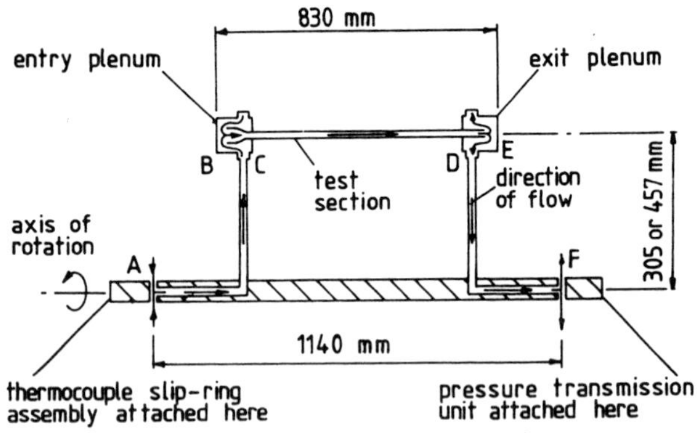

Fig 1b Schematic layout of the experimental flow circuit

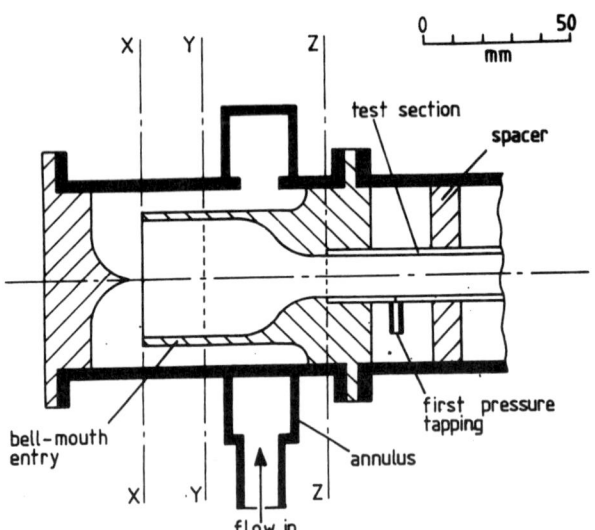

Fig 1c Test section inlet plenum chamber

Inlet configuration details

	Inlet Type			
	A	B	C	D
Gauze P at section XX	No	Yes	Yes	Yes
Flow straightener honeycomb between sections XX and YY	No	Yes	Yes	Yes
Gauze Q at section ZZ	No	No	Yes	No
Gauze R at section ZZ	No	No	No	Yes

Gauze details

Gauze	Wire diameter, mm	Wire pitch, mm	Flow area, %
P	0.23	0.64	42
Q	0.17	0.42	35
R	0.13	0.32	34

Honeycomb flow straightener details

Tube outside diameter = 4.25 mm
Tube thickness   = 0.15 mm
Tube length   = 18 mm
Tube configuration - touching with centres on a square grid

Fig 1 Geometric and experimental configurations for the circular tube rotating about a parallel axis.

Fig 2 Typical numerical prediction for the development and decay of the axisymmetric swirling velocity for laminar flow. (Reynolds number = 973, rotational Reynolds number = 800)

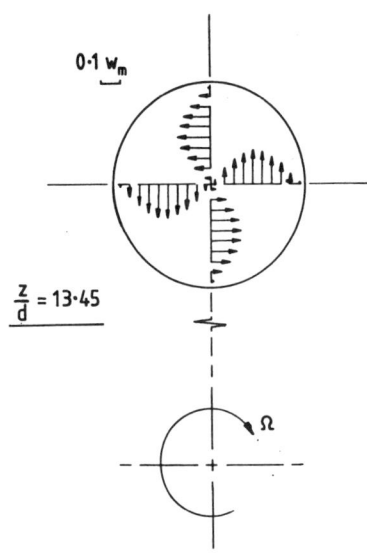

Fig 2a Axisymmetric swirling velocity

Fig 2b Development and decay of axisymmetric swirling velocity profile

Fig 2c Development and decay of the peak swirl velocity

Fig 3 Correlated laminar swirl velocity for all combinations of Reynolds numbers (486, 973, 1460, 1946) and rotational Reynolds numbers (400, 800, 1400) investigated numerically

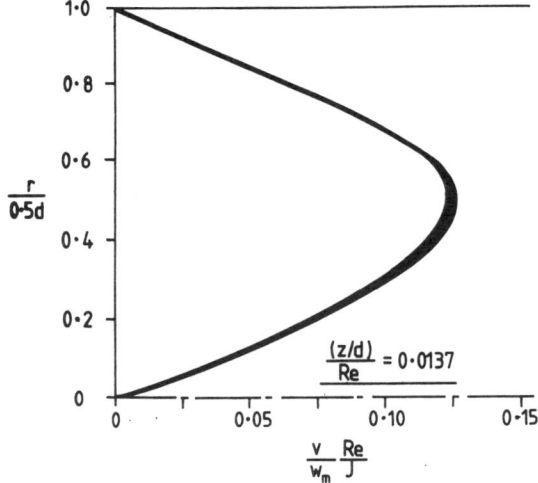

Fig 3a Correlated swirling velocity profile

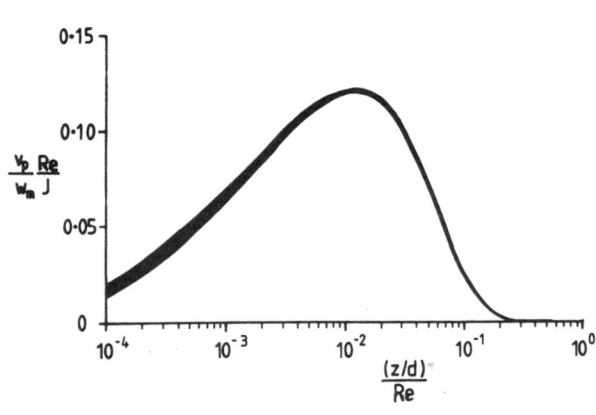

Fig 3b Correlated development and decay of the peak swirl velocity

Fig 4 Typical effect of rotation on the numerically predicted development of the axial velocity and friction factor for laminar flow (Reynolds number = 973)

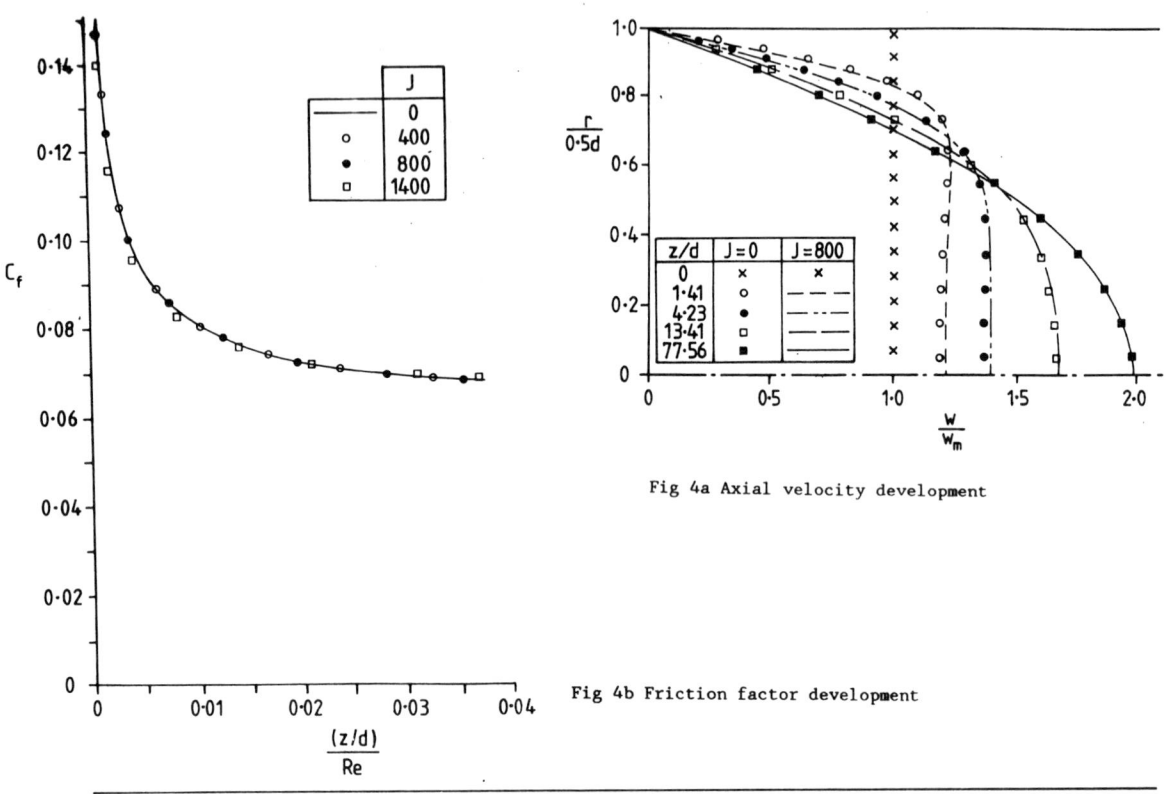

Fig 4a Axial velocity development

Fig 4b Friction factor development

Fig 5 Typical numerical prediction for the development and decay of the axisymmetric swirling velocity for turbulent flow.

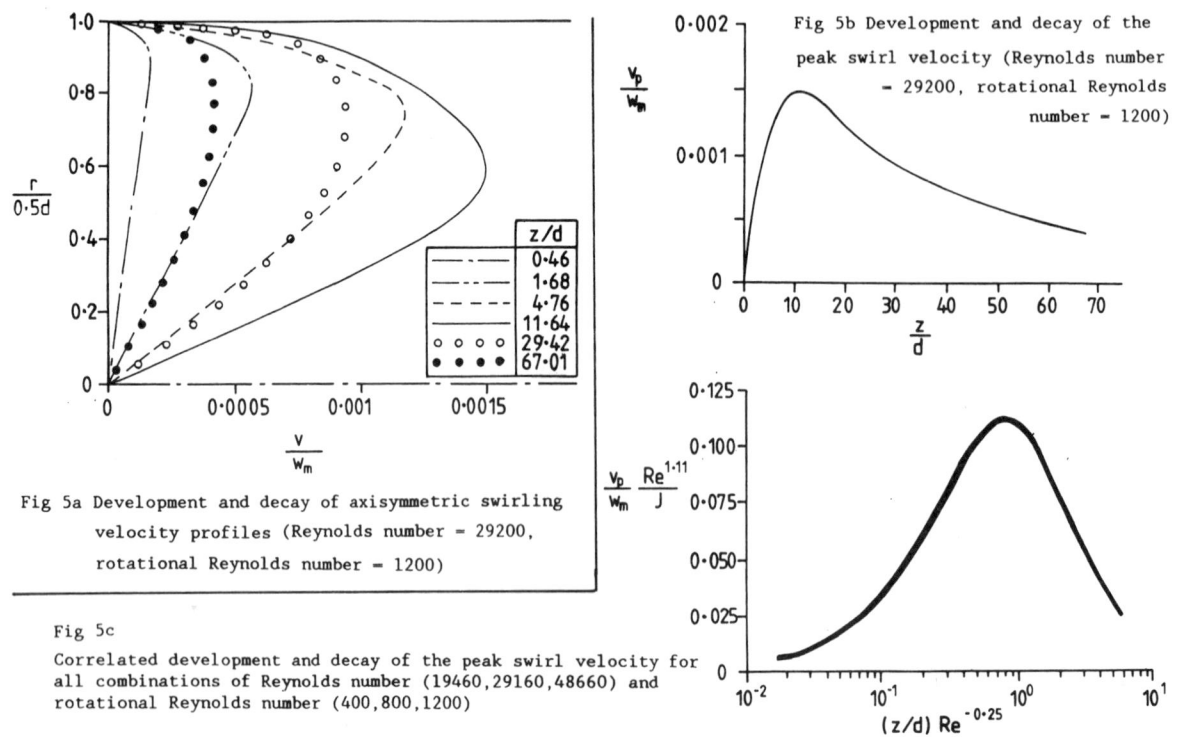

Fig 5a Development and decay of axisymmetric swirling velocity profiles (Reynolds number = 29200, rotational Reynolds number = 1200)

Fig 5b Development and decay of the peak swirl velocity (Reynolds number = 29200, rotational Reynolds number = 1200)

Fig 5c Correlated development and decay of the peak swirl velocity for all combinations of Reynolds number (19460, 29160, 48660) and rotational Reynolds number (400, 800, 1200)

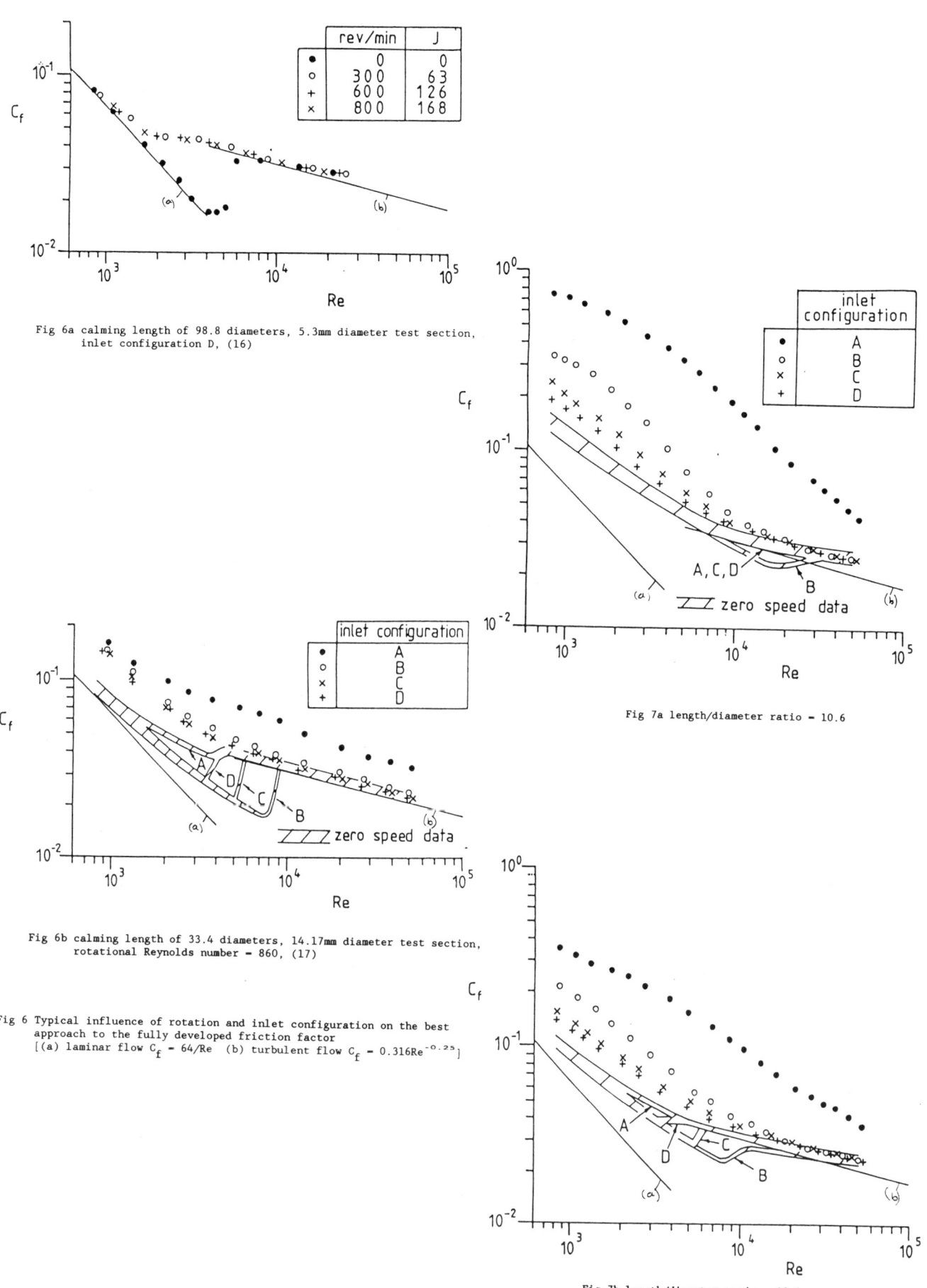

Fig 6a calming length of 98.8 diameters, 5.3mm diameter test section, inlet configuration D, (16)

Fig 6b calming length of 33.4 diameters, 14.17mm diameter test section, rotational Reynolds number = 860, (17)

Fig 6 Typical influence of rotation and inlet configuration on the best approach to the fully developed friction factor
[(a) laminar flow $C_f = 64/Re$ (b) turbulent flow $C_f = 0.316 Re^{-0.25}$]

Fig 7a length/diameter ratio = 10.6

Fig 7b length/diameter ratio = 42.3

Fig 7 Typical influence of rotation, inlet configuration and length/diameter ratio on the friction factor (14.17mm diameter test section, rotational Reynolds number = 860), (17)
[(a) laminar flow $C_f = 64/Re$ (b) turbulent flow $C_f = 0.316 Re^{-0.25}$]

Fig 8 Typical influence of eccentricity ratio on the friction factor (rotational Reynolds number = 300) [(a) laminar flow $C_f = 64/Re$ (b) turbulent flow $C_f = 0.316Re^{-0.25}$]

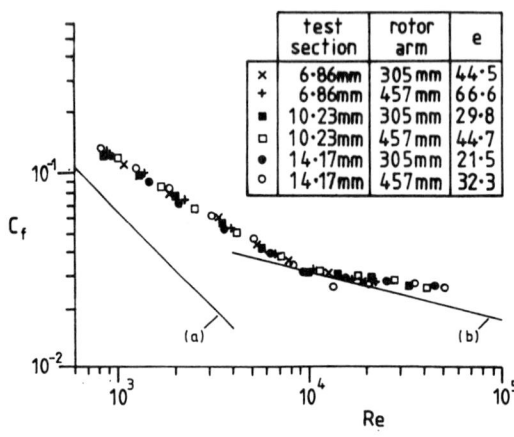

Fig 8a length/diameter ratio = 10.6

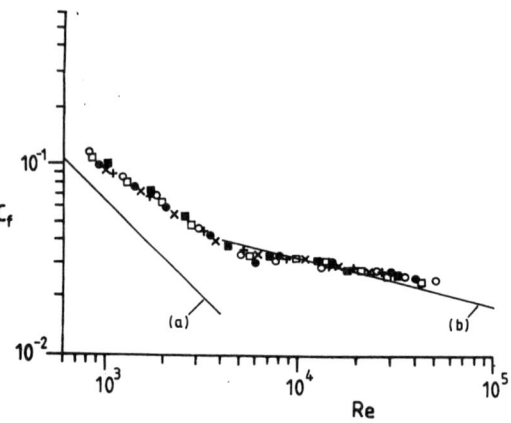

Fig 8b length/diameter ratio = 31.8

Fig 9 Typical influence of rotation on the friction factor (14.17mm diameter test section, inlet configuration D) [(a) laminar flow $C_f = 64/Re$ (b) turbulent flow $C_f = 0.316Re^{-0.25}$]

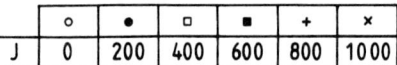

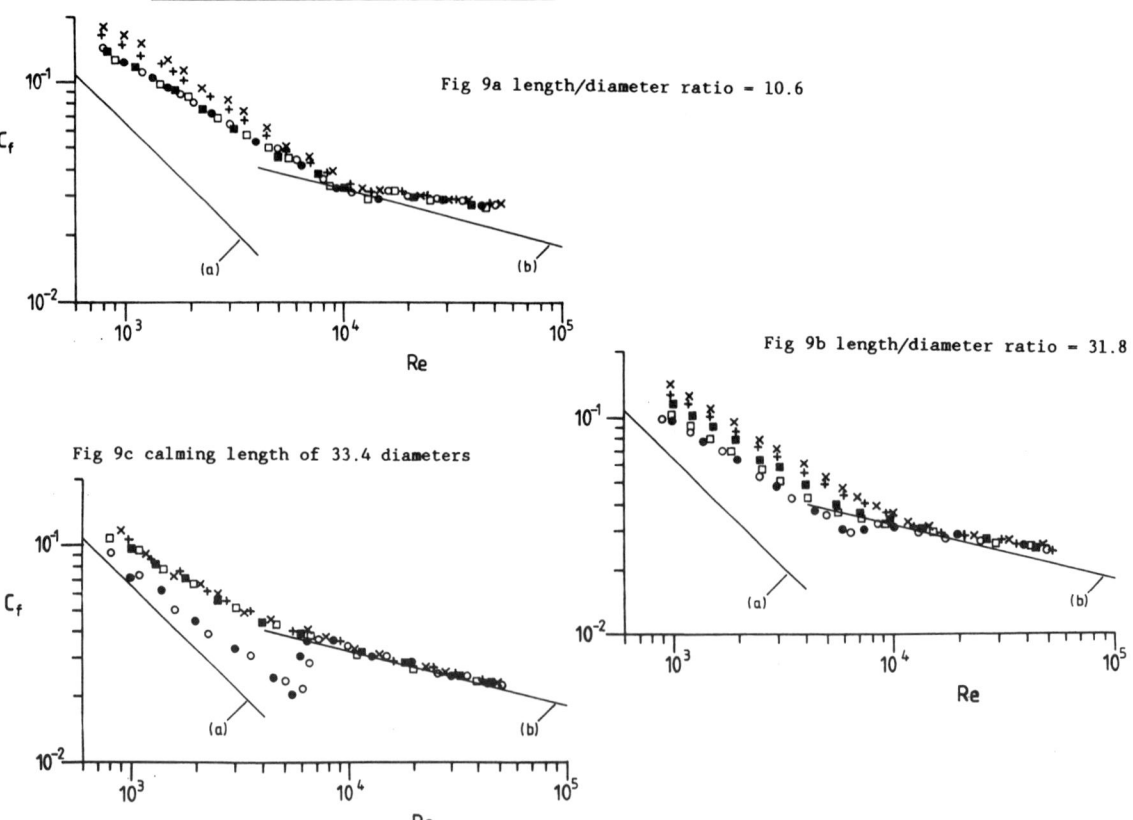

Fig 9a length/diameter ratio = 10.6

Fig 9b length/diameter ratio = 31.8

Fig 9c calming length of 33.4 diameters

Fig 10 Typical correlations for the friction factor (14.17mm diameter test section, length/diameter ratio = 31.8, inlet configuration D) [(a) laminar flow $C_f = 64/Re$ (b) turbulent flow $C_f = 0.316Re^{-0.25}$]

	○	●	□	■	+
J	200	400	600	800	1000

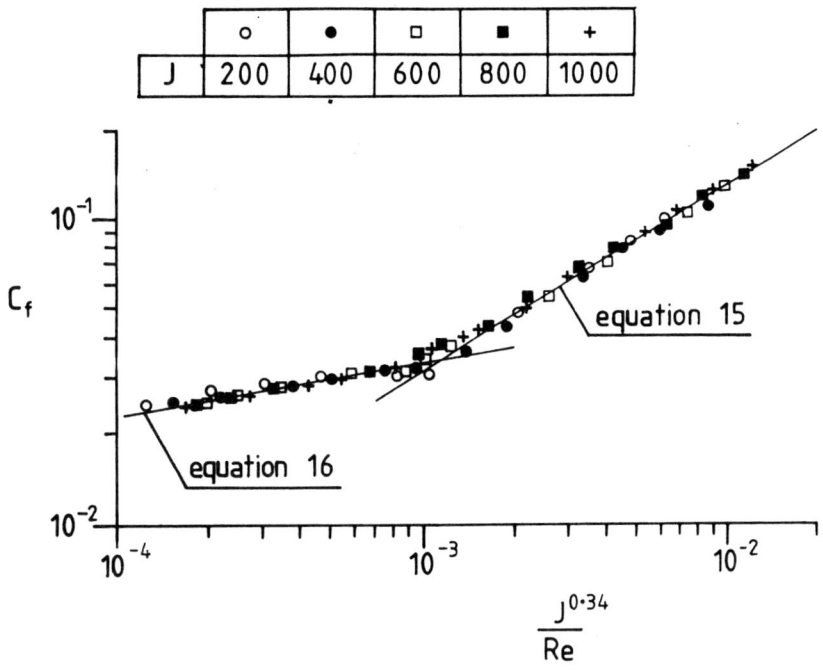

Fig 10a Correlations for the effect of rotation on the friction factor for all combinations of Reynolds number (406 to 41000) and rotational Reynolds number (200 to 1000)

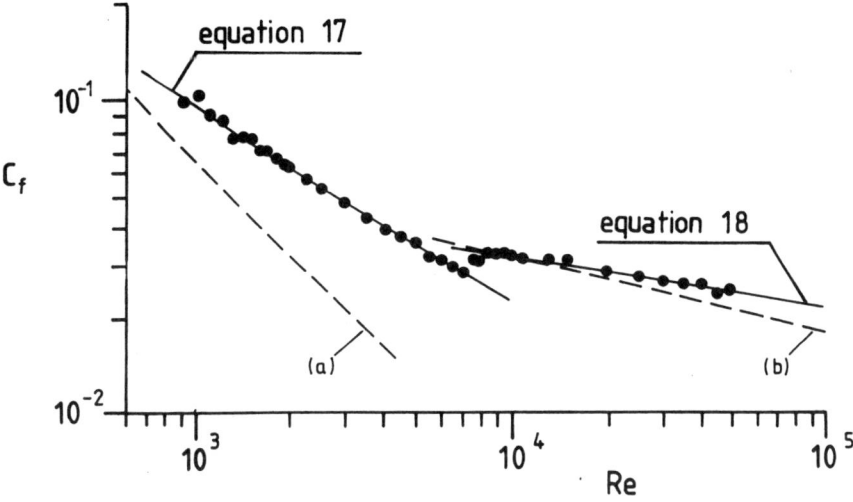

Fig 10b Correlations for the friction factor for a stationary tube.

C413/003

Biofilm formation and control in flowing aqueous systems

T R BOTT, BSc, PhD, DSc, CEng, FIChemE
School of Chemical Engineering, University of Birmingham

SYNOPSIS The paper describes research undertaken to investigate the factors influencing attachment and removal of micro-organisms at surfaces, with particular reference to industrial cooling water. Velocity and temperature are important variables. A 5°C rise in temperature can result in a 70% increase in biofilm thickness, and a velocity of about 1 ms^{-1} in tubes, similar in diameter to those found in shell and tube exchangers produces a relatively thick biofilm. Other factors of importance are material of construction and surface finish, cell concentration and nutrient availability. Ozone is an effective biocide in comparison with chlorine, and is capable of biofilm removal.

1 INTRODUCTION

The contamination of surfaces in industrial equipment by living matter is often referred to as biofouling. The phenomenon is essentially an adaption of a natural occurrence. Creatures such as barnacles and mussels or algae for instance, find it beneficial to cling to rocks in contact with sea water, and if the water is used industrially the adhesion is transferred to the associated equipment in contact with the water. If living matter of a size that makes it readily identifiable such as shell fish, is observed on surfaces the contamination is usually called macrofouling. On the other hand if the accumulation of living matter consists of algae, fungi or bacteria, it is usually referred to as microfouling, or microbial fouling.

The occurrence of either category of biofouling usually in combination with other deposition processes such as particulate deposition or crystallisation, represents a serious hindrance to the efficient operation of any industrial equipment involved. The problem is very prevalent in cooling water systems, that often although not always, use 'fresh' water from rivers, canals or lakes. Extensive accumulation of living matter on the inside of pipework and flow channels represents a restriction to flow and increased roughness in heat exchangers, e.g. the condensers on a power station, and can result in loss of heat transfer efficiency and hence an increase in the unit cost of production.

The potential for biofouling in water systems is essentially dependent upon three factors:

1. The origin and quality of the water in the system.
2. The water temperature. Cooling water systems for instance, often operate at temperatures conducive to biofilm formation.
3. Nutrient availability in the water.

The increased costs associated with fouling include:

1. Additional energy requirements for pumping due to the restrictions to flow and roughened surfaces in contact with the flowing water, and to make up for the shortfall in heat recovery associated with heat exchanger inefficiency.
2. Additional maintenance costs, e.g. increased incidence of leaks and joint failures due to the increased back pressure in the system brought about by the presence of fouling.
3. Lost production when the equipment is out of service for cleaning or maintenance.
4. The cost of cleaning in terms of materials and manpower.
5. The cost of techniques to control the problem, e.g. the use of chemical additives or biocides.
6. The loss of employee morale that may occur as a result of intractable fouling problems probably manifest in product quality.

In addition, because the likely problems of fouling are recognised at the design stage, increased heat transfer area say, or larger diameter pipes, are installed to offset the effects of fouling with consequent increased capital cost.

This paper is concerned with the problem of microbial fouling in 'fresh' water systems and summarises research work that has been undertaken over the past few years, in the School of Chemical Engineering at the University of Birmingham carried out to provide a better understanding of the problem as it affects industrial operations.

2 BIOFILM DEVELOPMENT

If a virgin clean surface is exposed to 'natural' water, in a very short time the surface will become contaminated with organic material. The initial contaminant is generally thought to be non-living macro molecules, that are essentially decomposition products of formerly living matter, ie. from decomposed or decomposing plant life, or dead animals or by-products from micro-organisms or animals. The material is held on solid surfaces by physico-chemical forces. The macro-molecular layer is generally firmly attached. It is via this 'conditioning' layer that it is generally supposed micro-organisms attach to the surface, either through some form of chemical reaction or through electrical or physical forces.

The mechanisms by which the micro-organisms arrive at the surface include: Brownian diffusion, eddy diffusion due to the fluid turbulence in the system, sedimentation in regions of low velocity, and motility or taxis of micro-organisms that enable them 'to swim' towards the surface.

It has been suggested [1] that ideally the attachment is a two stage process.

1. A 'reversible' stage where the organism is loosely bound to the surface by weak physico-chemico forces. If sufficient force is imposed on the organism under these conditions, due to fluid turbulence, there is the possibility of removal of the organism from the surface.

2. After a lapse of time, an 'irreversible' stage, that is likely to involve the production of extracellular polymers (eg. polysaccharides) that can more firmly bond to the surface. The 'sticky' character of these polymers may also be responsible for the 'capture' of other organisms and particulate matter, that may be present in the system, including nutrients.

Once firmly bound to the surface the development of the biofilm will be dependent on the availability of nutrients and trace elements essential to the life and well being of the living cells.

There is a great deal of uncertainty about the true nature of the adhesion of micro-organisms to surfaces in industrial systems. Under the complex conditions of a system containing different organisms and in the presence of debris such as mineral particles, scaling, and corrosion products, an exact description of the adhesion mechanism is unlikely to be established. In the face of these difficulties a more empirical approach is necessary.

In a flowing system the development of a biofilm over a period of time can be ideally visualised as having three stages.

1. An initiation period where the conditioning of the surface takes place and the colonisation of the surface begins.

2. A period of relatively rapid growth dependent upon the prevailing conditions.

3. A 'steady state' or plateau region, that can last indefinitely, where the extent of the biofilm remains more or less constant but oscillating about a mean thickness.

From this brief discussion it will be apparent that biofilm development in an industrial environment will be associated with two aspects of the total system:
1. The metabolism of the micro-organisms.
2. The engineering design and operation of the system.

It is difficult to classify all the variables affecting the system into the above categories because of complex interactions but recent research has given insights into the effects of modifying the variables. In turn this has given indications of how to control biofouling to minimise its impact in industrial systems with particular reference to cooling water.

3 THE EFFECTS OF CHANGING VARIABLES

Although there is considerable interaction between the effects of different variables it is useful to consider each of the variables independently so that their separate influences may be assessed. The discussion is primarily concerned with cooling water but the implications for other operations can be readily appreciated.

3.1 <u>Flow velocity</u>

Increase in water velocity for a given geometry increases turbulence and produces two major effects:

1. The viscous sub layer, near the wall (the inside surface of tubes in a shell and tube heat exchanger for example) becomes thinner. The thickness of this viscous sub layer is largely responsible for resistance to mass transfer. It will affect the transport of micro-organisms to the surface for initial colonisation, the subsequent mass transfer of nutrients and trace elements to the organisms residing on the surface, and the removal of waste materials from the biofilm.

2. The shear forces at the interface become larger. The increase in the shearing action may increase removal by assisting the shedding or sloughing process, or by suppressing further attachment.

Increasing velocity therefore has two opposing effects. On the one hand mass transfer resistance is reduced thereby promoting enhanced conditions for microbial growth, and on the other an increase in biofilm removal. An optimum velocity at which maximum growth can occur is therefore possible.

For economic reasons it is generally accepted that the operating velocity should be greater than the velocity where maximum growth occurs. A 'rule of thumb' often quoted in connection with cooling water systems, is that velocities should be in excess of 1 ms^{-1}. Fig. 1 shows results of some experimental work using simulated cooling water containing *Pseudomonas fluorescens* flowing through tubes having diameters similar to those found in shell and tube heat exchangers [2]. The results tend to confirm that 1 ms^{-1} appears to be the critical velocity. It is anticipated that other species of bacteria would behave in a similar way.

In the light of the earlier discussion however, in connection with turbulence levels it has to be borne in mind that velocity is only one component in the Reynolds number, the usual measure of turbulence. If the characteristic dimension, diameter in the case of a tube, is very different from those encountered in heat exchangers, the critical velocity for operation could well be quite different.

3.2 Temperature effects

The effect of temperature on the growth of micro-organisms very much depends upon the type of organism involved. In general, as the temperature of the environment in which the micro-organism resides is increased above say $0°C$, the chemical and enzyme reactions, responsible for growth and reproduction, are increased and hence the rate of biofilm development will accelerate. There is however, a temperature range within which a particular species may grow, outside this range metabolic rates will be suppressed and in the extremes the cells will die. For most micro-organisms there is an optimum temperature at which growth is a maximum in a nutrient rich environment. The optimum temperature for growth of a particular organism is generally nearer the maximum permissible temperature for sustained growth rather than the lower limit of temperature. It is the temperature of the biofilm that is important, as opposed to the bulk water temperature remote from the surface.

The temperature of the biofilm is dependent on the temperature distribution between the hot fluid (condensing steam in the case of power generation condensers) and the bulk cooling water, which in turn is controlled by the thermal resistances operating in the system and in particular the biofilm thermal resistance. If the microbial layer has any appreciable thickness then the temperature at the metal/biofilm interface is likely to be quite different from the water/biofilm interface, giving rise to differences in metabolic rates at different points through the biofilm thickness.

Even small differences in temperature can have a pronounced effect on biofilm development as illustrated by Fig. 2: An increase of $5°C$ has the effect of increasing the biofilm thickness by almost 70% |3|. The consequences for heat exchanger efficiency can be readily appreciated. These data were obtained using the *Escherichia coli* microbe under isothermal conditions, i.e. when there was no heat transfer taking place.

3.3 Concentration of micro-organisms and nutrient availability

The rate of mass transfer depends on concentration differences. Since the initial colonisation of the surface is a mass transfer operation (the micro-organisms may be regarded as behaving like particulate matter in this respect), the concentration of cells in the system will affect the initial accumulation of the film on the surface. The further development of the biofilm however, not only depends on concentration differences but the reproduction of cells already on the surface.

For an active biofilm nutrients are required, this includes not only sources of carbon and nitrogen, and oxygen for aerobic species, but trace elements essential for growth. In some experiments using *Pseudomonas fluorescens* |4|, the colonisation process was delayed till the trace concentrations of iron had been introduced into the system. Again since mass transfer depends on concentration differences, high concentrations of dissolved nutrients may result in extensive biofilms. As the biofilm grows in thickness diffusion through the biofilm will be necessary to supply nutrients to the cells in the layers near the solid/biofilm interface. As the transfer through the biofilm takes place some of the nutrients will be consumed by the upper layers of cells, and may result in little or no nutrient availability for cells in the lower levels. The outcome may be for instance, that anaerobic conditions are produced near the metal wall, or that the cells in this region are starved and die. These changing conditions as the film develops, may lead to sloughing and biofilm removal.

Research has shown |5| that oxygen concentrations at the biofilm/metal interface fall rapidly as the biofilm grows in thickness. In one particular experiment the oxygen concentration was zero at the interface after only 14 days operation (water velocity 1 ms^{-1}) and when the biofilm thickness had reached about 250 μm.

It has been demonstrated |6| that once colonisation has taken place, it is the availability of nutrients that influences the continued growth of the biofilm; the living biofilm is self generating by cell reproduction. Figs. 3 and 4 based on research using *Pseudomonas fluorescens* under isothermal conditions, illustrates these comments. Once the surface is contaminated restriction of the nutrient supply inhibits growth whereas removing cells from the flowing liquid does not appear to affect the biofilm development. For any system of biofilm control this fact has to be taken into account. It is of little value to make the water aseptic for instance, leaving living cells intact on the surface to reproduce and develop.

3.4 The effect of pH

Under natural environmental conditions the overall pH of the system will be reasonably constant at around a value of 7. Large deviations from this figure will affect the cell metabolism, and hence the potential for the biofilm to develop. It is to be expected however, that due to the rather artificial conditions in a cooling water system, variations in pH could occur. The pH of the system will be dependent not only upon any water treatment that has been necessary, e.g. to control corrosion, but also the effects of concentrating dissolved salts due to evaporation in the cooling tower or spray pond, the presence of contaminants due to process leaks, and the uptake of carbon dioxide from the atmosphere.

Although the pH does affect biofilm formation, pH adjustment is not generally practised as a method for biofilm control. The reason for this is that other problems may accrue from this approach such as corrosion, or changes in scaling potential, and in any case other methods of biofilm control are likely to be more cost effective.

The metabolism of the cells within a biofilm is likely to affect the pH, say by the production of acidic waste products. It will be this local pH rather than the bulk pH, that will affect the residual cells, but clearly the bulk solution pH will influence the level of the local value. For these reasons the pH at the biofilm/metal interface, may be quite different from the bulk pH and in some cases causes corrosion of the metal through chemical reaction or galvinic action.

The effects of turbulence, discussed earlier, on the removal of waste products, and the replacement of the adjacent water in the vicinity of the cells, will affect this local pH.

3.5 Material of construction

A wide range of materials of construction is available for the manufacture of heat exchangers, and the different surface characteristics that these materials display can affect the extent of the biofilm on the surface. The major influence occurs during the colonisation process, so that the character of the surface, and the cells, e.g. hydrophobicity or hydrophilicity will be important factors in the process. During colonisation, as already discussed, a complex range of interacting effects will combine to give the overall result.

Once the surface is contaminated and covered with developing biofilm under the preliminary conditions, the interface presented to the flowing water is not the original material of construction but the outer layers of biofilm. Nevertheless the physico-chemical interaction between the colonising cells and the heat exchanger or duct surfaces will influence the retention of the biofilm. For instance, if the bonding is weak it is more than likely that, under the influence of the hydrodynamic conditions, the biofilm will slough off the surface when it has reached a certain critical depth. On the other hand if the colonisation bond is strong, there is a greater opportunity for the biofilm to be stable and develop to a substantial thickness before removal mechanisms become apparent.

Some preliminary experiments |7| on the effect of surface suggested that biofilms grown on rough surfaces were more extensive than biofilms grown on smooth surfaces under otherwise similar conditions. Other work |8| comparing the differences between materials was carried out using an experimental flow system with *Pseudomonas fluorescens* as the contaminant. Some of the results are shown on Fig. 5. The considerable difference between the electropolished 316 stainless steel tube and a similar tube used in the 'as received' (rough) condition is quite pronounced. The figure also illustrates the difference in fouling propensity between the 'as received' stainless steel tube, and tubes fabricated from glass and a polypropylene polymer.

Electropolishing stainless steel is a relatively costly process that is additional to the cost of the material in the first place. An alternative approach is to coat the heat exchanger surface with a monomolecular layer of say, a polymer, that changes the surface characteristics making it less hospitable to biofilm growth. Development work on such polymers has shown that they may be effective in reducing the adhesion of biofilms |9|.

An economic balance has to be struck between the increased first cost and the reduced operating costs resulting from more effective biofilm removal. At the present time the majority of cooling water operators prefer to use chemical treatment, rather than using special materials or the modification of traditional materials of construction, as being more cost effective.

3.6 Other factors

Observations |10| on an industrial cooling water system, employing forced draught cooling towers for temperature restoration, have shown that seasonal and climatic conditions affect biofilm growth. The greatest change appears to be from summer to winter operation. Although the effects are complex the essential element would appear to be the seasonal temperature changes and may be related to the discussion in Section 3.2 In general the higher summer temperatures produced more extensive biofilms. It is also possible that in particular locations, prevailing wind direction may be a factor. Wind blowing across say an effluent treatment plant, may carry nutrients (or micro-organisms) to an 'open' cooling water circuit.

If the recirculating water contains particles, e.g. sand, it is possible that the abrasive conditions produced will control the extent of the biofilm |11|. Although the deliberate introduction of particles is practised in some operations, e.g. a fluidised bed heat exchanger, it is not usual in cooling water systems due to potential erosion problems and associated high maintenance charges.

3.7 Biofilm development - general observations

The foregoing discussion indicates that biofilm initiation and growth in cooling water circuits is a complex phenomenon, with a number of interacting factors contributing to the overall result. Of major importance are the velocity of flow of the bulk water across the biofilm and the temperature. Low velocity and relatively high temperatures favour biofilm development. Some measure of control can be exercised by attention to these two variables by the process operator. Investigations of the origin of biological nutrients in the system may also be helpful in reducing the incidence of biofouling, by providing an opportunity to reduce or even eliminate nutrient availability. At the design stage careful choice of materials of construction may also help to reduce biofouling problems.

4 BIOFOULING CONTROL

Despite attention to the points emphasised in Section 3.7, it is likely that an open, 'once through' or recirculating cooling water system will require some 'on line' means of coping with the potential fouling problem. Essentially two broad opportunities are available, namely mechanical and chemical techniques.

4.1 Mechanical methods

Mechanical control of biofilms usually involves the use of some 'projectile' that moves under the influence of the water flow wiping the surface free of residual biofilm.

One such technique, the Taprogge system, involves the circulation of sponge rubber balls on the inside of the tubes of a shell and tube heat exchanger through which the water flows. The soft rubber balls, a little larger in diameter than the tube diameter, are slightly compressed in the tube so that they effectively wipe the tube surface. After passing through the exchanger, the balls are collected and recirculated through the heat exchanger. The technique is particularly suitable for power station use where very large heat exchangers (the condensers) are employed and the tubes are of uniform size. It is claimed that the technique is very cost effective, and leads to improvements in plant operation |12|.

An alternative approach is to use what is known as a cage and brush technique. In this method a brush is fitted in each exchanger tube so that it can travel the length of the tube under the influence of the water flow and is caught by a cage at the other end.

After a period of time the water flow is reversed and the brush travels back to its original position to be restrained by a cage at that end. Intermittent flow reversal maintains clean conditions in the tubes. The repetitive reversal of flow may interfere with the smooth operation of the process, and is likely to require elaborate (and costly) valve arrangements.

4.2 The use of biocides

One of the common methods for control of biofilms is the use of biocides, particularly where the heat exchange equipment does not lend itself to the use of mechanical techniques for example, on a large chemical complex. In such an installation the heat exchangers using cooling water are generally isolated from one another on individual process plants, and the tube size may vary from exchanger to exchanger. The usual technique employed to control biofilms under these circumstances, is to dose the circulating water with a biocide.

A favoured biocide for many years has been the strong oxidising agent, elemental chlorine, but its continued application represents a number of hazards, not least for the environment. Briefly they include:

1. Elemental chlorine is chemically persistent. In most large scale 'open' cooling water systems, the water after use, is returned to the environment, i.e. to rivers, canals and lakes, and at least some of the added chlorine will be contained in this discharge. Its presence can be hazardous to aquatic life in the natural environment. It is also capable of reaction with organic material in the environment (see 2 below).

2. Because of its chemical nature chlorine, in contact with organic material contained in the feed water to the system, may produce haloforms that will eventually be discharged to the environment. These chlorine compounds are generally regarded as hazardous to health.

3. Large cooling systems will require that large volumes of chlorine are kept in storage in readiness for use. Three problems associated with the environment and local populations, accrue from this practice:

(a) Large quantities of chlorine have to be transported to the site of application with the risk of accidental discharge.

(b) There is a risk of plant failure or accidental discharge.

(c) The action of terrorists.

For these reasons the storage of elemental chlorine is being discouraged.

The alternative is to generate chlorine in situ by say electrolysis of brine or to use sodium hypochlorite in solution.

Both methods are likely to be more expensive to operate than the purchase of elemental chlorine from an appropriate supplier.

In general very low concentrations of free chlorine are effective in reducing biofilm formation, or if used intermittently in removing at least part, of a biofilm from a heat exchanger surface. Low levels of chlorine of the order of 0.2 ppm, are capable of controlling biofilm development |13|. Concentrations of chlorine for the maintenance of domestic water disinfection in mains supply lines, are often maintained in this range.

In some experiments |14| to examine the effectiveness of various concentrations of chlorine derived from sodium hypochlorite on established biofilms it was demonstrated that it was difficult to remove completely a biofilm once it had been established. In the concentration range of 1 - 4 ppm, with a flow velocity of 1.2 ms^{-1} operating in the system, it was possible to reduce the quantity of biofilm on the inside of a tube surface. The micro-organism adhering to the surface was *Pseudomonas fluorescens*. At the low concentration of 1 ppm after a period of around 8 hours 60% (by weight) of the biofilm remained on the surface. The indications were that 40% removal was unlikely to be improved for these conditions. At the higher concentration of 4 ppm 80% of the biofilm was removed in about 5 hours. Little improvement in removal appeared to be possible after this time with a final retention of about 20%.

In the light of the potential problems in the use of chlorine for biocidal treatment of cooling water, a promising alternative is ozone, a strong oxidising agent, that does not persist in the environment and is considered to produce fewer noxious compounds in water compared to chlorine. Ozone is generally produced by subjecting dry air, oxygen or air oxygen mixtures with an electric discharge. It is a relatively unstable allotrope of oxygen, particularly in solution in water. For this reason ozone cannot be stored and has to be generated on site, as required. It could be argued that the additional cost of the ozone generating equipment and its operation would be prohibitive. It has to be borne in mind however, that the changed thinking in respect of the safety in the use of chlorine, is likely to make the cost of ozone more acceptable. It is also not without significance that the UK water companies are seriously considering ozone as an alternative to chlorine at least in the treatment works.

Some recent work reports |15| that even very small concentrations of ozone are effective in the total removal of biofilms. Fig. 6 illustrates the effectiveness of ozone in removing a biofilm of *Pseudomonas fluorescens*. A concentration as low as 0.2 ppm with water velocity in excess of 1.0 ms^{-1} as shown to remove 80% (by weight) of the biofilm. Fig. 6 also demonstrates the importance of velocity in the control of biofilms, and shows that relatively high levels of turbulence associated with higher velocities are necessary for the ozone to be effective. This is clearly a mass transfer effect for a given concentration of ozone.

5 CONCLUSION

The deposition of organisms and growth of biofilms on engineering surfaces (biofouling) is a complex combination of microbial and engineering factors. Their control and potential removal are also related to cell metabolism and the operating conditions of the equipment, i.e. the use of biocides and high velocities, and temperatures as low as possible.

6 ACKNOWLEDGEMENTS

The author would like to record his sincere thanks to all the postgraduate students who have contributed to this paper through their work and whose names appear in the references.

7 REFERENCES

(1) MARSHALL, K.C., STOUT, R. and MITCHELL, R. Selective sorption of bacterial from seawater. *Can. J. of Microbiol.* 1971, 17, 1413.

(2) MILLER, P.C. Biological fouling film formation and destruction. PhD Thesis, 1982, University of Birmingham.

(3) BOTT, T.R. and PINHEIRO, M.V.P.S. Biological fouling - velocity and temperature effects. *Can. J. Chem. Eng.*, 1977, 55, 473.

(4) SANTOS, R. Private communication, 1990.

(5) PATEL, T.D. Film structure and composition in biofouling systems. PhD Thesis, 1986, University of Birmingham.

(6) BOTT, T.R. and MILLER, P.C. Mechanisms of biofilm formation on aluminium tubes. *J. Chem. Tech. Biotechnol.* 1983, 33B, 177.

(7) HARTY, D.W.S. Microbial fouling of a simulated heat transfer surface. PhD Thesis, 1980, University of Birmingham.

(8) MOTT, I.E.C. and BOTT, T.R. The adhesion of biofilms to selected materials of construction for heat exchangers. Paper presented at the 9th Int. Heat Trans. Conf., 1990, Jerusalem.

(9) SANTOS, R. Internal report on co-operative research project 'Novel polymers for the prevention of the adhesion of biofilms to surfaces', 1990, University of Birmingham.

(10) BOTT, T.R., MILLER, P.C. and PATEL, T.D. Biofouling in an industrial cooling water system. *Process Biochem.* Jan/Feb 1983, 10.

(11) LOWE, M.J. The effect of inorganic particulate materials on the development of biological fouling films. PhD Thesis, 1988, University of Birmingham.

(12) BOTT, T.R. Fouling Notebook. *Inst. Chem. Engnrs.* Rugby, 1990.

(13) KAUR, K. Ozone as a biocide in cooling water systems. PhD Thesis, 1990, University of Birmingham.

(14) NESARATNAM, R.N. and BOTT, T.R. Effects of velocity and sodium hypochlorite derived chlorine concentration on biofilm removal from aluminium tubes. *Heating and Ventilating Engineer*, Feb/Mar 1984, 5.

(15) KAUR, K., LEADBEATER, B.S.C. and BOTT, T.R. Ozone as a biocide in industrial cooling water treatment. Paper presented at the Conference Wasser Berlin '89. International Ozone Assn. 1989, Berlin.

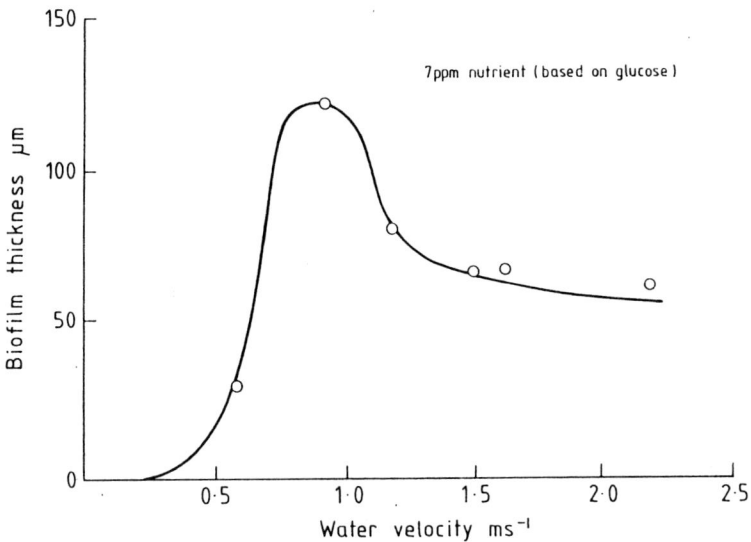

Fig 1 The variation of biofilm (*Pseudomonas fluorescens*) thickness with water velocity through the test section

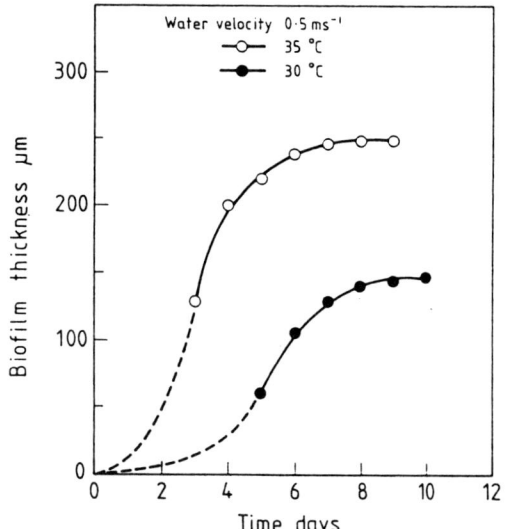

Fig 2 The effect of temperature on slime formation (*Escherichia coli*)

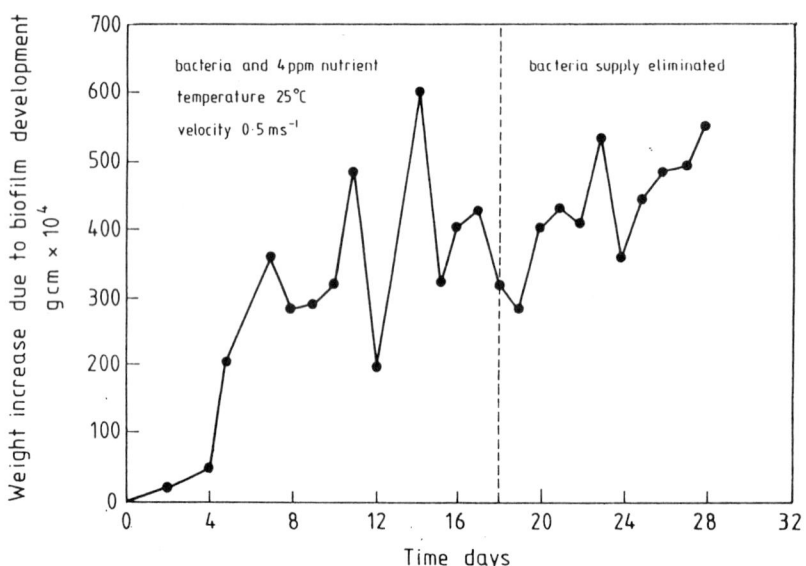

Fig 3 The effect of eliminating the supply of bacteria (*Pseudomonas fluorescens*) on the development of a biofilm

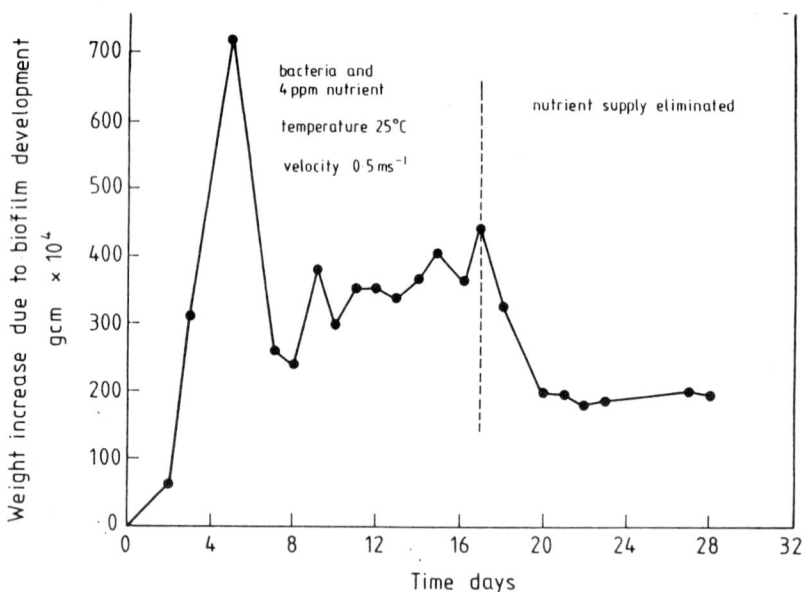

Fig 4 The effect of eliminating the supply of nutrient on the development of a biofilm (*Pseudomonas fluorescens*)

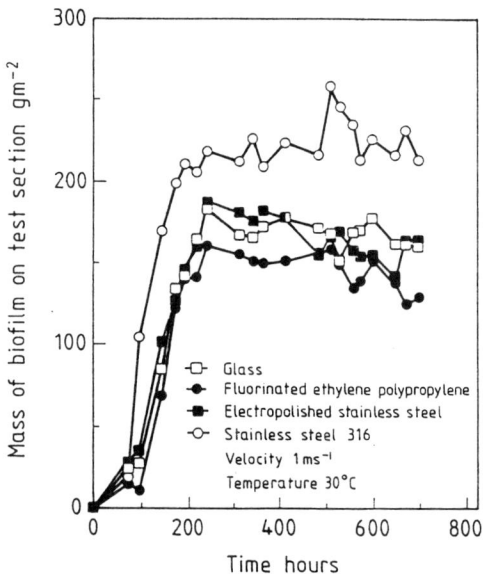

Fig 5 Biofilm (*Pseudomonas fluorescens*) development on various surfaces

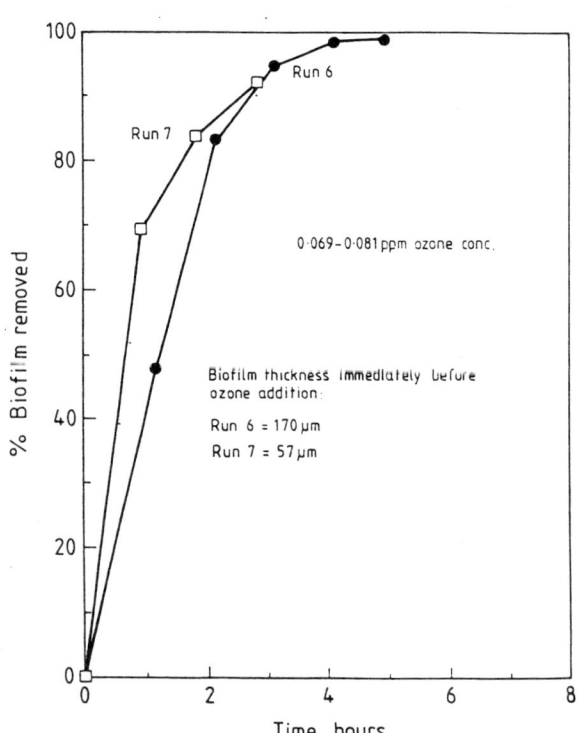

Fig 6 The effect of ozone on biofilm of different thicknesses (*Pseudomonas fluorescens*)

C413/040

Liquid crystal mapping of local heat transfer in crossed corrugated geometrical elements for air heat exchangers

J STASIEK, MSc, PhD and M W COLLINS, MA, PhD, DSc, CEng, MIMechE, MIEE, MINucE
Thermo-Fluids Engineering Research Centre, The City University, London
P E CHEW, BSc,
PowerGen, Marchwood, Southampton

SYNOPSIS The geometrical elements used in rotary regenerators for fossil-fuelled power stations are of a diamond-shaped configuration. For performance reasons it is desirable to know the detailed flow and thermal characteristics. In parallel with computational fluid dynamics (CFD) predictions, experimental work involves applying optical methods to a number of elements in a heated wind-tunnel. Thermochromic liquid crystals (TLC) are used to provide complete local surface temperature data. The methods used to obtain reliable quantitative information are explained. It is concluded that it would also be desirable to use image and automatic data processing.

1 INTRODUCTION

The rotary air regenerators used in fossil fuel power stations perform the essential function of recovering low grade heat from the combustion product gases. They carry out the last stage of transfer of heat from the flue gas, prior to the gas being exhausted to the stack, and reduce the gas temperature to a level well below that which can be economically utilised in water heating and steam raising. The thermal energy transferred from the gas is also used to preheat air supplied for combustion. This preheating is important in its own right to ensure efficient combustion of the low grade fuels employed and, in pulverised coal fired boilers, to enable drying of the coal for milling and transport to the furnace.

Rotary regenerators are employed, in preference to recuperators, for air preheating on all modern generating sets. These heat exchangers rely for their operation on the thermal storage capacity of a central matrix which is periodically exposed to the flue gas and combustion air. The rotary air heater offers advantages in terms of smaller size and lower capital cost, these advantages stemming from the ability to utilise compact heat transfer surfaces which are also readily removable for replacement or maintenance. The two main types of rotary heater are the Ljungstrom and Rothemuhle designs, manufactured in the U.K. by James Howden & Company and Davidson & Company respectively. While functionally similar they differ in the method adopted for exposing the matrix to the air and gas streams. The Ljungstrom type has a rotating matrix and stationary headers while the Rothermuhle design employs a stationary matrix and a system of rotating headers (Fig. 1). The need to minimise powerplant development costs implies that component design methods must be as accurate as possible [5]. These methods, of course, require detailed knowledge of heat transfer and pressure drop performance.

This paper presents the use of liquid crystal mapping in the study of convective heat transfer in crossed corrugated geometrical elements for air rotary regenerators. The rotary regenerators used by PowerGen are heat exchangers in which high rates of heat transfer are extracted from large air-flows. While the exchangers themselves are substantial pieces of engineering equipment, they are composed of large numbers of nominally identical fairly small geometrical elements. Various complex designs of these are used, and they have been developed on an empirical basis.

Ljungstrom type heaters are used and an optimum design involves factors such as the compromise between heat transfer and pressure drop, effect of fouling, and minimising the thermal margin. This last is the empirical reduction encountered in heat transfer rates on site compared with those measured in small scale laboratory tests.

PowerGen has good laboratory performance data in the form of overall heat transfer and pressure drop, but the effects which cause these are not known. It is, therefore, of considerable value to have reliable local data and to be able, by integrating the heat transfer and obtaining drag factors, to relate these to the available bulk measurements. More specifically, the fluid flow patterns (e.g. shear layers) and regimes (laminar/transitional/turbulent) of these three-dimensional swirling flows need to be identified, together with local surface heat transfer values.

In the current study, the liquid crystal thermographic technique has been used to map

---
*On leave of absence from Technical University of Gdansk, Poland.
+Powergen, Ratcliff-on-Soar, U.K.

temperature and local convective heat transfer on a uniformly cooled surface plate and corrugated sheets, coated with thermochromic liquid crystal (TLC)

Of course, heat transfer by convection is closely related to fluid flow conditions, and, whilst not the direct subject of this paper, the TLC approach is being combined with fluid measurements in the form of particle image velocimetry [16] and detailed pressure drop knowledge.

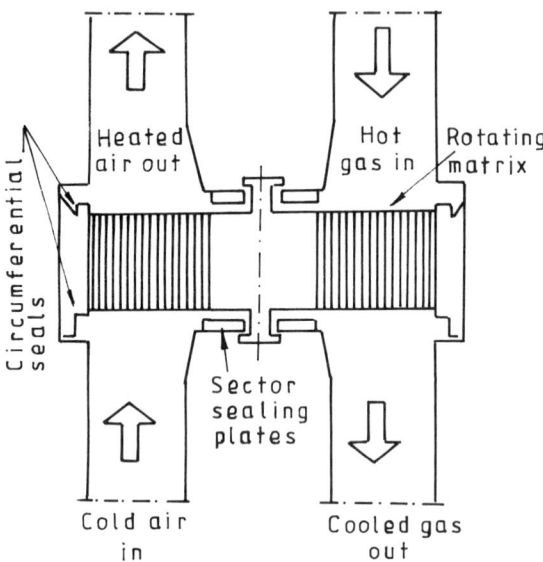

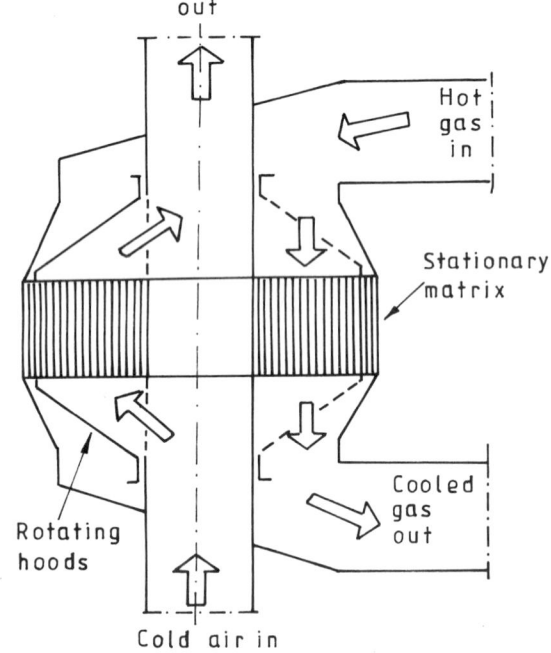

Fig. 1 Power Station rotary air heaters
(a) - Ljungstrom type, (b) Rothemuhle type

2 LIQUID CRYSTALS: A BACKGROUND SURVEY

Some liquid crystal materials (cholesteric or chiral-nematic type) exhibit attractive colours with a visible spectrum when they are heated to the specified temperature ranges. Since the colour change is reversible and repeatable, they can be calibrated accurately with proper care and used as temperature indicators. Thermosensitive liquid crystals have been employed in a number of interesting applications to heat transfer studies for several years now. Fergason [6] extensively reviewed the chemical, optical and thermal properties, and the techniques important to the thermal mapping for non-destructive material testing. Recently, these thermochronic liquid crystals have become commercially available in film or paint forms, in which liquid crystals are stabilised by microencapsulation. They can be easily attached or painted on a heat transfer surface, and two-dimensional temperature distributions can be directly visualised as a coloured pattern.

Raad and Myers [13] used liquid crystals to observe nucleation sites in a study of pool boiling. Cooper et al [3] [4] employed liquid crystals on thin mylar sheets to observe the temperature fields produced by resistively heated, radio-frequency, and cryosurgical cannulas, respectively.

The use of cholesteric liquid crystals in wind tunnel experiments was first investigated by Klein [8] [9] in 1968. Klein employed unencapsulated cholesteric liquid crystals as the surface temperature sensor. Goldstein and Timmers [7] determined the local heat transfer coefficient to an impinging jet from a heated surface. To evaluate temperature distributions quantitatively, they made use of only one temperature, corresponding to the colour boundary between green and blue, which was the sharpest to the naked eye. In this way the isothermal line was determined almost completely whilst avoiding ambiguous judgment of colour by human sensation. Kasagi et al [10] and Simonich and Moffat [15] improved the method in a study of film-cooling heat transfer and provided the first careful examination of the frequency response of the liquid crystal techniques. The monochromatic light source was used selectively to illuminate the single isothermal line. New transient techniques using microencapsulated chiral nematic liquid crystals have been recently developed by Ireland and Jones [11] [12] to measure local heat transfer coefficients in gas turbine blade geometries.

Akino et al [1] [2] have developed a new method by which they can determine isothermal maps at a single heating condition, without the help of human sensation. The method is based on the use of a set of sharp band-pass optical filters, one of which is attached to a black and white video camera to take a monochromatic image having a specified colour. From the image an isothermal line was drawn with the aid of a digital image processing technique that excludes human colour sensation.

Current work of the authors involves the use of filters and cameras only. However, it is intended to commence in the near future work with a fully automated system using image processing. (See discussion on Fig. 4 below).

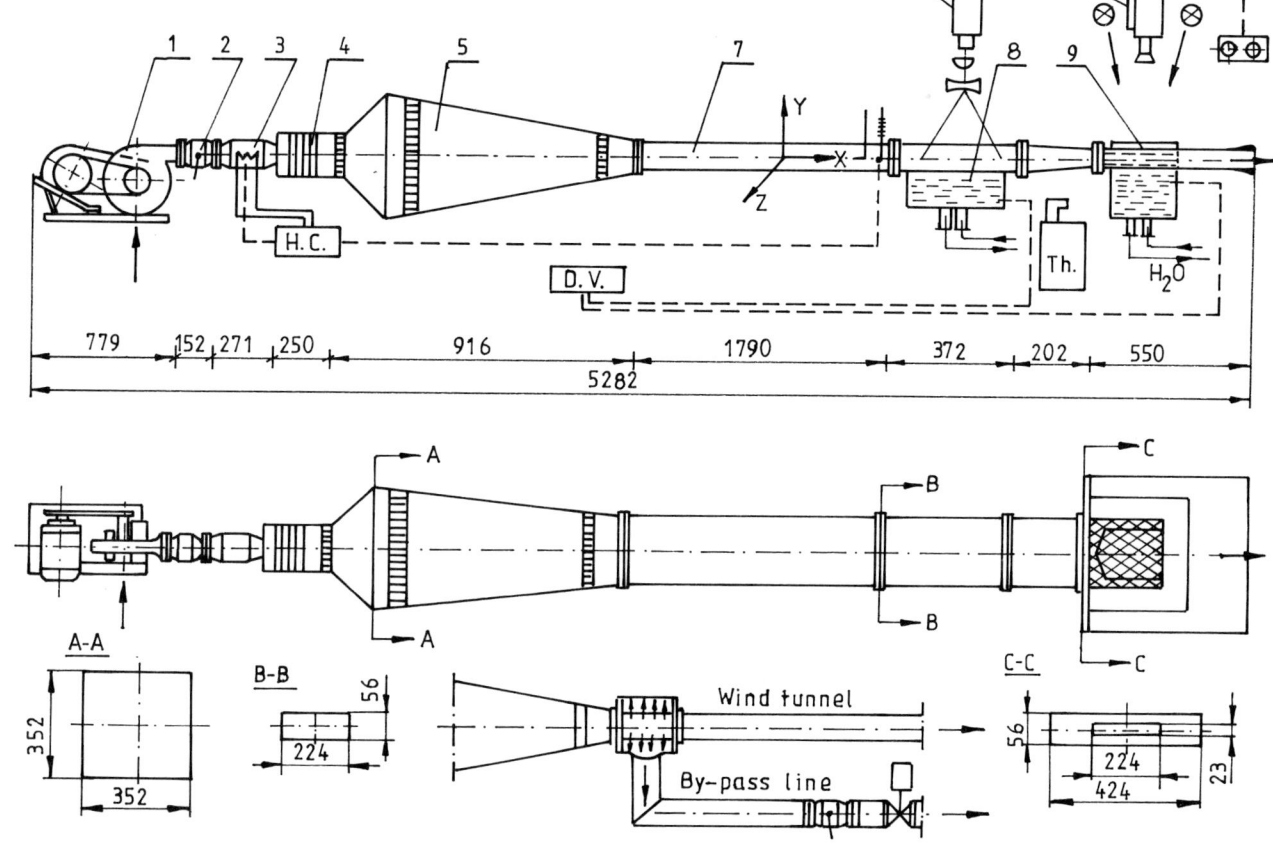

Fig. 2 Open low-speed wind tunnel. 1. Fan, 2 - Valves, 3 - Heater, 4 - Hose, 5 - Large settling chambers. 6 - Alternative arrangement with by-pass 7 - Settling length 8 - Mapping section, 9 - Working section, 10 - Particle Image Velocimetry. 11 - Video-camera.

3 EXPERIMENTAL ARRANGEMENT

The experimental study is carried out using a blow through open low-speed wind tunnel with dimensions as shown in Fig. 2. It can be seen that the wind tunnel consisting of entrance section (with fan and heaters), large settling chambers with diffusing screens and honeycomb, and the mapping and working sections. Air is drawn through the tunnel using a fan able to generate Reynolds numbers of between 1000 and 10000 in the working section. The working air temperatures in the rig range between $25°C$ to $50°C$ produced by the heater positioned just downstream of the inlet. The mapping section is a channel of rectangular cross-section 56 mm x 224 mm, 280 mm long working plate surface. The working section consists of crossed perspex corrugations twice the size of the actual corrugations. Both sides and end of the working section void into a large box. The downstream end of the working section consists of seven diamonds, each of pitch 31.8 mm, amplitude 9.53 mm and cross-sectional area of 145 mm^2. The lower side of the mapping and working sections are covered by commercially available liquid crystal sheets (Fig. 3b) with an event temperature range $21°C$ - $23°C$ and $27.0°C$ - $29.6°C$. Photographs are taken using a video-camera with high bandwidth detector and band-pass filters.

Velocity vector data were obtained by particle image velocimetry (PIV) where a double-pulse sheet illumination of seeded flow is photographed. [15] [16].

It is important to compare wall static pressure distributions in different flow structures with computational predictions. Therefore, the static pressure drop along the peak and trough were measured using 6 pressure tappings 1mm in diameter. A reference pressure tapping is located in the outlet box. As can be seen from the plan view of Figure 2 the double corrugated sheets form a series of identical basic diamond-style geometries, and Figure 3a gives a general view of the base geometry. The upper and lower corrugations have inlet flows at V_{1I} and V_{1V} respectively, with mixing across the plane of the diamond. Locations of pressure orifices are also shown.

Before the quantitative determination of the isothermal lines, the peak intensity temperatures for each band-pass filter must be found by a calibration test. A schematic view of the calibration apparatus and a diagram of the image processing system are shown in Fig. 4. [1] [2] [15]. The calibration plate was made of brass, with length of 200 mm, width of 165 mm and thickness of 8 mm. One end is heated electrically and the other is cooled by water. Consequently, a linear temperature distribution is established within a length of 140 mm and measured by five thermocouples inserted in the plate. In this study, 7 interferential band-pass filters are used. Their central wavelengths are in the range of 400 - 700 nm and the full width of the half maximum (F.W.H.M.) of transmittance is 10(± 2) nm.

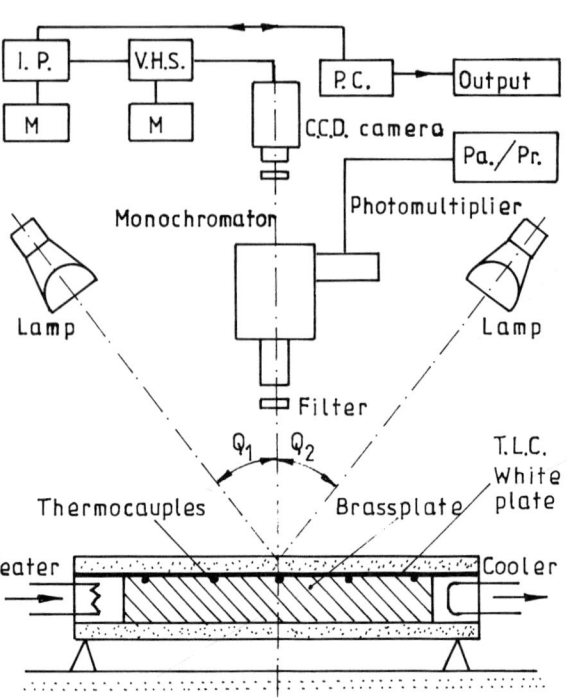

Fig. 4. Schematic view of the calibration apparatus

4. EXPERIMENTAL RESULTS

Outlet mean velocity profiles (for seven ducts and for both lower or trough, and upper or peak, corrugated plates) were measured using a Pitot pressure tube mounted 45 mm inside each duct. Fig 5 show the outlet mean velocity distribution for the corrugated working section (the shaded and clear columns corresponding respectively to the upper and lower walls. It is evident from the graphs that the flow is not everywhere strictly uniform. However, for the central ducts III and V respectively, the velocities satisfactorily agree to within ±0.7 % and ± 1.2 % for both graphs as a whole.

Wall static pressure drops along peak and trough were measured at the wall tappings by means of a micromanometer model MDC FC002/Furness Controls Ltd. Bexhill U.K. with two ranges:

1: from 0 to 19,999 mm H_2O, and

2: from 0 to 1,999 mm H_2O

Fig 6 shows the peak and trough wall static pressure variations along the flow duct No. II and V together with mean inlet velocity V_i.

It is evident from the graphs that there is a significant variation in mean pressure drop along the peak and trough. Due to the streamwise and vertical velocity components of the turbulent flow the pressure drops for the upper corrugation (peak) are higher than for the lower corrugation (trough). For example, referring to Fig. 6 the static pressure drop $\Delta P_{st, 3a-5a}$ = 30 Pa and

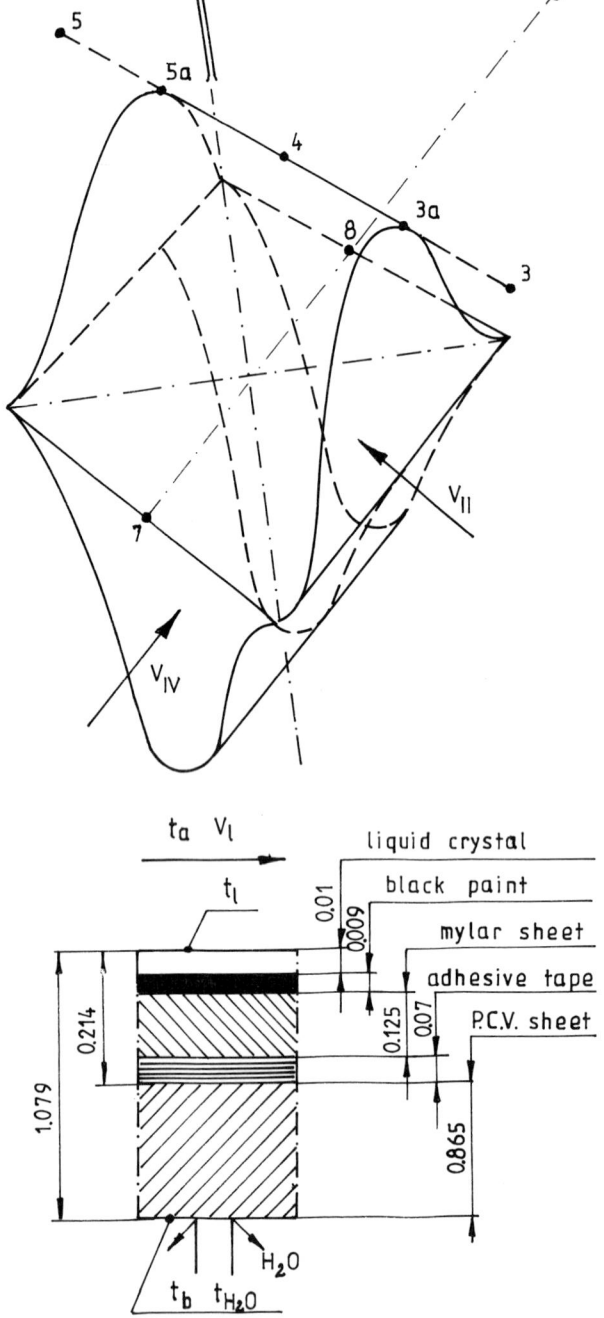

Fig 3. A. General view of three-dimensional diamond geometry and the locations of pressure orifices along the peak and trough
B. Liquid crystal package with component thicknesses.

$\Delta P_{st,7-8} = 10 Pa$ for $V_i = 15.9$ m/s, for upper and lower respectively.

The local convective-heat transfer coefficient α in "ℓ" follows from the fact that the conductive heat flux q_λ in the corrugated plate is equal to the convective heat flux from the surface to the air in the stationery state.

$$\alpha_\ell = \frac{\lambda}{\delta} \frac{(t_\ell - t_b)}{(t_a - t_\ell)} \qquad (1)$$

where

λ - mean conductivity of the liquid crystal package and corrugated plate,

δ - thickness of the liquid crystal package and corrugated plate ($\delta = 1.079$mm),

t_ℓ - temperature of surface

t_b - temperature of water

t_a - temperature of air

The conduction parallel to the plate can in our case be neglected. The temperature drop at the water-wetted side of the plate can also be neglected. From (1) we can see that for the determination of the local heat transfer coefficient α, only an accurate measurement of t_ℓ, together with the conductivity of the component liquid crystal sheet and corrugated plate are important. The use of TLC makes it easier to satisfy accurately these criteria, rather than alternative usually invasive methods.

Fig. 7 shows the distribution of local Nusselt number for a uniform wall temperature. The contours of constant Nusselt number were obtained using the red colour to define a specific isotherm. The liquid crystal colour temperature used is 27.7°C, some 17.6°C below the air temperature for these experiments. The maximum measured local Nusselt number or heat-transfer coefficient was several times higher than the average Nusselt numbers calculated according to the Sieder-Tate and Dittus-Boelter equations for laminar and turbulent flow respectively.

5. CONCLUSIONS AND FUTURE WORK

It is shown in this paper how TLC can be used to give quantitative local heat transfer data in a complex geometry. In this application the information will be used for CFD code validation, and checked also against bulk empirical thermal and fluid data for a section of regenerator geometry. However both the quantity and the human sensitivity involved in processing the results indicate that image and automatic data processing would be very desirable.

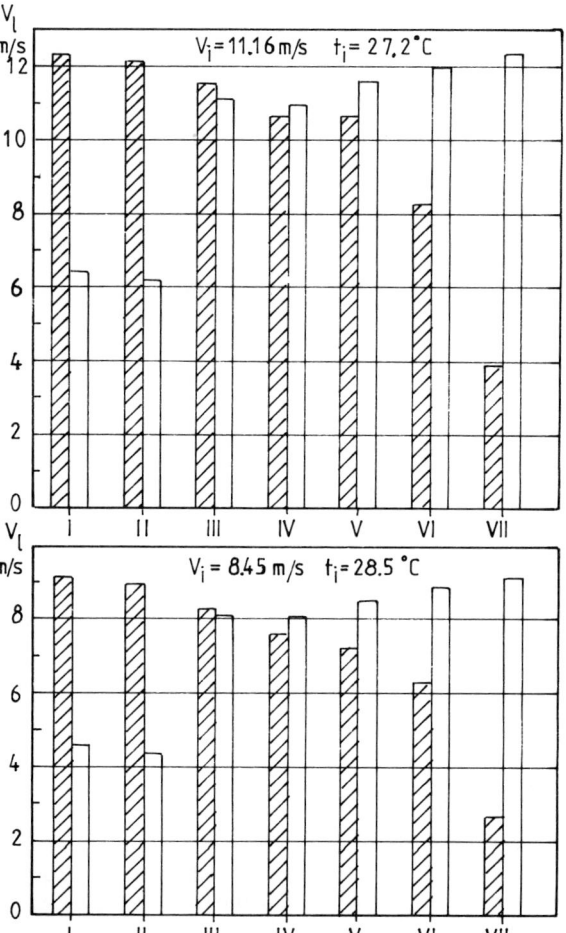

Fig. 5 Outlet velocity profiles: the shaded and clear columns correspond respectively to the upper and lower walls.

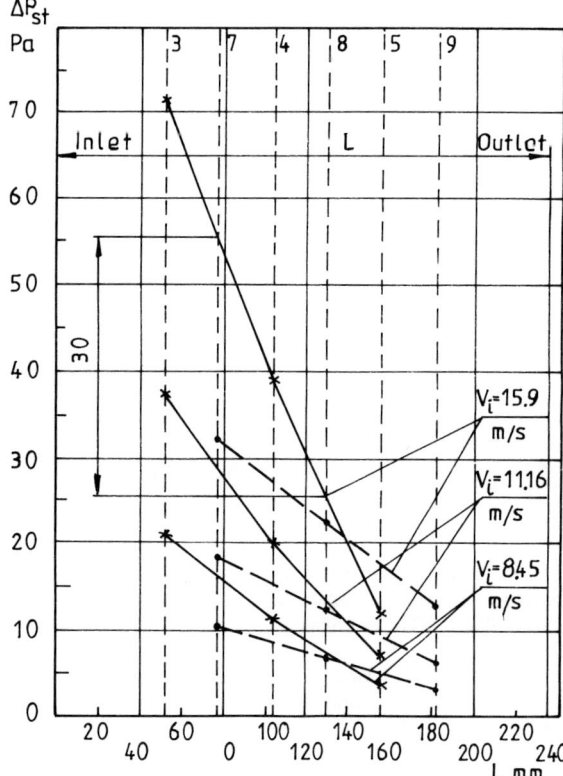

Fig. 6 Wall static pressure distribution along the flow ducts (* - the peak tappings, • - the trough tappings).

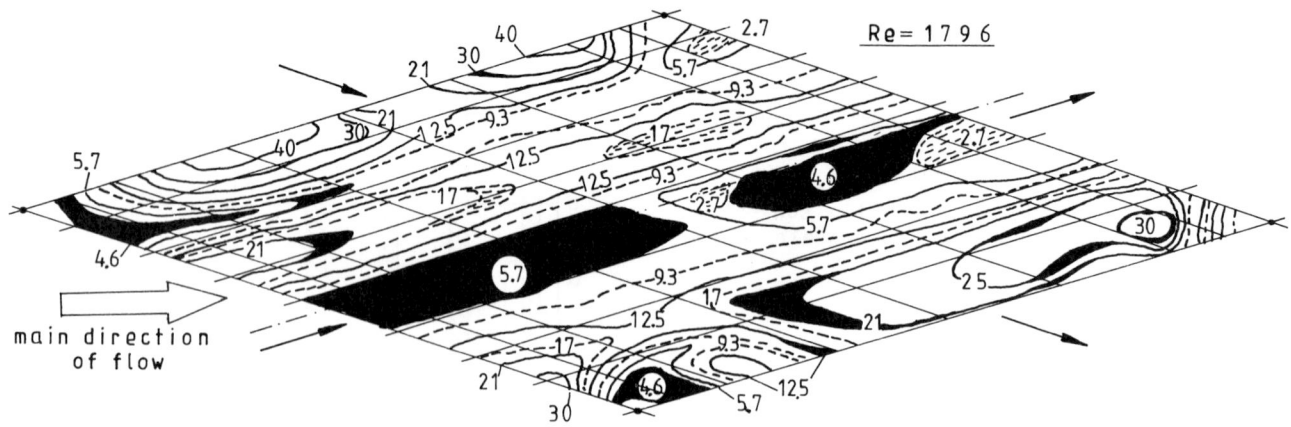

Fig. 7. Distribution of local Nusselt numbers for a uniform wall temperature

ACKNOWLEDGEMENTS

The authors are grateful to PowerGen (Central Electricity Generating Board) for their sponsorship of this work.

REFERENCES

1. Akino. N, Kunugi,. T., Ichimiya K, Matsushiro. K, and Ueda M: Improved Liquid Crystal Thermometry Excluding Human colour Sensation. J. Heat Transfer 1989, 111, 558-565.

2. Akino N, Kunugi T, Ueda M and Kurosawa A: A Study on Thermo-Camera Using a Liquid Crystal. N.H.T.C., Philadephia, USA, 1989

3. Cooper T.E. and Petrowie W.K. An Experimental Investigation of the Temperature Field Produced by a Cryosurgical Cannula.. J. Heat Transfer, 1974, 96, 415-420.

4. Cooper T.E., Field R.J. and Meyer J.F.: Liquid Crystal Thermography and its Application to the Study of Convective Heat Transfer. J. Heat Transfer, 1975, 97, 447-450

5. Chew P. E.: Rotary Air Preheaters on Power Station Boilers. Institute of Energy Symposium "Waste Heat Recovery and Utilisation". Portsmouth. September 1985.

6. Fergason J.L.: Liquid Crystals in Nondestructive Testing. Appl Optics, 1968, 7, 1729-1737.

7. Goldstein R.J. and Timmers J.F.: Visualisation of Heat Transfer from Arrays of Impinging Jets. Int. J. Heat Mass Transfer, 1982, 25, 1875 - 1868.

8. Klein E.J. Liquid Crystals in Aerodynamic Testing. Astronautics and Aeronautics, 1968, 6, 70-72.

9. Klein E.J.: Application of Liquid Crystals to Boundary Layer Flow Visualisation. AIAA 3rd Aerodynamic Testing Conference. AIAA Paper 68-376, 1968

10. Kasagi N, Hirata M and Kumada M: Studies of Full-Converge Film Cooling: Part 1, Cooling Effectiveness of Thermally Conductive Wall. ASME Paper No. 81-GT-37, 1981.

11. Ireland P.T. and Jones T.V. Detailed Measurements of Heat Transfer on and Around a Pedestal in Fully Developed flow. Proc. of 8th Int Heat Transfer Conf. San Francisco. Hemisphere Publishing Corp. 1986, 3, 975-980.

12. Ireland P.T. and Jones T.V.: The Response Time of Surface Thermometer Employing Encapsulated Thermochronic Liquid Crystals. J. Phys. E. 1987, 20, 1195 - 1199.

13. Raad T and Myer J.E. Nucleation Studies in Pool Boiling on Thin Plates Using Liquid Crystals. AICHE Journal, 1971, 5, 1260 - 1261.

14. Simonich J.C. and Moffat R.J. New Technique for Mapping Heat Transfer Coefficient Contours. Rev. Sci Instrum., 1982, 53, 678-683

15. Stasiek J and Collins M.W. Local Heat Transfer and Fluid Flow Fields in Crossed Corrugated Geometrical Elements for Rotary Heat Exchangers. Report No. 1 and No. 2, City University, London , 1989, 1990

16. Hunter J.C. and Collins M.W. Processing of data from both flow measurement methods (Particle Image Velocimetry) and Large Eddy Simulation to investigate coherent structures. Int. J. of Optoelectronics, 1990, 4,

© Copyright PowerGen 1991

C413/061

Simulation: a tool for improving the design process

B MURPHY, BSc, M A P MURRAY, BTech, PhD, CEng, MInstE and V I HANBY, BSc, PhD, CEng, MInstE
IDC Limited, Stratford upon Avon, Warwickshire

SYNOPSIS : The design process of mechanical services systems is discussed and the role of system simulation is identified. The development of a component based performance simulation technique is described, including the types of component models utilised and the difficulties in obtaining part load performance data from manufacturers. The structure of a graphical user interface is described which aims to make the simulation tool easy to use by practising designers. Developments in the numerical methods used to determine the operating point of the system are described. The practical application of the technique to design problems is described.

1. INTRODUCTION TO SIMULATION

1.1 The Design Process

The design of mechanical services systems (such as thermofluid and heating ventilating and air-conditioned (HVAC) systems) relies upon the analysis of the performance requirements of the system at fixed, agreed design parameters. This usually determines that system load characteristics are analysed at peak loads, the appropriate peak design criteria being established by historical experience.

At this stage the designer would sketch a schematic arrangement of plant to represent the system. The peak system performance requirements are then matched to individual items of plant, each selected to give optimum performance at design conditions.

Alternative arrangements of plant may be considered but are rarely analysed ; the selection between alternative systems being based upon qualitative analysis by the designer, or at best a brief consideration of the likely performance of the systems.

A detailed analysis of the alternatives is generally impractical due to the pressure of design time which reflects the low capital value of HVAC systems compared with the design costs incurred. On high capital cost projects the design costs incurred for iterative refinement of a design may be offset by the capital savings on an optimum design.

Designers usually avoid the need for an iterative design process by the adoption of 'standard' system design criteria based upon historical precedent and peak load conditions. The selection of a cooling coil, for example, would usually be based upon arbitrarily selected chilled water flow and return temperatures. These parameters represent 'best practice' as a compromise between, for example, pump power input and chiller performance. The selection of water chillers is linked to the chilled water design temperatures and external ambient conditions for water or air cooled condensers.

It would be impractical using manual analysis to consider the inter-relationship of these various design parameters to select an optimum set. In all but the most rudimentary of systems, manual analysis of part load performance is too complex or time consuming for practical use in design. Computer simulation or modelling of systems is the only practical alternative to the construction of prototypes.

1.2 System Simulation

Several approaches to computer simulation of systems have been developed over the last fifteen years. Simulation of 'standard' systems in which certain criteria may be adjusted to reflect the scope, (such as zoning), of the application have been widely used (Ref. 1). This approach suffers, however, from the restriction of an ever expanding list of standard systems and the inability to analyse hybrid or innovative designs.

Component based simulation techniques have distinct advantages in being able to reflect the actual proposed system design. The computer model of the system is built-up by selection from a menu of components in a manner akin to the actual schematic design process. Whilst offering flexibility in system definition these simulation procedures have in practice been limited to energy accounting simulations. The individual components are simply represented as energy consuming or producing models linked by a limited set of system variables (Ref. 2).

Many of these energy simulation programs have been developed to compliment existing building fabric analysis suites. The systems analysis tends to be directly linked to the building analysis, including parameters, such as dynamic response, which may not be of interest to designers.

1.3 Component Based Performance Simulation

The procedure discussed in this paper has been developed as a designers' tool with the aim of putting system simulation into the hands of designers.

The system is component based, thus mirroring the design process via the engineering schematic. The component models are based upon actual performance data relating to true systems variables (such as flow rate, pressure and temperature).

Individual component models are linked together into a network representation of the system via a graphical user-interface. This gives the designer complete control over the definition of the system or sub-system to be analysed.

The system operating point is established for each set of operating conditions using a solution routine based upon numerical methods. The system variables can then be analysed directly or used to determine performance parameters such as efficiency or energy consumption.

The key aspects of this process are discussed in detail in the following sections together with a review of the practical application of system simulation to design.

2. COMPONENT MODELS

The component models used by the solution routine have several important features;

a) They are in neutral format. This is to say that the equations are not deterministic (ie. input - output) but are cast as residual equations (Figure 1). This enables the equations to be solved for any viable set of system variables.

b) They are data abstract. Any data used in equations is passed in the subroutine call : only the algorithm is fixed.

c) They contain a results routine to calculate performance parameters, required by the user which require interpretation of system variables (eg. total duties, efficiencies).

Two distinct approaches have been considered in formulating component models; fundamental analysis and empirical data. In practice most models represent a combination of these with the fundamental laws of thermodynamics being applied alongside manufacturers experimental performance data.

2.1 Data Types

In general, component models utilise three types of data;

a) Variables data - this is data which represents items such as mass flows and temperatures. This data will be changed by the solution routines, subject to certain conditions, in order to establish the operating point of the system.

b) Constants data - this is data associated with the particular model being simulated. It only changes when a new model is selected from the component data library.

c) Properties data - this is data associated with a particular component model but may be changed between operating points. For example, it could be the throttling range of a controller.

The implications of the above items can ne seen in considering two different type of models. A centrifugal fan (Figure 2) and a pipework component (Figure 3).

If the models are examined, it can be seen that in order to affect the performance characteristics of the fan it is necessary to select different component model data (ie. change the fan). With pipework, the performance characteristics can be changed by selecting new component model data (ie. change the pipework type) or by altering the properties data (eg. internal diameter). It should also be seen that this can be done without changing the code, proving that the models are in fact data abstract.

2.2 Data Acquisition

In order to simulate 'real' components, the models use 'real' data. This data can sometimes be obtained from standard references, such as heat exchanger data from Kays & London (Ref. 3). If this is the case then the models may be written based on this data.

The usual case however, is that the data required can only be obtained from test results from the manufacturers of the components. The following problems have been experienced;

a) The manufacturers may not have the performance data which is required.

b) The manufacturers maybe unwilling to release proprietary data for fear of commercial misuse.

c) The data available is represented in different formats by different manufacturers.

With further use of simulation as an integral part of the design process manufacturers will be under increasing pressure to publish the information required.

2.3 Component Library

A pragmatic approach has been adopted to the development of the range of component models, with research and development effort concentrated on the components necessary to prove the viability of the technique and to serve the immediate practical application of simulation to specific design tasks. The current models are limited to water, air and controls systems although the formulation of the simulation program will allow extension to cover two phase fluids, such as refrigeration and steam, without the need for any re-coding.

3. USER INTERFACE

The user interface is the link between the system designer and the simulation procedure. It is the design of this interface which makes the simulation program unique, in that it has the same degree of flexibility as the actual process of producing a design schematic.

3.1 System Definition

The designer builds up the computer model of the system by placing component items from a menu of components on the screen and defining appropriate links between these components to represent the system.

It should be emphasised that the same degree of engineering judgement should be applied to the selection of component models for a simulation exercise as would be used in live project design. Poorly designed systems, with mismatched components have proved difficult or impossible to simulate.

3.2 Linking Components

The links between components are formed with a degree of automation, in that variable 'sets' can be linked (e.g. air outputs to air inputs) with the system designer able to override the automatic process to define links between variables explicitly if required.

When a link between two components is made by the designer there are automatic checks made in order to validate the link. The system utilises the component data for each component to ensure the following;

a) The link is not a link back to the component itself.

b) The link is correct in terms of its sense (ie. output to input and visa-versa).

c) The link is correct in terms of the fluids (ie. air to air, water to water).

d) Each of the variables can be connected to an equivalent variable within the other component (ie. temperature to temperature).

If this validation fails at stage (d) then a manual method can be utilised by the designer to complete the link or abort it. This is sometimes inevitable, for example when specifying a controlled variable as an input to a controller.

3.3 Component Model Data Storage

Component model data files for each component are selected from a database of manufacturers information. Although the designer manipulates the system and data graphically, this data is held by the system on disk and accessed during system definition. There are currently thirty-five component models available with upwards of ten sets of model data each for each manufacturer. In order to handle this large amount of data, database methods have been employed. This database is manipulated via library routines but once finalised, the data tends to be constant and only needs to be updated to reflect changes in manufacturers data.

The solution technique requires an initial estimate of each of the system variables. These initial estimates are currently set-up from the component model data files although operational experience suggests that there are benefits in terms of speed of solution in adjusting the initial estimates. Techniques for automating this process are the subject of further research at Loughborough University of Technology (LUT).

3.4 Identification of System Variables

The system variables are identified automatically by the simulation program which allocates a unique identification reference to each variable. This allocation process recognises a variable which passes through a series of components unchanged and allocates a single number for such a variable. The total number of variables will exceed the total number of residual equations in the system ; the excess variables represent the 'input' variables to the simulation of a particular operating point. By fixing the values of these input variables a set of equations in a matched set of unknowns is described which can be solved as described in section 4.

The system network defined is checked to ensure that a viable system has been defined (see section 4.2) and the simulation initiated to identify the operating point. Multiple operating points may be simulated using a load profile to define a series of conditions such as hourly or seasonal values of design parameters.

3.5 Analysis of Simulation Results

The graphical user interface can be used to view the results of a simulation, either by examining the system variables or by reviewing component performance parameters which are calculated from the variables at the operating point.

4. SOLUTION TECHNIQUE

4.1 Numerical Formulation

At the completion of the input process, a system model is formed by a set of equations generated by each component model. The links between the components, originally defined by the user, are now transposed into the presence of system variables in more than one component equation.

As the component model is represented in residual form, the set of system equations takes the following form :

$$F_1 = f(X_1, X_2, \ldots)$$
$$F_n = f(\ldots, X_{n-1}, X_n)$$

Where X_1 to X_n are the system variables and F is the vector of residual values. The equations are solved by starting with initial estimates of the system variables and employing a non-linear optimization algorithm to adjust the values of the estimates such that the vector of residuals is forced to a near-zero value.

A number of algorithms have been evaluated for this purpose and the current method uses an implementation of the Generalised Reduced Gradient method (GRG2) devised by Lasdon et al (Ref. 4). This is a non-linear optimization method which handles non-linear constraints. This algorithm has proved generally satisfactory on terms of speed and robustness in a research environment.

4.2 Solution Method

In the early phase of development of the simulation program the equations were simply treated as a simultaneous set and solved accordingly. When development as a design tool was undertaken two major shortcomings of this approach was recognised.

Firstly, it was apparent that some dialogue with the user was required to ensure that a viable network had been defined. In particular, it was possible to assign fixed variables in such a way as to produce an invalid set of system equations and unknowns. In addition, some 'intelligent' direction of the solution process was required as the solution algorithm could, under some circumstances, fail to find a solution to a valid equation set.

An additional stage in the simulation process was interposed between the problem definition and the equation solver (a solution manager), incorporating knowledge derived both from external sources and from the topology of the network. The key element of the latter is an adjacency matrix - an array in which each component (node) is assigned a row and column. A non-zero element in this matrix means that the two nodes denoted by the respective row and column are connected. The network description held in this matrix is called the system graph.

An example of a system graph described by an adjacency matrix is shown in Figure 4. The schematic shows part of an air conditioning system. The principal components, with their identifying numbers, are a cooling coil, supply air fan, duct and the occupied space (zone). The cooling duty of the coil is controlled by a room thermostat modulating the mass flow rate of chilled water, adding a further three components to the sub-system.

The resulting adjacency matrix describes the connectivity between these components. There are two points concerning this matrix which are worthy of note. Firstly, the diagonal elements of the matrix are set to 1, i.e. each node is show as being connected to itself. Secondly, the complete matrix is shown, although strictly speaking only a triangular half-matrix is needed. Several of the graph algorithms used are more straightforward to code if the diagonals are non-zero and if the full matrix is stored as an array.

This graph is processed by a number of algorithms which both provide an effective dialogue with the user and improve the effectiveness and robustness of the equation solver. At its current stage of development, the graph algorithms :-

* check that all components are connected i.e. that only one system has been defined.

* ensure that a valid set of fixed variables has been chosen.

* identify linear 'chains' of components which can be solved deterministically and pass these through to the equation solver as self contained blocks.

The intermediate solution manager has proved effective in the development of the simulation into a design tool. The use of these techniques is still at an early stage and further research is being undertaken to exploit the potential of this concept.

5. APPLICATION IN THE DESIGN PROCESS

The practical application of system simulation to the design process requires an understanding by the designer of the potential, and limitations, of component based simulation.

The introduction of simulation into the design practice has been based upon several key studies of particular design problems, together with providing a quick response to designer led queries. The latter effort has resulted in the need for rapid development of additional component models together with expansion of the component model database. The ability to respond to these requests has reinforced the flexibility of the computer suite design which allows additions to, and development of, component models without the need for re-coding of the program.

The simulation technique has been applied to several live design projects, to verify conceptual design, in the analysis and selection of energy efficiency measures, and to validate field test results against design criteria.

5.1 Conceptual Design of Containment Systems

One project design problem included the provision of fume cupboards for product containment within a pharmaceutical production facility. This consisted of thirteen walk-in fume cupboards plus a standard bench top fume cupboard. The design problem was to provide a fume extract system to serve up to four of the walk-in cupboards and the bench cupboard in the open position, with the remaining cupboards closed, whilst maintaining the containment pressure regime as negative to other production areas, positive to outside.

The extract system configuration is shown in Figure 5 and consists of a constant volume extract fan, extract duct header, branch connections to each fume cupboard including manual regulation damper and automatic control damper, and relief damper.

The extract header duct was designed as a plenum to ensure that the total system pressure characteristic is unaffected by which combination of fume cupboards are operational. The relief damper is automatically controlled to maintain a constant pressure within the extract header duct, and thus a constant volume at the fan.

The control damper in each sub-branch was designed as two-position to ensure that the air-flow is either maximum or minimum depending on the position of the fume cupboard sash.

The system simulation was carried out in two phases, firstly to analyse a single branch sub-system and secondly to analyse the performance of the whole extract system.

The initial simulation provided confirmation of the required commissioning setting of the manual regulation damper and of the control authority of the branch control dampers.

Simulation of the whole extract system enabled the performance of the system in all operating states to be simulated. Because the extract plenum was considered to be at constant pressure a total of nine operating states are possible if no more than four cupboards are open at one time, these states are summarised in Table 1.

The simulation indicated a variation in extract volume of approximately ±0.5% of design across the range of operation, confirming the assumption of the constant pressure in the header duct. The branch flows are maintained at constant rates across the throttling range of the controllers.

Table 1 Operating States for Fume Cupboard System

NO.	BENCH CUPBOARD	WALK-IN UP	WALK-IN DOWN
1	DOWN	0	13
2	DOWN	1	12
3	DOWN	2	11
4	DOWN	3	10
5	DOWN	4	9
6	UP	0	13
7	UP	1	12
8	UP	2	11
9	UP	3	10

The two position control operation of the branch control dampers was confirmed as was the effectiveness of the relief damper operation. The simulation results did indicate that the operating range of the relief damper could be widened by an improvement in the authority through a reduction in size of this component.

The simulation of this system confirmed that the installation would achieve the design objectives provided that the damper control characteristics installed corresponded to the design data used in the simulation. The simulation also indicated the initial commissioning setting required for the various manual balancing dampers.

The installation was completed as designed and the commissioning results confirmed that the system performs as predicted by the simulation.

5.2 Validation of Field Test Results

Field test results of the performance of cooling coils and heater batteries are normally limited to the measurement of water and air flow rates, and recording of the water and air temperatures on and off-coil. The water flow rate temperatures can usually be controlled at design conditions but the air on-coil conditions are dependent upon the ambient conditions ate the time of the test. It is not usually possible to ensure that the performance tests take place at or near design conditions. Comparison of measured performance against design performance is limited by the availability to analyse coil performance in detail.

On several pharmaceutical projects, where coil performances are critical, the measured performances have been compared with design criteria by utilising the component based simulation program.

The constructional details of the actual coils installed were used to define a coil model which was then simulated at test conditions to verify the model accuracy against field test results. The performance of the coil was then simulated to confirm the coil performance at the design operating point.

Given a cursory glance this 'problem' appears simple, however the part load operation of coils, especially cooling coils, would require very detailed and tedious manual calculations. An approximation to the coil performance could be obtained by manufacturers' performing multiple runs on computer selection programs. Since these programs are designed for equipment selection at peak design conditions, this approach is not academically rigorous and is not an acceptable form of analysis for validation.

5.3 Conclusions

The examples given of the application of component based simulation illustrate the scope for practical application of this technique to real design problems.

Further work is in hand both at IDC and LUT, to develop further component models, improve the numerical solution techniques and to provide a knowledge based system for system performance diagnostics.

The most practical application of component based system performance is as an aid in the optimum selection of components to form a system and in giving designers additional confidence in the system performance at off-peak design conditions.

The simulation technique described is forming a useful additional design tool for use directly by designers which should result in improved system design.

REFERENCES

(1) Stoecker. W. (editor), 'Procedures for Simulating the Performance of Components and Systems for Energy Calculations', ASHRAE, New York, 1975.

(2) Murray. M.A.P., ' Component Based Performance Simulation of HVAC Systems', Doctoral Thesis, University of Technology, Loughborough, 1984.

(3) Kays. W.M. and London. A.L., 'Compact Heat Exchangers',, The National Press, Palo Alto.

(4) Lasdon. L.S., Warren. A.D., Jain. A. and Ratner. M., 'Design and Testing of a Generalised Reduced Gradient Code for non-linear Programming', ACM Transactions on Mathematical Software, Vol 4 No. 1, 1978, pp 34 onwards.

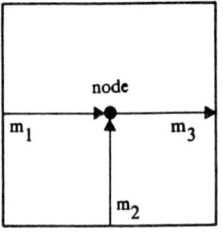

For nodal balance:
$m_1 + m_2 = m_3$

In residual format:
$F = m_3 - (m_1 + m_2)$

Figure 1 - Residual equation example

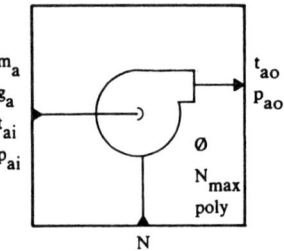

Variables:

m_a — air mass flow rate
g_a — air moisture content
t_{ai} — inlet air temperature
t_{ao} — outlet air pressure
P_{ai} — inlet air pressure
P_{ao} — outlet air pressure
N — fan speed

Constants:

$\emptyset$ — diameter
N_{max} — max fan speed
poly — polynomial coefficients

Residual Equations:

1 — pressure relationship
2 — temperature/power relationship

Figure 2 - Centrifugal fan definition

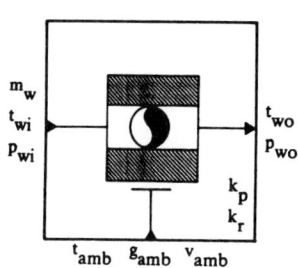

Variables:

m_w	-	water mass flow rate
t_{wi}	-	inlet water temperature
t_{wo}	-	outlet water temperature
p_{wi}	-	inlet water pressure
p_{wo}	-	outlet water pressure
t_{amb}	-	ambient air temperature
g_{amb}	-	ambient air moisture content
v_{amb}	-	ambient air velocity

Constants:

k_p	-	pipe conductivity
k_r	-	pipe roughness

Properties:

$\emptyset_i$	-	pipe internal diameter
k_t	-	pipe wall thickness
i_i	-	insulation thickness
i_k	-	insulation conductivity

Residual Equations:

1 - heat transfer
2 - pressure relationship

Figure 3 - Pipework definition

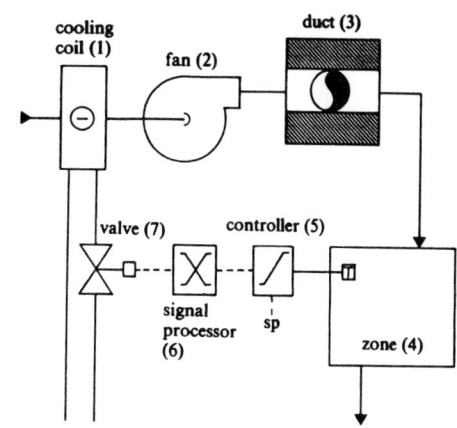

Schematic of example subsystem

NODE	1	2	3	4	5	6	7
1	1	1	0	0	0	0	1
2	1	1	1	0	0	0	0
3	0	1	1	1	0	0	0
4	0	0	1	1	1	0	0
5	0	0	0	1	1	1	0
6	0	0	0	0	1	1	1
7	1	0	0	0	0	1	1

Adjacency matrix for example subsystem

Figure 4 - Example subsystem

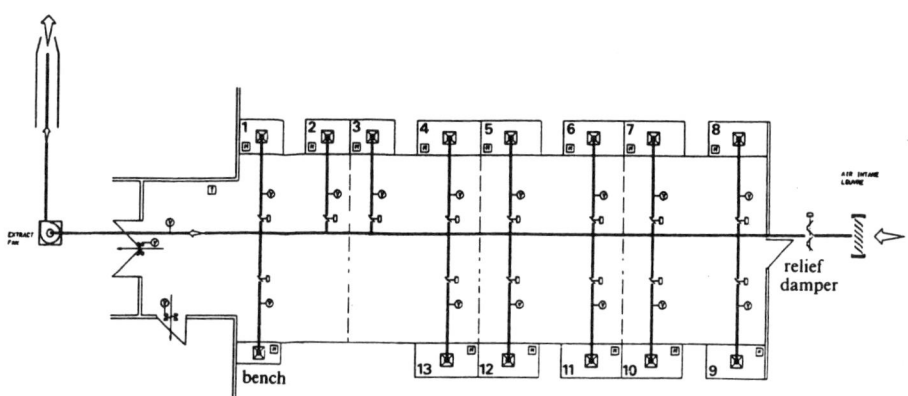

Figure 5 - Fume cupboard schematic

C413/011

A study of flow oscillation and heating in a Hartmann-Sprenger tube – a literature survey

J IWAMOTO, PhD, MASME, MJSME, MSAE
Tokyo Denki University, Tokyo, Japan

SYNOPSIS A Hartmann-Sprenger tube consists of a nozzle and a tube with one end open and the other closed. When an underexpanded jet through the nozzle is directed towards the open end of the tube, the air in the tube is excited into an oscillation and high temperature is obtained at the closed end. The study of the mechanism of flow oscillation and heat production in the H-S tube is reviewed and some of the studies in an attempt to apply this phenomenon to practical use are presented.

1 INTRODUCTION

The phenomenon known as a thermal effect of Hartmann-Sprenger tube was first discovered by Sprenger(1) in 1954 at E. T. H. and since then the study has been made by some researchers(2,3) to establish the mechanism of heat production in the tube.

As shown in Figure 1(a), the H-S tube consists of the nozzle and the tube with one end open and the other closed. When the underexpanded jet is directed towards the open end of the tube, the flow in the tube is excited into a violent oscillation at certain nozzle-tube spacings and the high temperature is produced at the closed end. The temperature obtained at the closed end depends on various parameters such as the stagnation pressure of the jet, nozzle-tube spacing, tube material, the geometry of nozzle and tube, and so on. Using the tube of 3 mm in diameter and 102 mm in length the temperature at the closed end can reach up to 400 degrees Celsius when the driving jet has the stagnation pressure of 5 atmospheres and the stagnation temperature of 20 degrees Celsius.

Prior to the discovery of thermal effect of H-S tube, Hartmann(4,5) found the generation of strong pressure oscillation and intense noise in a short tube as shown in Fig. 1(b). The tube has been named Hartmann whistle since many papers dealing with this problem were published by Hartmann. With regards the Hartmann whistle only the temperature rise of 7 degrees Celsius was measured.

In the present paper the papers on the study of H-S tube published so far are reviewed, and the mechanism of the initiation and maintenance of the self-excited oscillation of the flow and the production of thermal effect as a result of the flow oscillation are described.

2. HEAT PRODUCTION IN THE TUBE

The temperatures at the closed end of the tube are shown against the nozzle-tube spacings in Fig. 2(6). The ordinate is the non-dimensional temperature ß which is the ratio of temperature increase of the air in the H-S tube to the temperature drop due to the isentropic expansion of the driving jet. Thus, ß = 1 when there is no flow oscillation and no temperature rise in the tube. As shown in the figure, the temperature depends on the nozzle-tube spacing. In order to understand the mechanism of such heat production the flow pattern in the tube must be described.

If the flow in the tube can be assumed to be one-dimensional, then different ways of calculating the flow pattern in the tube are available. The wave diagram prepared by the method of characteristics is shown in Fig. 3, which describes in detail what is happening in the tube for one cycle of oscillation. The ordinate is the nondimensional time τ (= $a_0 t/L$, where t is the time, a_0 the stagnation speed of sound, and L the length of the tube) and the abscissa is the position X (=x/L, where x is the distance measured from the open end of the tube). For simplicity inviscid flow and complete insulation of the tube are assumed, and the entropy increase due to shock wave is neglected. It is known that the pressure changes of the flow in the tube obtained from this wave diagram are qualitatively in good agreement with those by experiments.

In this figure the dashed line shows the contact surface which divides the fluid in the tube into two; i.e. the indigenous fluid which remains always within the tube and the extraneous fluid. The former fluid cannot go out of the tube and if the tube is completely insulated, its entropy always increases due to wall friction and shock wave as it is expanded and compressed. On the other hand the extraneous fluid goes out of the tube and comes into the tube with the new value of entropy during one cycle of oscillation.

If the indigenous fluid changes its state isentropically, then its temperature goes up when it is compressed and comes down to the original temperature when expanded, and thus the average temperature does not change. But actually the entropy of the indigenous fluid increases. So, when it is compressed, the temperature goes up and when expanded it comes down to the temperature which is higher than the original temperature because the entropy is also higher now. This is repeated a number of times until the maximum temperature is attained when the entropy increase is balanced by the entropy decrease due to the incomplete insulation, diffusion at the contact surface and so on. Therefore, any means of increasing the entropy of the indigenous fluid results in the increase of maximum temperature. Tapering the tube (see Fig. 1(c)) is one of the means of increasing the entropy because the shock wave becomes strong at the closed end.

It is known that very high temperature rise is obtained in the tapered tube and that the nozzle-tube spacings at which the heating effect is produced are limited to a certain degree(7). It was shown by Rakowsky et al(8) that even higher temperature is obtained in the stepped tube shown in Fig. 1(d) The logarithmic spiral tube where a planar shock produces a stable implosion at the spiral focal point was also used to obtain the high temperature(9).

3. SELF-EXCITED OSCILLATION OF FLOW

How the heat is produced and high temperature is attained in the tube are now well-known and are described in the previous section. In order to generate heat the flow in the tube must be excited into a periodic oscillation. The mechanism of the initiation and maintenance of flow oscillation in the tube can be clarified by describing in detail the flow field in the space between the nozzle and the open end of the tube(10). Figure 4(a) shows the shadowgraph pictures during one cycle of oscillation of the flow fields in the space between the nozzle and the open end of the tube. τ shown under each picture is the nondimensional time ($\tau = a_0 t/L$, where t is the time measured from the instant when the shock wave reaches the closed end of the tube, a_0 the stagnation speed of sound and L the length of the tube). The period of one cycle is $\tau = 4.6$. The shadowgraph picture at $\tau = 0$ shows the inflowing pattern where almost all the jet from the nozzle is going into the tube. At the lip of the tube mouth there is an oblique shock wave. This flow pattern is relatively stable and continues for about 20 per cent of the period of one cycle. Then, there is a transient phase as shown in the picture at $\tau = 1.20$ where the flow into the tube is changing into an outflow. The outlow starts at the instant when the shock wave comes out of the tube and the expansion wave is reflected from the open end. The picture at $\tau = 2.00$ shows the almost steady flow field where the jets from the nozzle and the tube collide each other. This flow pattern lasts for about 30 per cent of the period of one cycle. Then, when the head of the expansion wave comes back to the open end, the velocity of the outflowing fluid from the tube decreases and gradually the outflow from the tube changes into inflow.

Figure 4(b) shows the steady flow pattern when the both ends of the tube are open. As can be seen this steady flow pattern is almost the same as that during the inflow at $\tau = 0$ in Fig. 4(a). The mechanism that initiates the flow oscillation, may be described in the following way.

As a starting point, the jet from the nozzle may be assumed to be directed against a tube that has both ends open and that the steady state flow pattern of Fig. 4(b) has been attained. If the one end of the open tube which is farther from the nozzle is now suddenly closed, the flow pattern in the space between the nozzle and the open end of the tube will change. What has to be considered here is whether, or not, the same flow pattern as that of the original steady flow can be restored after the lapse of a certain time. If the flow pattern reverts to that of the steady state, then a periodic oscillation of the flow could result, since quasi-steady inflow could be one phase of a possible periodic oscillation.

When the end of the tube farther from the nozzle is suddenly closed while the tube with both ends open are steadily blown by the jet, hammer shock is generated at the closed end and moves towards the open end closer to the nozzle. And the expansion wave is reflected from the open end and moves towards the closed end. When this reflected expansion wave returns to the open end, the shock wave is reflected and the inflow pattern is established. However, after the hammer wave is generated, the expansion wave and shock wave must be strong enough to maintain oscillation. Otherwise, the oscillation dies down. In Fig. 5 the inflow pattern is schematically shown. As shown the oblique shock is attached at the lip of the open end of the tube. This shock reaches the jet boundary which is not well defined in the shadowgraph picture and expansion wave is reflected. Behind the reflected expansion wave the pressure is lower than the atmospheric pressure. Therefore, when the shock wave reaches the open end during the inflow phase after reflected from the closed end and it comes out of the tube, the oblique shock moves upstream together with the outflowing shock. So, the region of low pressure outside the tube comes closer to the lip of the open end and the expansion wave is reflected from the open end on the basis of this low pressure, which makes the expansion wave strong enough to maintain the oscillation. On the other hand in order for the shock wave to come out of the tube the pressure gradient in the jet near the open end must be positive. These two conditions are necessary for the initiation and maintenance of flow oscillation and the existence of the region of low pressure and positive pressure gradient was confirmed experimentally, although the experiment was made with rectangular tube(11).

Since there is an analogy between the shallow water flow and compressible flow, the experiments using water table were carried out to visualize the flow pattern in and out of the tube(12-14). Qualitatively the similar results described above are obtained.

4. APPLICATION OF THERMAL EFFECT OF H-S TUBE

Since the high temperature can easily be generated with the simple device of nozzle and tube, some researchers studied this phenomenon of thermal effect of H-S tube in an attempt to apply it to some practical use(15-17).

In the period between late 1960s and 1970s the research was carried out at NASA in the United States to apply this phenomenon to the igniter of the rocket engine. In the experiments at NASA it was demonstrated that ignition of a hydrogen-oxygen mixture can be achieved by using the H-S tube principle(15,16). The feasibility of applying the H-S tube to the rocket igniter was shown by Marchese(18), Rosen(19) and so on, but it is not actually in use yet.

Although the relatively high temperature is obtained in the H-S tube, the thermal efficiency is very low. In the paper by Thompson(2), it is shown that the heating over one cycle is of the order of 1% of kinetic energy discharge at the nozzle. So, the application is limited to the use where the thermal efficiency is not important.

As at present the mechanism of heat generation is established and we know how to increase the temperature, it is hoped that this interesting phenomenon can be applied to some industrial use.

5. CONCLUSION

The studies on thermal effect and self-excited flow oscillation of the H-S tube made so far were described.

The maximum attainable temperature in the H-S tube was predicted by Brocher(20). But practically since the heat capacity generated in the tube is very small, the maximum temperature depends very sensitively on how well the tube is insulated and so it is quite difficult to predict the maximum temperature.

Since the mechanism whereby the heat is generated is now almost established, the applied research to some industrial use is hoped to be made besides the igniter of the rocket engine described above.

REFERENCES

(1) SPRENGER, H. Ueber thermische Effekte in Resonanzrohren. Mitt. Inst. Aerodynamik, E. T. H., 1954, 21, 18.
(2) THOMPSON, P. A. Jet-Driven Resonance Tube. AIAA J., 1964, 2-7, 1230.
(3) MANNING, J. R. Computerized Method of Characteristics Calculations for Unsteady Pneumatic Line Flows. Trans. ASME, J. Basic Eng., June 1965, 231.
(4) HARTMANN, J. On the Production of Acoustic Waves by means of an Air-jet of a Velocity exceeding that of Sound. Phil. Mag. and J. Sci., 1931, XI, 926.
(5) HARTMANN, J. and TROLLE, B. Construction, Performance and Design of the Acoustic Air-Jet Generator. J. Sci. Instr. 4
(6) IWAMOTO, J. and WATANABE, I. Investigation Concerning a Resonance Tube. Trans. JSME, 1968, 34-262, 1094 (in Japanese).
(7) IWAMOTO, J. Oscillatory Flow and Thermal Effect in a Tapered Hartmann-Sprenger Tube. Trans. JSME, 1986, 52-482, 3422 (in Japanese).
(8) RAKOWSKY, E. L. et al Fluidic Explosive Initiator. Fluidics Quarterly, 1974, 13.
(9) WU, J. H. T. et al Experimental Investigations of a Hartmann-Sprenger Logarithmic-Spiral Tube. Proc. 7th Canadian Congress of Appl. Mech., 1979, 669.
(10) IWAMOTO, J. Necessary Conditions for Starting and Maintaining a Stable Oscillatory Flow in a Hartmann-Sprenger Tube. Flow Visualization, 1986, IV, 507.
(11) IWAMOTO, J. Experimental Study of Flow Oscillation in a Rectangular Jet-Driven Tube. Trans. ASME, 1990, 112-1, 23.
(12) HARTEMBAUM, B. Hydraulic Analog Investigation of the Resonance Tube. MSc Thesis, 1960, Department of Mechanical Engineering, MIT.
(13) IWAMOTO, J. and DECKKER, B. E. L. A study of the Hartmann-Sprenger tube using the hydraulic analogy. Experiments in Fluids, 1985, 3, 245.
(14) SAROHIA, V. and BACK, L. H. Experimental investigation of flow and heating in a resonance tube. J. Fluid. Mech., 1979, 94-4, 649.
(15) CONRAD, E. W. and PAVLI, A. J. A Resonance-Tube Igniter for Hydrogen-Oxygen Rocket Engines. NASA TM X-1460, 1967.
(16) PHILLIPS, B. R. and PAVLI, A. J. Resonance Tube Ignition of Hydrogen-Oxygen mixtures. NASA TN D-6354, 1971.
(17) ROUNTREE, M. E. and KRANC S. C. Combustion Driven Resonance Tubes. Trans. ASME, J. Basic Eng., 1972, 94-2, 509.
(18) MARCHESE, V. P. Flueric Sounding Rocket Motor Ignition. Singer Co. Report, 1971, Contract NAS-1-10680.
(19) ROSEN, S. G. Investigation of the Hartmann-Sprenger Tube for Thermal Ignition of Hydrazine-based Propellants. Euromech Colloquium 73, 1976.
(20) BROCHER, E. On the Maximum Attainable Temperature in Hartmann-Sprenger tubes. Proc. 12th in. Symp. Shock Tubes and Waves, 1979, 161.

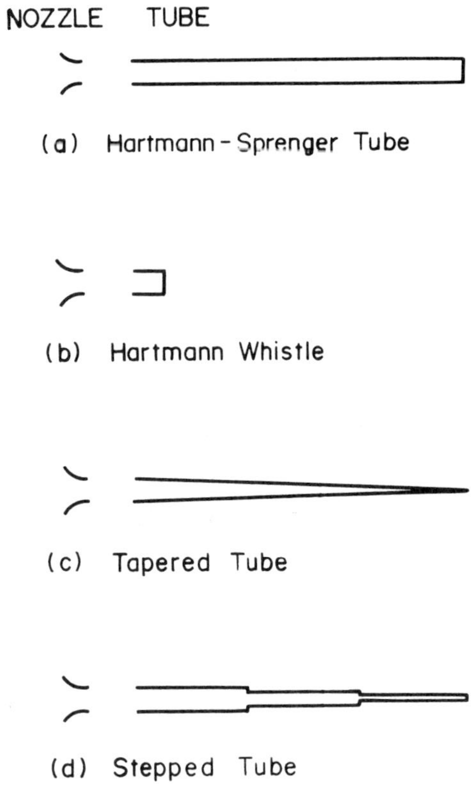

Fig 1 Hartmann-Sprenger tubes

(a) Hartmann-Sprenger Tube

(b) Hartmann Whistle

(c) Tapered Tube

(d) Stepped Tube

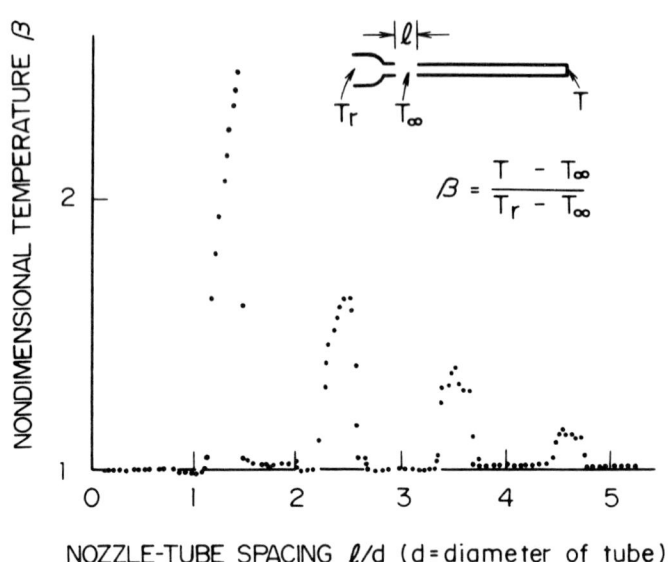

Fig 2 Relation between temperture at the closed end and nozzle-tube spacing

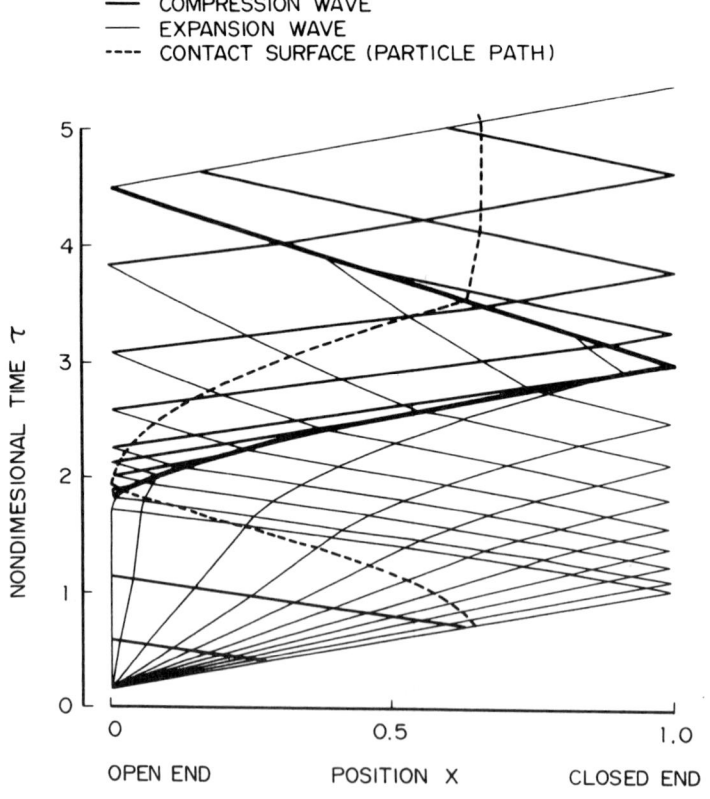

Fig 3 Wave diagram for flow in the tube

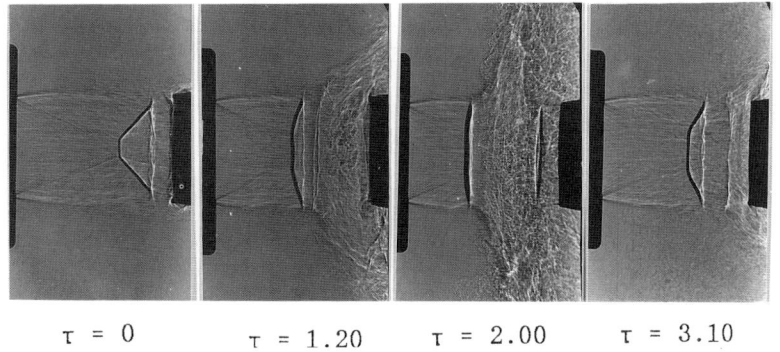

(a) SHADOWGRAPH PICTURES FOR ONE CYCLE OF OSCILLATION

(b) SHADOWGRAPH PICTURES OF STEADY FLOW PATTERN FOR THE TUBE WITH BOTH ENDS OPEN

Fig 4 Shadowgraph pictures of the flow field in the space between the nozzle and the open end of the tube

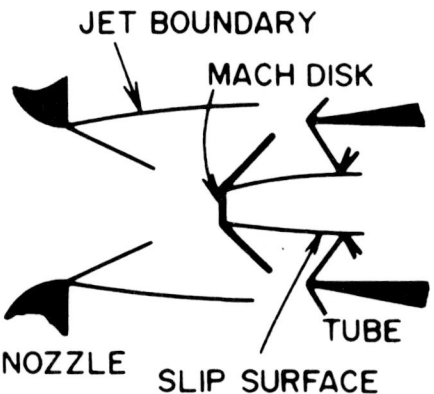

Fig 5 Sketch of the inflow pattern

C413/017

Thermal insulation for optimum design of annular cavity industrial chimneys

M M A SHAHIN, PhD, CEng, MIMechE, FESME
College of Engineering, University of Qatar

SYNOPSIS The characteristics of the major types of chimneys, are reviewed. Atmospheric, mechanical and chemical kinds of attacks on chimneys are then discussed. Next, the effect on environment and the corrosion, due to acid smut production, are examined. Following that, the factors affecting the design and performance of these chimneys, such as thermal insulation and flue gas properties, are analyzed. The effects of separation dimension of different air-filled vertical annular cavities, on maximum thermal resistance, hence, minimum heat transfer across the cavity, were studied. A sophisticated test rig was built and an extensive test programme was carried out. The results appear to indicate that the use of a 17 mm air-filled annular cavity gave the optimum thermal insulation. However, further measurements are needed to reveal the dependency of this value upon other factors, such as temperature, diameter and height of chimney.

1 INTRODUCTION

The main objective, when designing an industrial chimney, is to produce the draught required to disperse the exhaust gases into the atmosphere with the least cross-sectional area and at the minimum cost. This involves the development of forced or induced furnace systems for which natural draught is not required.

Depending on the fuel used and the type of combustion, considerable damage to life and property in the neighborhood of the chimney can occur as a result of the acidic products of combustion and other harmful materials being deposited. There has, therefore, been a shift in the emphasis during the designing procedure. While the draught criteria is still a major consideration, greater emphasis is now being placed on reducing pollution.

With recent trend, from coal-fired boilers to more efficient oil-fired ones, which are often used in conjunction with steel chimneys, problems such as acid smut production and the following corrosion have increased. The work described in this paper is, thus, to review and analyze the factors affecting the previously mentioned problems, with a view to trying to define the ideal thermal insulation conditions, for optimum design.

Despite the extensive work done on heat transfer across vertical rectangular air-filled cavities [1], not much is known about the case of vertical annular air-filled cavities. The paper published by Thomas and V. Davis [2] is, therefore, of great interest. The paper describes a numerical study of vertical annular cavities, having their inner surfaces at higher temperatures than the outer surfaces, similar to that for a chimney where the flue gas passes through the inner flue.

The heat fluxes, across the air-filled annular cavity, consist of conductive, convective and radiative components. There are several methods for obtaining the values of these components, hence, the total value of the heat fluxes transferred across the cavity. However, this requires the knowledge of the emissivities of the surfaces, the thermal contact resistance between such surfaces, and the dependence of these factors on temperature difference. In such cases, assumed values of emissivities and thermal contact resistance are generally employed. Consequently, the validity of results are somewhat suspect. This is further complicated by the variation in values during operation. An experimental approach is, therefore, recommended. The objective of this test programme is to determine the optimal separation dimension of an air-filled annular cavity to provide maximum heat resistance and hence minimum heat transfer, in order to reduce the acid smut production and the subsequent corrosion due to the following chemical reactions.

2 ENVIRONMENTAL POLLUTION

The increasing efficiency in combustion processes have resulted in lowering the flue gas temperature. Fuels, including coal, oil and gas, may contain as much as 4% sulphur, which oxidizes during combustion to SO_2. A smaller percentage further oxidizes, in the presence of excess air, to SO_3. If these gases come into contact with a surface at a temperature between that of the water vapour and the acid dewpoint (at which condensation occurs) a certain film of concentrated sulphuric acid results. Whilst sulphuric acid dewpoint is about 132°C, the maximum corrosion rate occurs at about 110°C [3].

Acid dewpoint, hence acid condensation, depends mostly on the quantity of the excess air present, the temperature of flue gas, and the sulphuric content. After condensation, the condensed water evaporates and the sulphuric acid dries solid in particles. These particles are carried out of the chimney in the flue stream, in the form of acid smuts, then fall around the chimney environment. When this gets moistened, it leaves brown stains, and causes damage to paint-works, vehicle bodies, clothing, etc. The distribution of these solid particles depends on atmospheric conditions such as wind velocity and direction, flue gas velocity, excess temperature over that of the surrounding atmosphere, and chimney height. The latter has, recently, been reduced due to the use of artificial draughts. Thus, the chances of more deposition of flue dust on chimney neighbourhood are enhanced.

3 TYPES OF CHIMNEYS

The majority of chimneys in current use are either brickwork, concrete or steel. Each of these chimneys has its unique features. However, all are subject to three types of attacks; atmospheric, mechanical and chemical.

In the brick chimneys, the brick thermal resistance is high. Accordingly, the temperature of the flue gas is also high so that it does not fall below the acid dewpoint. To protect the main brick structure from this high temperature, it is common to line the inner surface with a sleeve of refractory brickwork. Due also to this high thermal resistance of brick, a high temperature difference between the inner and outer surfaces exists and different expansions occur, giving rise to stresses that can lead to cracks in the shell. The refractory lining is also susceptible to erosion by grit particles carried out by the stream of the flue gas. However, a low rate of this type of attack upon the brick structure can be obtained by using high density and low porosity bricks.

Concrete chimneys are subjected to similar thermal and chemical attacks, as in brick chimneys. There are two types of these chimneys; monolithic and precast chimneys. The surface of the monolithic is often crazed with minute hair cracks. If moisture gets in as far as the reinforcing steel bars, it will cause the steel to corrode. The precast concrete chimneys suffer also from the same type of attack, but in a different way. The wind causes the chimney to bend at the joints. Consequently, hair cracks appear and the steel reinforcement may also be attacked.

Steel chimneys are usually of riveted plate construction. In this form, they are particularly prone to abrasion by grit particles which are common in the flue gas, emerging from the oil-fired boiler installations. Frequently, rivet heads on the inner walls of the chimney become completely eroded. Complete erosion of steel chimney shells by grit is particularly rapid due to the relatively thin walls required in the chimneys of this type. Although abrasion by grit particles has disappeared with the introduction of oil-fired economic boilers, other serious problems have resulted such as, emission of acid smuts and reduction of chimney life by acid attacks.

Bare steel surfaces corrode rapidly when exposed to atmosphere, and mechanical properties are consequently deteriorated. Wind loading does not usually cause serious problems to steel chimneys, as steel structures have an inherent elasticity, and small deflections can be accommodated. However, tall and self-supported steel chimneys may be subjected to wind-excited oscillations. These are produced by Von Karman vortices from each side of chimney. A solution for this problem was produced in the form of helical stacks, wound round the upper part of the chimney. The regular formation of vortices is, therefore, offset, and a small damping force is also provided.

Performance of chimneys is dependent, in general, on design and operation criteria. These criteria are: least acid condensation, least smut emission and, finally, least 'downwash' which is sometimes observed when exhaust gases leave the chimney and are caused by wind to swirl down the upper part of the stack on the leeward side. The main factors affecting these criteria are flue gas properties (velocity, temperature and composition) thermal insulation, characteristics of wind and fuel being burned.

4 PROPERTIES OF FLUE GAS

Flue gas velocity is a crucial parameter governing the performance of chimneys, especially the steel ones. Etoc and Gills [4] showed that the overall thermal resistance of a chimney decreases with the increase in flue gas velocity, as shown in Fig.1, and that the gas outlet temperature increases with higher mass flow, as appears in Fig.2. In general, the flue gas velocity is a variable quantity, and the turn-down ratio is the governing factor for the amount of that variation. The turn-down ratio is defined as the ratio of the mass of the steam produced in unit time by the boiler, when on maximum load, to that produced in unit time, when on minimum load. To reduce the problem of turn-down ratios, the modern trend is to employ one flue for each boiler, and recent designs have accommodated that by allowing several flues to be incorporated in one chimney.

Work by Beaumont et al. [5] has yielded results, as given in Fig.3, by which the required overall thermal resistance of a chimney of given dimensions can be determined, in order to avoid acid condensation, for any assumed minimum flue gas entry velocity. They also derived a relation which gives the flue gas temperature at any point in the chimney, to avoid acid condensation at the top.

5 INSULATION OF CHIMNEYS

Modern economic boilers have been designed to keep the temperature of the exhaust gases at relatively low value. To counter the problem of smut emission and the subsequent corrosion, it is necessary to retain as much of the heat as possible in the flue gas during its passage through the chimney so that the temperature does not fall at any point below the dewpoint. One way of satisfying this condition is by thermally insulating the chimney surface, thus, keeping the temperature of the flue gas high enough, to prevent the condensation of acid on the inner surface of the chimney.

Two methods for thermal insulation of chimneys are in common use. A combination of the two of them is recommended for better performance. These methods are:

(a) Cladding the chimney with a high thermal resistance material such as mineral wool applied to the outside surface of the chimney shell in strips of appropriate thickness. This is then covered with a skin of 16 gauge bright aluminium sheet. This method is known as material insulation.

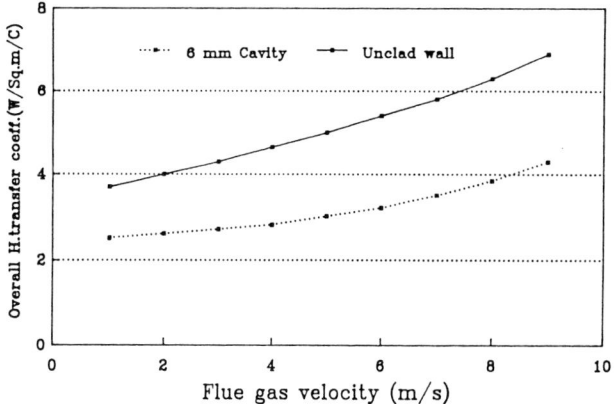

Fig.1 Effect of gas vel. on heat Coeff.

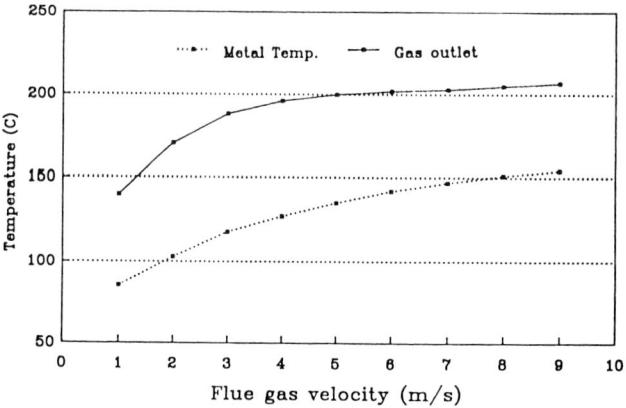

Fig.2 Effect of gas vel. on temperatures

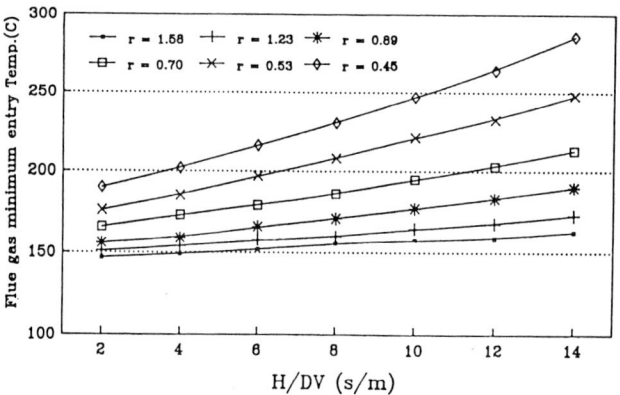

Fig.3 Gas Temp. and acid condensation

(b) Cladding the chimney with bright aluminium sheet as before, with an annular air-filled gap of 6 mm between that sheet and the steel shell. The trapped air inside that gap acts as an insulant. Its effectiveness depends on the heat transfer properties and the physical dimensions of the cavity. These dimensions are critical, as greater air space increases convection heat losses, whilst smaller air space increases conduction heat losses. This method of insulation, known as aluminium clad annular cavity, is generally accepted as one of the simplest, most efficient and durable form of thermal insulation to chimneys.

The second method (aluminium clad annular cavity) can be used in installations having flue gas temperatures exceeding 260°C, without the risk of smut emission and corrosion. But with lower flue gas temperatures, it becomes necessary to have a layer of insulating material between the aluminium cladding and the steel flue. This conclusion was drawn on the basis of experimenting with aluminium cladding insulation, keeping an air gap of 6 mm [6], as recommended in common practice. However, there is a strong doubt whether this 6 mm air gap provides the optimal separation for maximum thermal resistance. This is not a surprise judgement after the recent findings of the optimal separation for the rectangular cavities [7] which appreciably differs from that annular ones. Also, heat transfer measurements by Etoc and Gills again [4] for 6 mm and 20 mm annular cavity chimneys show that heat losses for the 20 mm air gap are 15 to 20% less than those measured for the 6 mm air gap, irrespective of flue gas velocity. The 20 mm air gap was arbitrarily chosen, and there was no suggestion that it represents the optimal separation value.

6 EXPERIMENTAL WORK

Different air gap thicknesses were studied in the experimental programme described in this paper. The air gap height was maintained constant, hence, different aspect ratios were examined. An outer cold cylinder was used throughout the test programme, and to vary the annular air gap thickness, an inner hot cylinder was changed for each test with different outer diameter.

6.1 Test Apparatus

The experimental rig, as shown in Fig.4, consists of two main sections, the pumping system and the test section. The pumping system is used to create a vacuum down to 10^{-3} torr in the test chamber, using a diffusion pump backed by a matched rotary pump and connected to it through a rubber tube to inhibit any vibration transmission. This pumping system is also used to introduce compressed air for leak testing. The test section consists of outer cold cylinder, inner hot cylinder, support table, and heating element. The outer cold cylinder is made of a mild steel pipe with a fixed 101.6 mm bore diameter and 114.3 mm outer diameter. The outer surface of this cylinder is wound spirally with a 3/8" copper tubing. There are a total of 17 turns of cooling pipe. They are held in position by brazing, so that a good thermal contact is achieved. Near the middle of the cylinder, a small hole is drilled and a copper pipe is also brazed to it, to take the gauge connections and the thermocouple lead to the inner hot cylinder. Copper constantine thermocouples are also used to measure the water inlet and outlet temperatures. To inhibit any thermal leak, the surface of this outer cold cylinder is wrapped with asbestos, while the bottom flange and the disc over the support table are insulated by glass wool. The inner hot cylinders are made of mild steel pipes of suitable outside diameters, to provide different air gap cavities. One end of each of these pipes is sealed, by brazing, with a circular mild steel disc of the same size as the inner diameter of the pipe. The other end is brazed with an annular disc with an inner diameter of the same size as the outer diameter, making a flanged end. A rectangular section groove is cut in the lower side of the flange to locate the vacuum seal. Thermocouple wires of copper-constantine are led through the vacuum chamber, and are of sufficient lengths to connect them, in five different positions at equal intervals along the length and at 90° angular displacement. The support table is made

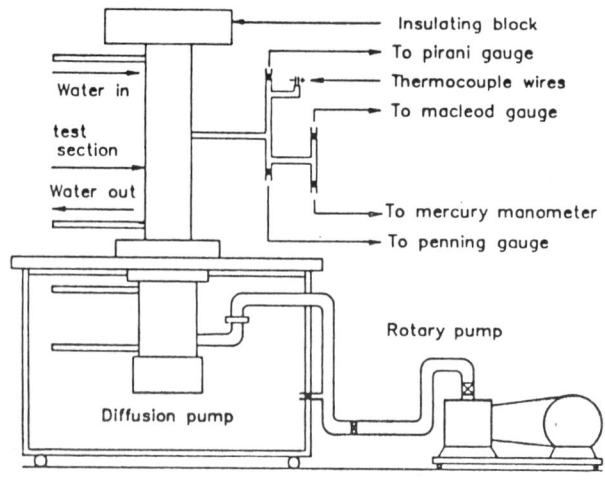

Fig.4 Schematic layout of test apparatus

of a wooden plank with a circular hole in the middle. The whole test assembly is placed vertically on top of this support table, with the bottom disc resting on the wooden surface and fixed in position with three clamps. The heating element is made of Ni-Cr wire, sheathed in a copper tube. A total length of 6 meters of this element is wound on another copper tube of 19 mm diameter and 445 mm length. To attain an isothermal condition along the whole length of the tube, more number of turns of the heating element per unit length are used at the bottom region of the tube. To restrict also the heat flow from bottom to top, in the high effective conductivity liquid surrounding the heating element, four brass annular rings are fixed at different heights, to divide the whole length into five sections. A stabilized d.c. power supply, with built in potentiometer, voltmeter and ammeter for control, is used for providing the power needed for heating.

6.2 Test Procedure

The inner hot cylinder is placed inside the cold outer one, and the heating oil is poured in position and covered. The system is then allowed to attain a steady state, before readings are taken. This is done by continuously heating the inner cylinder and cooling the outer one for about 10-12 hours, so that no further changes would occur while experimenting. After the experiment at atmospheric pressure is completed, the oil diffusion pump is switched on, and its cooling water is opened. Sets of experimental data, at several environmental pressures down to 10^{-3} torr are then taken. This process is repeated with all inner hot cylinders which provide the different air gap cavities. In each case, the steady state rate of heat transfer across the air filled annular cavity is calculated as the product of the flow rate of the cooling water, its specific heat, and the temperature rise as a result of its passage through the welded pipe to the outer cylinder.

7 RESULTS AND DISCUSSION

Fig.5 gives the total steady state rate of the heat transfer across the air-filled gap versus the annular cavity separation dimension for different environmental pressures. From this figure, it appears that the optimum separation dimension for annular cavity wall chimneys with average temperature about 125°C and temperature difference across the annular about 200°C, is at a cavity width of 17 mm rather than 6 mm as recommended in common practice. At this separation dimension, the heat resistance is maximum and the heat transfer across the cavity is minimum.

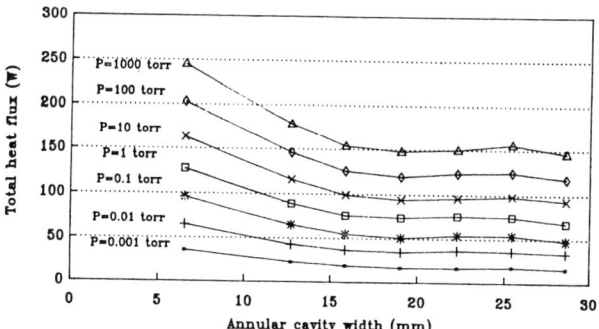

Fig.5 Effect of cavity width on H. flux

8 CONCLUSIONS

From the work described in this paper, the following points can be concluded:

(a) The majority of chimneys, in current use, are either brickwork, concrete or steel chimneys. Each of these chimneys has its unique features. However, all are subjected to three different types of attacks; atmospheric, mechanical and chemical.

(b) Operation criteria upon which the performance of industrial chimneys is judged are: least acid condensation, least smut emission and least 'downwash'. The factors affecting these operation criteria are: flue gas properties, thermal insulation, characteristics of wind and fuel being burned.

(c) The effect of flue gas velocity is crucial on performance of chimneys, especially the steel ones. It decreases the overall thermal resistance of chimney and increases the gas outlet temperature with the increase in velocity.

(d) Aluminium clad annular cavities can be used in installations having flue gas temperature exceeding 260°C, but with lower temperature, it is necessary to have a layer of insulating material between the cladding and the steel flue.

(e) The 6 mm annular air gap, recommended in common practice, is apparently misleading and it is better, from a thermal insulation point of view, to use a separation of nearer 17 mm. However, further measurements are needed to reveal the dependency of the optimum separation dimension upon other factors, such as temperature, diameter and height of chimney.

REFERENCES

(1) BROOKS, R. and PROBERT, S., An Interferometric Study of Heat Transfer Across a Rectangular Cavity, J. Mech. Eng. Soc. Vol. 14, No. 2, 1972.

(2) THOMAS, R. and DAVIS, D., Natural Convection in Annular and Rectangular Cavities - a Numerical Study, 4th Int. Heat Transfer Conference, Paris, Versailles, 1970.

(3) BOSANQUET et al., Dust Deposition from Chimney Stack, Proc. Ins. of Mech. Eng., Vol.162, No. 3, pp 335, 1950.

(4) ETOC, P. and GILLS, B., Design Performance of Aluminium Clad Chimneys, Journal of Inst. of Fuel Vol. 42, pp 104, 1969.

(5) BEAUMONT et al., Comparative Observations on the Performance of Three Steel Chimneys, IHVE Journal, pp 345-351, 1970.

(6) GUNLIMEDE, Aluminium Insulation Cladding of Chimneys, Oil Firing, Vol. 7, pp 34, April 1969.

(7) ROBERT et al., Heat Transfer Across Rectangular Cavities, CPE Heat Transfer Survey, pp 35-42, 1970.

C413/028

The design and testing of heat exchangers for a Stirling cycle heat pump

D H RIX, BSc, PhD, CEng, MIMechE
Department of Engineering Science, University of Cambridge

SYNOPSIS A theoretical model of the Stirling cycle heat pump is outlined. Its application to the design optimisation of new heat exchangers for an existing machine is described, with regard to such features as tube length, diameter and number. The resultant heat exchangers were then manufactured. Initial testing has shown a favourable improvement in heat pump COP and good agreement between predicted and measured performance parameters.

1 INTRODUCTION

Considerable efforts have been devoted to the development of Stirling engines and reversed Stirling cycle cryocoolers, but by comparison, the reversed Stirling cycle heat pump has received scant attention. The full title is stressed in order to distinguish it from the more common arrangement of a Stirling engine driving a Rankine cycle heat pump. This is something which is currently undergoing intensive development in Japan and elsewhere. In the case of the Stirling cycle heat pump, no particular prime mover is implied. The machine discussed below was driven by an electric motor, whilst Sunpower and Kawasaki (1,2) the previous main workers in this field, used a Stirling engine.

Some mention needs to be made of applications, in order to understand what follows. The Sunpower/Kawasaki work and a recent detailed theoretical study by Lundqvist (3) have both been centred on the use of such a heat pump for domestic space heating and hot water supply. Though the domestic market, by its very size, must be of considerable interest, it has always been considered that in terms of efficiency, or COP, the Stirling cycle heat pump would be inferior to Rankine cycle machines. Therefore it is significant that Lundqvist calculates annually averaged COP values for the Stockholm climate as close as 2.3 and 2.1, for Rankine and Stirling type machines respectively.

This author has argued (4) that a more favourable application might be the industrial waste heat recovery market. The temperatures involved are generally too great for the conventional type of working fluids used by Rankine cycle machines i.e. above 100°C. Further, relatively large temperature lifts are required i.e. around 100°C, though Lundqvist suggests that even in the (Scandinavian) domestic heating market this is a problem. Recent moves to phase out CFC working fluids may be another reason to give more consideration to the Stirling cycle heat pump.

As part of a programme of work to demonstrate the potential of the "high temperature" Stirling cycle heat pump, a small scale prototype was constructed by this author (4). A notable feature of the design process was the use of a theoretical model to optimise the design of the heat exchangers. This was something which posed a problem due to the lack of established design data and practice. This model has subsequently been considerably further developed. It is significant for its use of a full three conservation law treatment of the working fluid behaviour and will be described in more detail.

Tests with the resultant heat pump demonstrated very poor values of COP, of around one, but they did demonstrate its ability to lift heat over large temperature differences - over 200°C. There were considered to be several factors to explain this poor performance -

1. The design was optimised for maximum heat output rather than COP. It will be shown later that these are often in an inverse relationship when a particular parameter is varied.
2. No account was taken of external heat transfer coefficients (oil-to-metal). Conventional wisdom has it that the limiting factor is the internal (gas-to-metal) heat transfer coefficient. Subsequent analysis showed that this was untrue.
3. The mechanical efficiency of the heat pump was very low (less than 50%) due to the novel type of crankdrive chosen and the disproportionately high seal and piston ring losses on a machine of small heat output.

On a more positive note, it was possible to extend the theoretical model to include external heat transfer and to verify it against the experimental results actually obtained. Use was also made of this model for a theoretical investigation into the potential of the Stirling cycle heat pump (4) - in terms of just what values of COP and specific output might be realised in a fully optimised machine. The results looked promising - for a temperature lift of 100°C, a COP of about 3.5 and a specific output approaching 1J/cycle cm^3 of piston displacement. The main objective of this paper is to report subsequent progress, since further experimental work was obviously required to substantiate such findings. Given the limited financial resources available, it was considered that work should be phased in two stages -

1. New heat exchangers for the existing machine.

2. A new crankdrive arrangement with better optimised parameters - piston phase angle, internal dead volume, dead volume distribution etc.

This paper deals with the first stage.

2 HEAT PUMP SYSTEM SPECIFICATION

The SM3 heat pump is illustrated in Fig. 1. It consists of expansion and compression spaces linked by associated heat exchangers and a regenerator. Volume variations in the two spaces are produced by a working piston in each - the so called "alpha" configuration. The pistons are actuated by a Ross crankdrive mechanism which produces a phase difference between the two pistons of about 90 degs. The working fluid is so by expanded in the expansion space, with a net take-up of heat and transferred through the regenerator, where heat is added, to the compression space. Here compression produces a net outflow of heat. Finally the working fluid is transferred back to the expansion space, with a removal of heat as it passes through the regenerator.

To minimise parasitic heat transfer losses, from hot to cold end of the machine, thin-wall construction is employed for the cylinder walls etc., and tall domes are fitted to the pistons. To minimise the crankdrive loading, the crankcase is pressurised to the mean working pressure of the machine - to form a "buffer" space and a standard pressure balanced seal is provided on the crankshaft for sealing. A high mean working pressure is needed to maximise the mass and hence heat transfer on each cycle. A brief specification of the machine is given below.

Configuration - 2 cylinder - alpha
Bore - 54 mm
Stroke - 25.4/25.7 mm
Maximum speed - 1900r/min
Maximum mean pressure - 5.0 MPa
Working fluid - helium.

The heat pump exchanges heat with a test system which comprises two hot oil loops - Fig.1. One of these, which exchanges 'heat' with the expansion space heat exchanger, represents a low temperature heat source - typically waste heat. The other, which exchanges heat with the compression space heat exchanger, represents a high temperature heat sink - for the useful heat output of the machine. Electric heating is provided for the oil in both loops, with a heat exchanger to dissipate surplus heat. The temperature of both may be set at any required value up to a maximum of 250°C.

3 THEORETICAL MODEL

3.1 Discretisation method

As shown in Fig. 2, the working fluid contained within the heat pump - the expansion and compression spaces, heat exchangers and regenerator, is divided into a system of discrete sub-masses. The average value of the gas properties within each sub-mass - density and temperature, is considered to apply, whilst velocity is defined at the sub-mass boundaries. The three one-dimensional conservation equations - mass, momentum and energy are applied to each sub-mass and are repeatedly solved at incremental advances in time (or crankangle) to yield new values of density, temperature and velocity. Details of the equations are given in reference (5).

Previous theoretical models of the Stirling engine have tended to use a fixed discretisation 'framework' i.e. the gas flows through the framework, as distinct from the Langrangian reference system used here, which moves with the gas particles. This preference has been due to the ease with which the various changes in cross-sectional flow area can be accommodated. However this potentially conflicts with the numerical methods commonly employed to solve the conservations equations. In simple terms this translates into temperature and pressure waves travelling through the machine at a speed set by the incremental time step used, rather than molecular and acoustic speed respectively. In Stirling cycle machines it is the former which seems to be particularly important. The Lagrangian method does have the problem of changes in the sub-mass cross-sectional area as it crosses area transitions. This is dealt with by finding the exact moment at which it occurs and treating this as an infinitely short time increment in which local values of temperature, density and velocity are unchanged. These can then be re-allocated to the new pattern of sub-masses.

3.2 Wall temperatures

It is assumed that the compression and expansion space walls and their associated heat exchangers are at the mean external heat transfer media temperature i.e. oil - Fig. 2.

In the case of the regenerator, it has been found that a rather more rigorous treatment is necessary and the matrix temperature distribution is derived by considering the heat flows to and from it. Again the use of a Lagrangian reference system creates a difficulty in that the regenerator sections, with which the sub-masses exchange heat, are continually changing. The regenerator matrix temperature distribution is therefore represented in a continuous fashion by a cubic spline function - Fig. 2. As in a real machine, this does mean that a steady state temperature distribution is not achieved until a relatively large number of crankshaft revolutions have been completed.

3.3 Heat transfer and friction factor correlations

Since for reasons of practicality, only a one-dimensional formulation of the continuity equations is used, it is necessary to use heat transfer and friction factor correlations in terms of non-dimensional terms such as Reynolds number. The established correlations were all derived under steady flow conditions, but little evidence has ever emerged to suggest that these might be inappropriate under the oscillating flow conditions existing at the internal gas/wall interfaces. As already noted, the external heat transfer media/wall interface, where steady flow conditions do exist, may prove equally important.

In the case of the friction factor, the only reservation is that this author has shown that steady flow correlations may break down when air or another high molecular weight working fluid is used (6).

3.4 Solution of the conservation equations

Conventional solution techniques require that the conservation equations are made linear i.e. terms such as x^a are converted to a linear form bx. The technique used is to re-express the variables in terms of the difference in their values between the new" and old' time steps. Then

$$(x")^a = (x' + \Delta x)^a$$
$$= 1/x' (1 + \Delta x/x')^a$$
$$= 1/x' (1 + b\Delta x/x' + \text{other terms which may be neglected.}$$

3.5 Calculation of heat pump performance

At each time step the various heat transfers to and from each sub-mass may be computed and by addition the net heat transfers in the expansion and compression spaces found. Likewise the work done by, or done on, the pistons may be calculated.

3.6 Additional losses

In a practical machine, there are a number of additional losses which are not covered by the formulation of the conservation equations used here. These include

1. longitudinal/radial heat transfer by solid wall conduction and surface convection
2. mechanical losses
3. piston ring leakage.

Parasitic heat transfer has been evaluated experimentally and found to be small. As such it will have minimal influence on thermodynamic design optimisation, but it is subtracted when predicted heat transfers are compared to those measured. Similar comments apply to mechanical losses, which only need to be considered when comparing net COP values.

Piston ring leakage is a potentially more serious problem, since it will affect the mass of gas flowing through the heat exchangers. The author has so far only studied the problem by means of a simplified theoretical model, with data gleaned from the literature. The results do look to be significant. None the less, as will be shown, measured and predicted results do seem to agree reasonably well, though there is room for improvement.

4 HEAT EXCHANGER OPTIMISATION AND DESIGN

The original heat exchangers were of multi-tubular design, with gas on the inside and oil flowing over the outside - Fig. 3. It has already been noted that these suffered from poor heat transfer between the oil and tubes. The compression space heat exchanger, which utilised curved tubes, was particularly bad due to -

1. Low oil velocity over the tubes.
2. Oil by-passing the tubes - some tubes were badly formed with no clearance around them.
3. Thick walled stainless steel tubes were used without finning - the choice of wall thickness was dictated by availability.

Experimental measurements showed that the resultant oil-to-tube heat transfer coefficients were often considerably less than 400W/m^{2}K.

Based on Kays and London (7) data, it appeared that this could be readily increased to over 1000 W/m^{2}K, so as to be comparable to that on the gas side. This would involve the use of smaller diameter finned copper tubes, multiple passes of oil over the tubes and improved baffling of the oil flow. At this juncture it became clear that the existing crankdrive arrangement with parallel cylinders - Fig. 1, was particularly unsuitable. This is due to the need to curve the heat exchanger tubes, to get from one cylinder to the other. This makes it difficult to fabricate tubes with fins and cross-baffles. The poor mechanical efficiency of the Ross crankdrive system has already been noted and previous work (4) suggested that the piston phase angle, which was fixed, was an important variable in optimising COP. Given that the resources were not available to construct a completely new machine, a compromise solution was to construct a pair of new heat exchangers for the existing machine, along the lines suggested above. If these were to have straight tubes, a semi-circular connecting duct would be needed to connect them. Such a duct was not without merit, since it was known that dead volume should be increased in order to increase COP, though it would be better to have the extra dead volume as clearance in the cylinders. In addition, it appeared that constructing these heat exchangers would present such difficulties, with the brazing of so many small tubes, fins and baffles, that they should be tackled first.

For thermodynamic design optimisation purposes, the following operating parameters were chosen -

Heat delivery temperature	200°C
Heat source temperature	100°C
Operating speed	950 r/min
Mean pressure	2.5 MPa
Working fluid	Helium
Oil - tube heat transfer coefficient	
- Compression space	1800W/m^{2}K
- Expansion space	1000W/m^{2}K.

The source and sink temperatures were chosen as typical of what might be required in an industrial waste heat recovery scheme, whilst speed and pressure were chosen at somewhat less than those previously used, in the interests of crankshaft seal reliability.

The variables which were investigated for optimisation were -

1. Length, diameter and number of the expansion space heat exchanger tubes.
2. The same for the compression space heat exchanger.
3. Length and packing (wire gauze thickness and pitch) of the regenerator.
4. Internal dead volume and its distribution between the expansion and compression sides of the machine.

In the case of the heat exchanger tubes, the optimum diameter was limited by considerations of the smallest practical size. In the event, a smaller tube diameter than that used before was specified - 3 mm compared to 4.8 mm. Fig. 4 shows sample design optimisation curves for the length and number of compression space heat exchanger tubes. It can be seen that in the case of length, whereas heat output and COP vary inversely, in the case of tube number they vary in proportion. Tube number would therefore seem to be a particularly important variable, in that it allows both these quantities to be maximised simultaneously. A similar effect was noted with the expansion space heat exchanger.

Fig. 5 shows further design optimisation curves for the regenerator. It is surprising that the regenerator design is not more critical than these curves appear to show. Not only is the regenerator the major source of pressure drop in the machine, but typically it will transfer several times more heat per cycle than the tubular heat exchangers do. Once again there is a small "trade-off" between heat output and COP as the length is varied, but it would seem that different wire meshes have surprisingly little effect. Wire diameters were chosen so as to keep porosity approximately constant.

Finally Fig. 6 shows design optimisation curves for increasing dead volume, expressed as a fraction of the original machine's dead volume. Some surprise has been expressed at the way COP significantly improves with increasing dead volume, albeit with a loss of heat output. Perhaps this surprise stems from Stirling engine design practice, where strenuous efforts are usually made to reduce dead volume to the minimum possible. So far as the heat pump is concerned, the explanation for this beneficial effect would seem to lie in the fact that as dead volume increases, the pressure and temperature swings within the machine decrease. Consequently the machine is operating much closer to the ideal isothermal case, which corresponds to the highest attainable value of COP.

Fig. 6 also shows that there is a further benefit in concentrating the dead volume in the hot compression side of the machine. This is evidently due to the fact that a smaller fraction of the working fluid mass is required to fill the space, leaving a relatively larger fraction to undergo the expansion and compression processes. It should be noted that, for simplicity, the theoretical model considers all dead volume to be lumped with either the expansion or compression space, whereas in reality some will lie between the regenerator and heat exchangers.

5 HEAT EXCHANGER DETAIL DESIGN

In the detail design of the new heat exchangers, the main innovation was the use of finned copper tubes - Fig. 3. Though some design features such as the bolted joints were carried over from the previous design and therefore required no fresh stress analysis, the strength of these tubes proved fairly critical at the required design pressure and temperature. That is, in meeting the stress criteria of BS 5500. Special tubing was drawn in CN 107 copper alloy (70% copper, 30% nickel). This was further strengthened by the use of 0.4 mm copper sheet fins, precision reamed for fit and brazed to the tubes.

Each heat exchanger was vacuum brazed as an assembly complete with stainless steel end sections. Braze paste was used. The components were assembled on a stainless steel jig, which has a similar coefficient of expansion to copper, to avoid any differential expansion problems in the brazing furnace. Brazing the fins appeared to pose the greatest problem, but this was solved by assembling the fins with spacers between them. It was found that once the brazing paste was dry, it held the fins sufficiently well to withdraw the spacers. The assembly was subsequently brazed horizontally to minimise the chances of the fins moving.

After pressure testing, a thin stainless steel sheet metal jacket was welded around the outside of the tubes for the oil to circulate through. Again the similar coefficient of expansion to copper avoided any problems with differential expansion. The heat exchanger design also included tappings for pressure transducers, pressure gauges and thermocouples.

6 REGENERATOR DETAIL DESIGN

As in the previous machine, the stainless steel regenerator 'can' was designed to double up as the pressure vessel, rather than mounting it inside a separate one. This meant that parasitic heat transfer area could be reduced as much as possible. The packing was formed from stacked discs of stainless steel woven wire gauze. These were precision cut with a die set, to avoid problems with leakage around the edges. After packing the regenerator, the entire assembly was 'sintered' solid in a vacuum furnace. This was done to avoid previous problems with screen movement and leakage. Though from the point of view of axial heat transfer loss, it would be preferable not to sinter the screens, previous measurements (4) had shown that the heat flow was still very small. No account is taken of it in the theoretical model.

7 EXPERIMENTAL MEASUREMENTS

At the present time it has not proved possible to do measurements other than at the design point for the machine, the conditions of which were specified above. These have comprised -

1. Pressure-time variation in the expansion and compression spaces. Near flush-mounted piezo electric type pressure transducers were used.
2. Heat transfer to and from the expansion and compression space heat exchangers respectively. The rest of the expansion and compression space surfaces were well lagged. These values were calculated from measurement of the volumetric flow rate and temperature variation of the oil flowing over the heat exchanger tubes. Though such measurements are generally considered unreliable, by careful calibration of the instrumentation, it proved possible to reduce unaccounted for heat losses to within 8% of the total energy input to the machine.

This is illustrated in Fig. 7, where experimentally measured and predicted heat flows are shown graphically. Examination of the former illustrates the poor mechanical efficiency of this particular machine, since a large proportion of the shaft power input to the machine is dissipated as losses. These comprise the sum of the measured heat flows to the cylinders and main gas seal coolants and so represent piston, bearing, and seal losses. Previous work had indicated that static heat conduction into the cylinder coolant would only be a few per cent of the total. Notwithstanding, comparison of the measured heat flows to and from the expansion and compression spaces does show reasonably good agreement. Certainly the two sets of figures are in the correct relationship to each other, given the various minor losses which are not included in the theoretical model. In terms of COP value, it is only realistic to consider the gross or indicated value, given the disproportionately high losses

i.e. $COP = Q_c / (Q_c - Q_e)$.

The figures shown in Fig. 7 give COP values of

2.8 (measured) and 2.9 (predicted) and as such represent a considerable improvement on the previous machine's performance, where a COP of about 1.0 was achieved.

Good agreement, to within better than 10% was also seen in comparisons of the measured and predicted pressure-time variations.

8 CONCLUSIONS

Details of a theoretical model for use in the thermodynamic design optimisation of Stirling cycle machines have been presented and its value in a practical heat exchanger design problem has been demonstrated. Taken together with previous work (4) this is now the second time that this theoretical model has been shown capable of making predictions which agree reasonably well with results measured experimentally. Though much more testing remains to be done, the results presented here must increase confidence that this is not a mere coincidence, but rather attributable to the basic correctness of the model's formulation. This is particularly so with regard to the use of a moving, or Langrangian, discretisation scheme. Further development is still needed if better agreement between predicted and measured results is to be achieved. In particular account needs to be taken of the various minor losses, such as piston ring leakage.

The results obtained also show that the reasons for the poor performance of the original design were correctly diagnosed and this was not an inherent defect of the Stirling cycle heat pump i.e. measured indicated COP was increased from 1.0 to 2.8. Though the latter value is not likely to be high enough to generate any commercial interest, there is scope for further improvement. Variables such as piston phase angle, dead volume distribution, swept volume ratio between the expansion and compression spaces, regenerator length to diameter ratio etc. are still to be optimised in the next phase of this work. Now that gas side heat transfer is the limiting factor, increases in this may be possible by the use of internally finned tubes or adaptation of 'printed circuit' heat exchanger design.

REFERENCES

(1) BEALE, W., et al. Duplex Stirling heat pump development. International Conference on Gas Research, London, 1983.

(2) INABE, M., et al. The development of free piston Stirling heat pumps. Proceedings of Third International Conference on Stirling engines, Rome, 1986.

(3) LUNDQVIST, P. G. Stirling cycle heat pump - an optimisation study. Advances in Heat Pumps - 1989, ASME, AES-Vol. 7, HTD-Vol. 125.

(4) RIX, D. H. The potential of the Stirling cycle heat pump, Proc. Instn Mech. Engrs, Part A, Vol. 203, 1989.

(5) RIX, D. H. A thermodynamic design simulation for Stirling cycle machines using a Lagrangian formulation. Proc. Instn Mech. Engrs, Part C, 1988, 202 (C2), 85-93.

(6) RIX, D. H. An enquiry into gas process asymmetry in Stirling cycle machines, PhD thesis, 1984, University of Cambridge.

(7) KAYS, W. M., LONDON, A. L. Compact heat exchangers, McGraw-Hill, 1955.

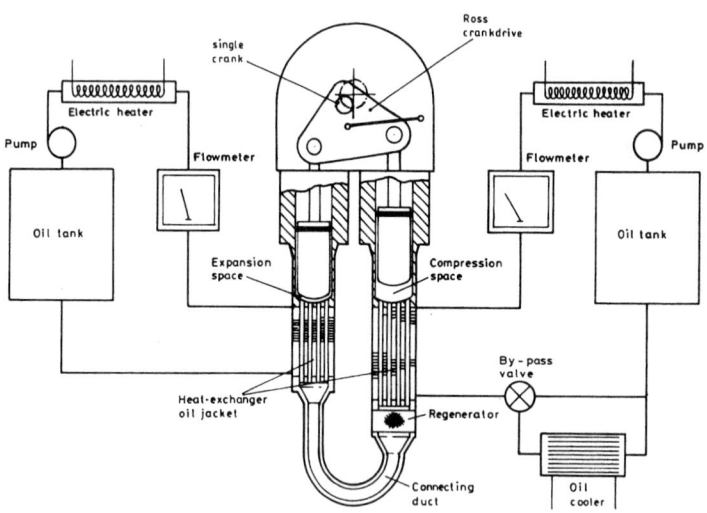

Fig. 1 Schematic arrangement of SM3 heat pump test system

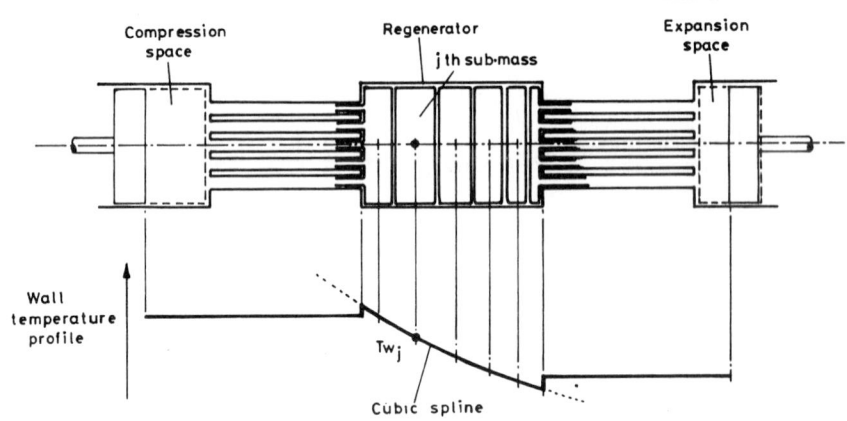

Fig. 2 Idealised Stirling cycle machine showing sample working fluid sub-masses

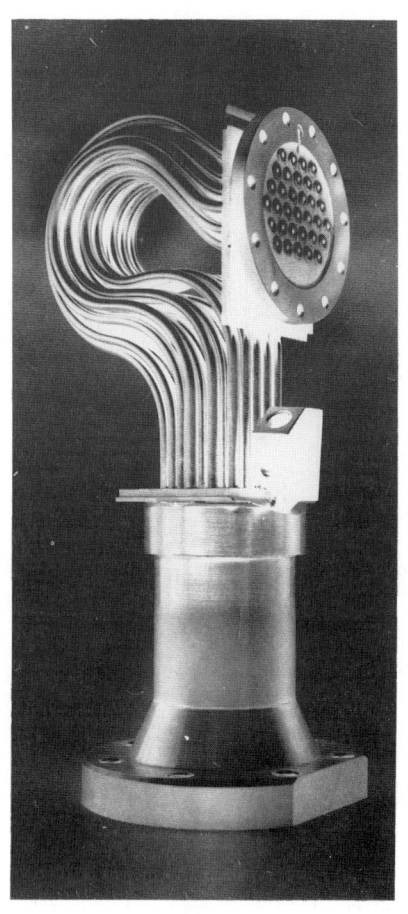 

Original design New design

Fig. 3 Old and new SM3 compression space heat exchangers, with oil circulation jackets removed.

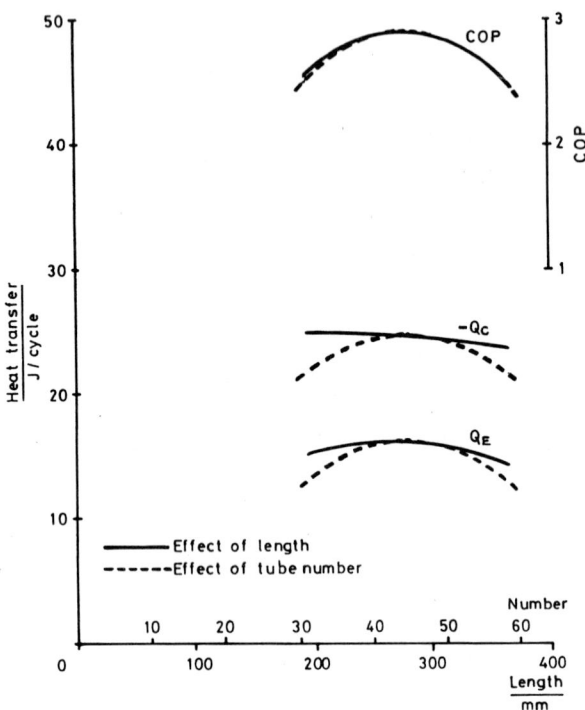

Fig. 4 Design optimisation curves for compression space heat exchanger, other variables at design point values

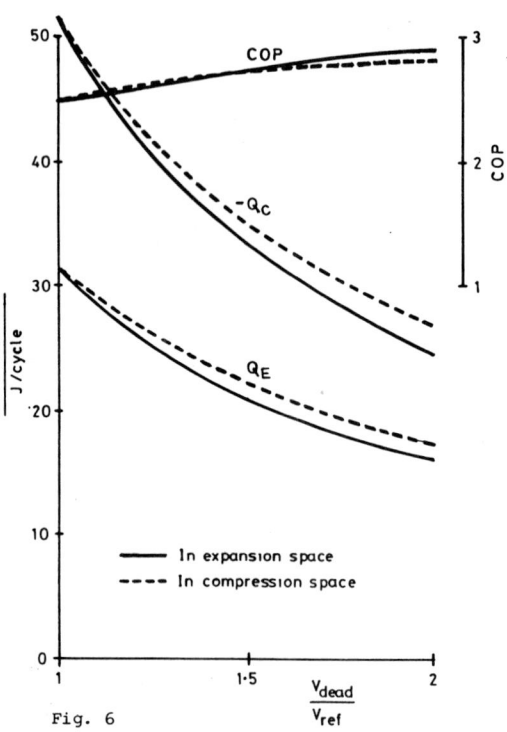

Fig. 6 Design optimisation curves for heat pump (dead volume/original dead volume) ratio, other variables at design point values

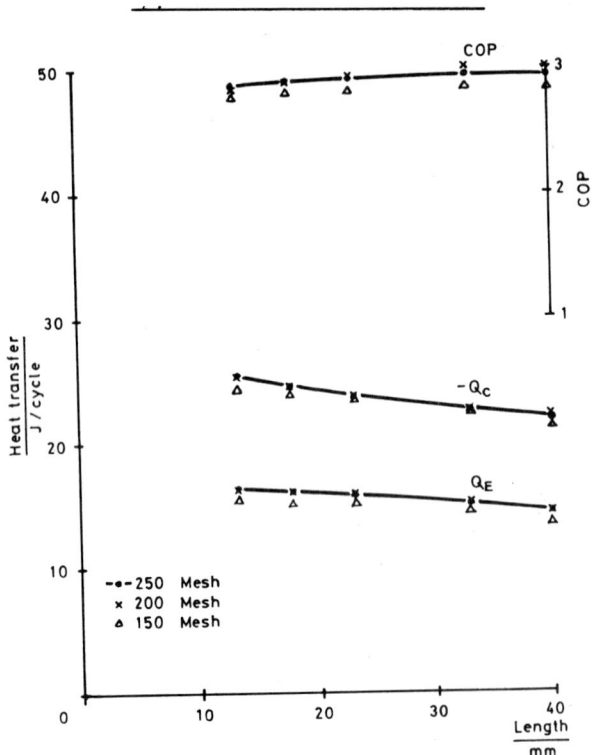

Fig. 5 Design optimisation curves for the regenerator, other variables at design point values

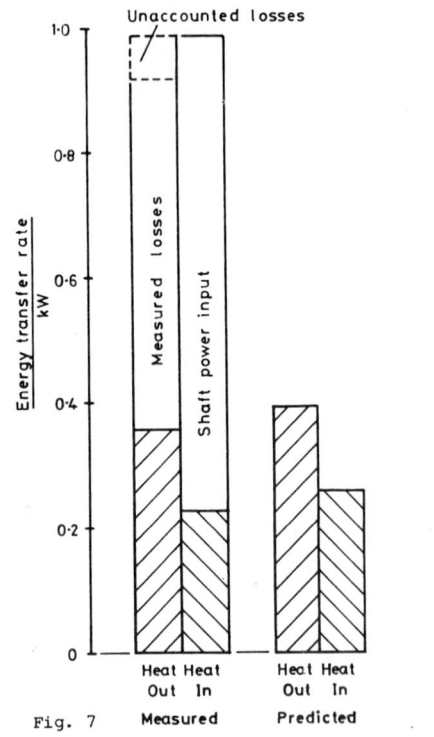

Fig. 7 Graphical representation of measured and predicted heat flows, at the design point